# Hydrodynamic Fluctuations, Broken Symmetry, and Correlation Functions

# Hydrodynamic Fluctuations, Broken Symmetry, and Correlation Functions

**DIETER FORSTER**

Temple University

This book was originally published as part of the Frontiers in Physics Series, edited by David Pines.

The Advanced Book Program

CRC Press
Taylor & Francis Group
Boca Raton London New York

CRC Press is an imprint of the
Taylor & Francis Group, an **informa** business
A CHAPMAN & HALL BOOK

First published 1975 by Perseus Books Group

Published 2018 by CRC Press
Taylor & Francis Group
6000 Broken Sound Parkway NW, Suite 300
Boca Raton, FL 33487-2742

*CRC Press is an imprint of the Taylor & Francis Group, an informa business*



Visit the Taylor & Francis Web site at
http://www.taylorandfrancis.com

and the CRC Press Web site at
http://www.crcpress.com

ISBN 13: 978-0-201-41049-5 (pbk)

# Publisher's Foreword

"Advanced Book Classics" is a reprint series which has come into being as a direct result of public demand for the individual volumes in this program. That was our initial criterion for launching the series. Additional criteria for selection of a book's inclusion in the series include:

- Its intrinsic value for the current scholarly buyer. It is not enough for the book to have some historic significance, but rather it must have a timeless quality attached to its content, as well. In a word, "uniqueness."
- The book's global appeal. A survey of our international markets revealed that readers of these volumes comprise a boundaryless, worldwide audience.
- The copyright date and imprint status of the book. Titles in the program are frequently fifteen to twenty years old. Many have gone out of print, some are about to go out of print. Our aim is to sustain the lifespan of these very special volumes.

We have devised an attractive design and trim-size for the "ABC" titles, giving the series a striking appearance, while lending the individual titles unifying identity as part of the "Advanced Book Classics" program. Since "classic" books demand a long-lasting binding, we have made them available in hardcover at an affordable price. We envision them being purchased by individuals for reference and research use, and for personal and public libraries. We also foresee their use as primary and recommended course materials for university level courses in the appropriate subject area.

The "Advanced Book Classics" program is not static. Titles will continue to be added to the series in ensuing years as works meet the criteria for inclusion which we've imposed. As the series grows, we naturally anticipate our book buying audience to grow with it. We welcome your support and your suggestions concerning future volumes in the program and invite you to communicate directly with us.

# Advanced Book Classics

V.I. Arnold and A. Avez, *Ergodic Problems of Classical Mechanics*

E. Artin and J. Tate, *Class Field Theory*

Michael F. Atiyah, *K-Theory*

David Bohm, *The Special Theory of Relativity*

P.C. Clemmow and J.P. Dougherty, *Electrodynamics of Particles and Plasmas*

Ronald C. Davidson, *Theory of Nonneutral Plasmas*

P.G. deGennes, *Superconductivity of Metals and Alloys*

Bernard d'Espagnat, *Conceptual Foundations of Quantum Mechanics, 2nd Edition*

Richard Feynman, *Photon-Hadron Interactions*

Dieter Forster, *Hydrodynamic Fluctuations, Broken Symmetry, and Correlation Functions*

William Fulton, *Algebraic Curves: An Introduction to Algebraic Geometry*

Kurt Gottfried, *Quantum Mechanics*

Leo Kadanoff and Gordon Baym, *Quantum Statistical Mechanics*

I.M. Khalatnikov, *An Introduction to the Theory of Superfluidity*

George W. Mackey, *Unitary Group Representations in Physics, Probability and Number Theory*

A.B. Migdal, *Qualitative Methods in Quantum Theory*

Phillipe Nozières and David Pines, *The Theory of Quantum Liquids, Volume II* - new material, 1989 copyright

David Pines and Phillipe Nozières, *The Theory of Quantum Liquids, Volume I: Normal Fermi Liquids*

F. Rohrlich, *Classical Charged Particles - Foundations of Their Theory*

David Ruelle, *Statistical Mechanics: Rigorous Results*

Julian Schwinger, *Particles, Sources and Fields, Volume I*

Julian Schwinger, *Particles, Sources and Fields, Volume II*

Julian Schwinger, *Particles, Sources and Fields, Volume III* - new material, 1989 copyright

Jean-Pierre Serre, *Abelian $\ell$-Adic Representations and Elliptic Curves*

R.F. Streater and A.S. Wightman, *PCT Spin and Statistics and All That*

René Thom, *Structural Stability and Morphogenesis*

# Vita

## Dieter Forster

Dieter Forster is a Professor of Physics at Temple University in Philadelphia. He was born in Southern Germany and received his "Diplom" in physics at the University of Stuttgart in 1964, working under Hermann Haken. He then moved to Harvard University where he completed his Ph.D. in 1969 under Paul C. Martin. He has worked at Columbia University, the University of Chicago, and as a Visiting Professor, at the Technical University of Munich, the University of Stuttgart, and the National University of Mexico. He has taught physics courses on many levels but takes a special interest in conceptual level courses designed for students not entering the physics profession. His research has focused on the fluctuation theory of liquid structure and on phase transitions.

# Special Preface

I am amazed, and delighted of course, to see that this book has remained useful for so long. When I wrote it I sought to provide an overview of work on response functions and the like and on ideas for their use. My instinctive orientation was microscopic: Hydrodynamic properties of correlation functions need to be firmly understood by those who compute by molecular dynamics the properties of simple liquids, for example. More recent work of this kind is perhaps best found in the book by Boon and Yip[1].

The continued vitality of my own book, however, seems to me also to stem from a reorientation in condensed matter physics which happened in the 1970's. The great success of Wilson's renormalization group ideas has led us to study systems of very great complexity. Ideas of scaling and universality allow us to *begin* that study on a mesoscopic level, using Landau-Wilson Hamiltonians and Langevin and Fokker-Planck equations of motion. Such work needs to be anchored in a few principles. The peculiar combination of ideas from phenomenology, statistical mechanics and microscopic dynamics which I have described in this book seems to have provided some workers with just such an anchor. It also seems to have given graduate students just enough of an introduction to fluctuations, correlations, and broken symmetry to get them started on their own more modern work.

Dieter Forster
August 1989

[1]J.P. Boon and S. Yip, *Molecular Hydrodynamics*, McGraw-Hill, New York, 1980.

To Martha and Willi Forster

# Editor's Foreword

The problem of communicating in a coherent fashion the recent developments in the most exciting and active fields of physics seems particularly pressing today. The enormous growth in the number of physicists has tended to make the familiar channels of communication considerably less effective. It has become increasingly difficult for experts in a given field to keep up with the current literature; the novice can only be confused. What is needed is both a consistent account of a field and the presentation of a definite "point of view" concerning it. Formal monographs cannot meet such a need in a rapidly developing field, and, perhaps more important, the review article seems to have fallen into disfavor. Indeed, it would seem that the people most actively engaged in developing a given field are the people least likely to write at length about it.

FRONTIERS IN PHYSICS has been conceived in an effort to improve the situation in several ways. First, to take advantage of the fact that the leading physicists today frequently give a series of lectures, a graduate seminar, or a graduate course in their special fields of interest. Such lectures serve to summarize the present status of a rapidly developing field and may well constitute the only coherent account available at the time. Often, notes on lectures exist (prepared by the lecturer himself, by graduate students, or by postdoctoral fellows) and are distributed in mimeographed form on a limited basis. One of the principal purposes of the FRONTIERS IN PHYSICS series is to make such notes available to a wider audience of physicists.

It should be emphasized that lecture notes are necessarily rough and informal, both in style and content, and those in the series will prove no exception. This is as it should be. The point of the series is to offer new, rapid, more informal, and it is hoped, more effective ways for physicists to teach one another. The point is lost if only elegant notes qualify.

Another way is to improve communication in very active fields of physics by the publication of collections of reprints of recent articles. Such collections are themselves useful to people working in the field. The value of the reprints would, however, seem much enhanced if the collection were accompanied by an introduction of moderate length, which would serve to tie the collection together and, necessarily, constitute a brief survey of the present status of the field. Again, it is appropriate that such an introduction be informal, in keeping with the active character of the field.

A third possibility for the Series might be called an informal monograph, to connote the fact that it represents an intermediate step between lecture notes and formal monographs. It would offer the author an opportunity to present his view of a field which has developed to the point where a summation might prove extraordinarily fruitful, but for which a formal monograph might not be feasible or desirable.

A fourth manner of presentation is the contemporary classics—papers of lectures which constitute a particularly valuable approach to the teaching and learning of physics today. Here one thinks of fields that lie at the heart of much of present-day research, but whose essentials are by now well understood, such as quantum electrodynamics or magnetic resonance. In such fields, some of the best pedagogical material is not readily available, either because it consists of papers long out of print or lectures that have never been published.

---

The above words, written in August 1961, continue to be applicable. During the past decade, condensed matter physics has undergone a period of substantial growth, as both experimentalists and theorists have made use of a wide variety of sophisticated techniques in dealing with a broad range of problems in the behavior of liquids and solids. Both experimentalists and theorists have come to realize that the language of correlation or response functions is a particularly useful one, in that it provides a simple way of describing what has actually been measured, and of specifying the content of a given approximation devised by the theorists. In this volume, Professor Forster, who has himself made significant contributions to our understanding of fluctuations and transport in systems as diverse as classical fluids, liquid crystals, and ferro- and antiferromagnets, presents an elementary account of this important language, with special emphasis on those ideas and theoretical techniques which are applicable to a wide variety of physical systems. His book is addressed to physics and chemistry graduate students who are beginning their studies, and has been written with the express aim of making the concepts of correlation, fluctuations, and broken symmetry readily accessible to experimentalists and theorists alike. I share his hope that, in addition to being a self-contained monograph for independent study, it may serve as well as a text for a one-semester course in non-equilibrium statistical mechanics at the first- or second-year graduate level.

David Pines

# Preface

In the past two decades or so it has become commonly appreciated among physicists and physical chemists that most of the experiments which they perform are conveniently discussed in terms of correlations functions, of one kind or another. Indeed, their use has become so pervasive that there is hardly a field of condensed matter physics that has not, at one time or another, been formulated in terms of correlation functions. Nor is their usefulness restricted to systems of many particles. Whenever one is interested in the response of a physical system to some external force, particularly in the linear response as is so often the case, correlation functions have proved to be a suitable language in which to express one's measurements. In statistical systems which are near thermal equilibrium, they give all desirable information about the intrinsic statistical fluctuations. They afford, in particular, the most convenient and transparent vehicle in terms of which one bridges the gap between a microscopic, reversible description and the irreversible macroscopic behavior of many-body systems.

For normal classical or quantum-mechanical systems there are several excellent texts which present the general theory of response functions and discuss specific examples (e.g., Pines and Nozières 1966, Martin 1968). The corresponding description of condensed systems, detailing the influence of a broken symmetry on the hydrodynamic fluctuations, is known to many experts but does not seem to have been simply presented in the book literature. For this reason, as well as for their intrinsic interest, beauty, and importance, I have devoted a large portion of this text to systems with broken continuous symmetries.

This book grew out of a graduate course which I taught at the University of Chicago in 1972 and again, in somewhat different form, at the University of Stuttgart in 1973. My aim in writing up these lecture in book form has been to provide a unified, elementary and easily accessible account of the language of correlation functions or response functions, and to discuss particularly those simple techniques whose usefulness is not restricted to one specific system. A reflection of that aim is the table of contents which covers a whole variety of apparently different physical systems. Some of those, like normal fluids or liquid crystals, are better known to physical chemists while others, especially superfluid Helium, have been of interest mostly to low temperature physicists. By treating all of these systems in a simple and common language I hope to make at least their most important properties and some of their specific fascinations more easily accessible to a larger audience.

Apart from being a self-contained monograph for independent study, the book can serve as a text for a one-semester course in non-equilibrium statistical mechanics, at the first or second year graduate level, as well as to supplement the more traditional courses in statistical mechanics or in solid state physics. The book requires no background which a first year graduate student of physics or chemistry would not have. At the same time I hope that there is enough material of interest to research workers in either field who would like to learn how to apply correlation function techniques to their particular problems, or how to transfer knowledge and experience gained in the study of superfluids to, say, the field of liquid crystals. The methods which are detailed are not very technical, and they certainly should not discourage experimentalists. It is for them, first and foremost, that this book has been written.

The bibliography is far from complete, and has been assembled largely with the aim of giving accessible references which further illustrate the points made in the text. For this reason I have usually quoted those papers whose purpose and approach is similar to that pursued here. I apologize to the many workers in this field whose contributions are not presented, or not adequately presented.

It is a pleasure to thank Mrs. Pauline Stockbauer for careful typing of the manuscript, and to acknowledge the many helpful discussions with my colleagues and students whose comments have improved the book.

Dieter Forster

# Contents

# Hydrodynamic Fluctuations, Broken Symmetry, and Correlation Functions

# CHAPTER I

# INTRODUCTION

Physics is a field that thrives on analogies. We can understand, partially at least, superfluid Helium on the basis of concepts borrowed from superconductors, superconductivity with techniques that were developed in the study of ferromagnetism, and we investigate fluctuations in liquid crystals by employing methods developed to understand spin waves in ferro- and antiferromagnets. The analogy is never complete, but it is almost always helpful.

The purpose of this book is to present and apply a language and to discuss methods which make it very convenient to exploit such analogies, and which are uniquely suited to describe and explain non-equilibrium phenomena in a rich variety of many-particle systems: the language of time correlation functions and linear response theory.

Texts about non-equilibrium statistical mechanics and irreversible processes have traditionally started with a series of ambitious questions which are very interesting, fundamental, and enormously difficult, and which have a long and colorful history of often controversial answers. To mention but one: how come a system of many interacting particles whose microscopic dynamics is invariant under the reversal of time, moves nevertheless irreversibly towards thermal equilibrium, and the same equilibrium state whatever that means, (almost) no matter from which initial state the system started? (Partial answers to this question began with L. Boltzmann's celebrated H-theorem which has just had its centenary, and they continue, in this century, with derivations of master equations by Pauli, Van Hove, Prigogine and many others.)

Such problems are very interesting philosophically, but they rarely concern the practical physicist. He knows, or believes, that many systems do come to equilibrium if given time, and not very much time as a rule. Optimistically, he takes yet fundamentally unresolved properties like ergodicity for granted, and he knows how to select ensembles for the statistical description of an equilibrium system, and thus how, in principle at least, to calculate measurable equilibrium properties for such a system. When he probes dynamical processes in many-particle systems, he usually employs as a probe an external force which disturbs the system only slightly from equilibrium, and he measures the linear time-dependent response to this force. A vast number of experimental methods falls in this category: lineshape studies of electronic, infrared, and Raman spectra; experiments in which light, neutrons, or electrons are inelastically scattered off a sample; lineshape studies in dielectric or spin relaxation experiments; studies of acoustic attenuation and transport behavior; and many others. In all these experiments, what he probes is the dynamical behavior of spontaneous fluctuations about the (presumably well-known) equilibrium state.

To explain these fluctuations, the requisite information is far less specific than that contained in the microscopic state or density matrix itself. According to linear response theory, these fluctuations may be rigorously described in terms of time-dependent correlation functions: products of pairs of certain dynamical variables $A(t)$ and $B(t')$ at different times $t$ and $t'$, which are averaged over a thermal ensemble,

$$S_{AB}(t,t') = \langle A(t)\, B(t')\rangle_{eq.} \,, \tag{1.1}$$

where the equilibrium average is indicated by the brackets $\langle \ldots \rangle_{eq.}$. These functions offer a powerful and versatile theoretical tool for the investigation of non-equilibrium behavior, and much of the great progress that has been achieved in the last two decades in the field of non-equilibrium statistical mechanics has been due to their study.

Of particular interest are fluctuations of measurable space and time dependent densities which describe, at long wavelength, the cooperative motion characteristic of many systems. From the particle density correlation function

$$S_{nn}(\vec{r}t, \vec{r}'t') = \langle n(\vec{r}t)\, n(\vec{r}'t')\rangle_{eq.} , \tag{1.2}$$

for example, one obtains the results of light, X-ray, and neutron scattering in, say, a normal liquid. Electron scattering measures the function $S_{\rho\rho}(\vec{r}t, \vec{r}'t')$ obtained by replacing the particle density $n(\vec{r}t)$ in (1.2) by the charge density, $\rho(\vec{r}t)$. In magnetic systems, neutron scattering measures the correlation function of the magnetization $M(\vec{r}t)$; and so forth.

It will thus be profitable to discuss the general properties of time correlation functions. These include various symmetries, among them time reversal symmetry which reemerges at a later stage to yield the Onsager relations between transport coefficients; positivity properties which are related to the dynamical stability of the system; fluctuation-dissipation theorems which connect, as the words indicate, spontaneous fluctuations and energy dissipation in a thermally equilibrated system; Kramers-Kronig relations which express causality, like the well-known one between the index of refraction and the absorption coefficient; Kubo relations which give the connection, in appropriate systems, between transport coefficients and correlation functions, relations which have proved extremely valuable in performing practical microscopic calculations. We will also discuss a host of sum rules which are an important input for approximate calculations since they are among the few quantitative statements about complicated many-body systems which can be derived and often calculated from first principles. Many of the methods for approximate calculations are based on rigorously valid dispersion relations which will be discussed in much detail. In particular, we will apply the projection operator techniques of Zwanzig and Mori to introduce memory functions, and motivate their approximate evaluation in a variety of specific examples.

The similarities among various systems are most apparent--and their differences most revealing--when one comes to study the highly cooperative modes of hydrodynamics: the collective fluctuations of large wavelength which involve coherent motion of a large number of particles, and whose lifetimes are very large, on a microscopic scale. The existence and structure of these collective modes manifest the most characteristic properties of a many-body system. And while this book is not exclusively reserved for their analysis, we shall pay them much attention. In particular, we will attempt to present our derivations in a fashion which makes the fundamental principles apparent on which the existence of collective modes rests.

In chapter 2 we will discuss at length the approach and illustrate many of the methods used throughout the book, on the simple example of spin diffusion in liquid $He^3$. General properties of time-dependent correlation functions are given and proved in chapter 3, and in chapter 4 we discuss the normal classical one-component fluid as a specific application. The treatment in these introductory chapters follows in large measure that of the fundamental paper by Kadanoff and Martin (1963).

An alternative and powerful technique is presented in chapter 5, namely the projection operator and memory function formalism, first given by Zwanzig (1961) and Mori (1965), which can be understood as a rigorous basis for the Langevin theory of thermal fluctuations. Chapter 6 contains a brief application to the Brownian motion of a heavy particle which is immersed in a fluid of light particles. Application of these methods to the normal fluid is easily distilled from our later treatment of a superfluid. Their real power becomes apparent in applications to systems with a broken symmetry.

As the temperature is lowered, many-particle systems generally undergo phase transitions to states which contain more order than do the completely chaotic high temperature phases. The existence of such order requires new quantities for the description of the state: order parameters. Below the transition, certain fluctuations are correlated over large distances, and additional long-wavelength collective modes appear. These concepts are introduced in chapter 7 where the general Bogoliubov inequalities and Goldstone theorems are proved on which the further analysis rests. The following chapters contain applications to several important physical systems. Long-ranged magnetic order and ferromagnetic and antiferromagnetic spin waves are discussed in chapters 8 and 9, broken gauge invariance and superfluidity in chapter 10. Chapter 11 is concerned with liquid crystals, particularly with the orientational light scattering modes in nematics. The final chapter 12 discusses, briefly, certain aspects of superconductors and of charged systems in general.

It is important, for a program of this magnitude, to clarify its limitations. Two particularly interesting fields of application are omitted here. Except for a brief section in chapter 12 I will not discuss in detail the changes which take place in a system of charged particles due to the long-range nature of the Coulomb force. These are analyzed, in a language very similar to the one given here, in the book by Pines and Nozieres (1966) and in the Les Houches lectures by Martin (1968). And I will not discuss the vast field of phase transitions and critical phenomena (Stanley 1971, Wilson and Kogut 1974, Ma 1975) which has received so much attention recently, and seen so much progress. These phenomena can be treated in the language presented here but require a whole array of techniques not covered here.

Moreover, this is not a text about Green's functions and the elaborate techniques usually combined under that label. (See Kadanoff and Baym 1962, Fetter and Walecka 1971.) Green's function methods are methods to perform detailed microscopic calculations, calculations which are always difficult, and often hard to check. It is useful therefore and desirable to obtain, from fundamental principles, constraints on such calculations. Most of the correlation functions which we will discuss here are in fact closely related to one- and two-particle Green's functions. And the results which we will derive -- sum rules, hydrodynamic limiting expressions and others -- provide restrictions which any fully microscopic theory must fulfil.

We will therefore, even when dealing with specific examples, concentrate on the structural aspects of the theory, aim at results which bridge the gap between microscopic equations of motion and macroscopically observable phenomena, and provide a common language in which experiments can be discussed that are performed on a wide variety of physical systems and with a similar variety of methods. A language which should be convenient for him or her who, when analyzing light scattering data from liquid crystals, would like to draw on knowledge of antiferromagnets or superfluids. The language of correlation functions is mathematical, of course, but a determined effort has been made to concentrate on the physical ideas involved, and tread lightly on questions of mathematical rigor. If I have succeeded, those who do the real work, the experimentalists, should find in this text a useful and intelligible bit of theory, and hopefully have some fun while reading it.

# CHAPTER 2

## A SIMPLE EXAMPLE-- SPIN DIFFUSION

As a simple example which illustrates many of the points we shall discuss, let us consider a fluid of uncharged particles with spin 1/2 (Kadanoff and Martin 1963). The essential assumption which makes this system so simple, is that the particles interact through a velocity- and spin-independent force. This situation is, in fact, realized to an excellent approximation in at least one real system, liquid $He^3$. (Much of the subsequent analysis will, however, also apply to the isotropic Heisenberg paramagnet, for example. See chapter 8 and e.g., Bennett and Martin [1965] and Lubensky [1970a]).

The spin of each particle can be taken to point either parallel (+) or antiparallel (-) to some arbitrary direction of quantization. In order to simplify things we will treat the spin as a scalar quantity; its vector character is of no importance for our purposes. The magnetization, $M(\vec{r}\ t)$, is then simply proportional to the difference in densities $n_+(\vec{r}\,t)$ and $n_-(\vec{r}\,t)$ at the space-time point $\vec{r},t$, i.e.,

$$M(\vec{r},t) = \mu[n_+(\vec{r},t) - n_-(\vec{r},t)], \tag{2.1}$$

where $\mu$ is the spin magnetic moment of a particle. A more microscopic way of writing this operator is

$$M(\vec{r},t) = \sum_\alpha 2\mu\, s^\alpha\, \delta(\vec{r}-\vec{r}^\alpha(t)), \tag{2.2}$$

where the a-th particle has the position $\vec{r}^{\alpha}(t)$ at time t and the spin $s^{\alpha}$ which is either 1/2 or −1/2. The δ-function sees to it that only those particles are counted which at time t are found at or near the point $\vec{r}$. The assumption of spin-independent forces is reflected in the fact that $s^{\alpha}$ is constant in time.

In thermal equilibrium, $n_+ = n_-$ on the average so that $M = 0$. Now assume that at some initial time, $t = 0$, there is a local imbalance at point $\vec{r}$ so that $M(\vec{r}, t=0) \neq 0$. We will be interested in the subsequent time development of $M(\vec{r}, t)$. In our simple model which neglects spin-flip processes, the time dependence of $M(\vec{r}, t)$ will be due to the fact that the particles move around, carrying their spin with them. In fact it does not much matter what they carry, i.e., what "spin up" and "spin down" mean. A system of green ($s^{\alpha} = \frac{1}{2}$) and red ($s^{\alpha} = -\frac{1}{2}$) particles would behave in the same way.

Since $\dot{\vec{r}}^{\alpha} = \vec{p}^{\alpha}/m$ where $\vec{p}^{\alpha}$ is the momentum of the a-th particle and m is its mass, we get from (2.2) the continuity equation

$$\partial_t M(\vec{r}, t) + \vec{\nabla} \cdot \vec{j}^M(\vec{r}, t) = 0 \, , \tag{2.3}$$

where $\vec{j}^M$ is the magnetization current. It can be written in the form

$$\vec{j}^M(\vec{r}, t) = \sum_{\alpha} (\mu\, s^{\alpha}/m) \left\{ \vec{p}^{\alpha}(t), \, \delta(\vec{r} - \vec{r}^{\alpha}(t)) \right\} \, . \tag{2.4}$$

We will always use curly brackets for the anticommutator

$$\{ A, B \} \equiv AB + BA \quad .$$

For classical particles, the symmetrization in (2.4) is, of course, unnecessary.

Eq. (2.3) expresses the fact that the total magnetization is conserved,

$$\frac{d}{dt} \int d\vec{r} \, M(\vec{r}, t) = 0 \, , \tag{2.5}$$

but it implies an important additional property: the current $\vec{j}^M(\vec{r}, t)$, as the magnetization $M(\vec{r},t)$, is a local density, dependent only on properties of particles which, at time t, are in some small neighborhood around the point $\vec{r}$. Differential conservation laws like (2.3) will play an important role in most of the processes which we shall discuss.

## 2.1 Hydrodynamic Description

The conservation law (2.3) is not a complete description; it just restricts the dynamics a little. To solve for $M(\vec{r},t)$, we need a second equation relating $\vec{j}^M$ to M. Now both $n_+(\vec{r},t)$ and $n_-(\vec{r},t)$ tend towards an equilibrium state in which they are spatially uniform. In other words, there is a net flow of magnetization from regions of large M to regions of small M. Phenomenologically,

$$<\vec{j}^M(\vec{r},t)> = -D\,\vec{\nabla} <M(\vec{r},t)> \,. \tag{2.6}$$

This is called a constitutive equation. The transport coefficient, D, is called the spin diffusion coefficient, and it is positive. Note that while (2.3) is microscopically exact, (2.6) can be true only on the average which is why we have put brackets < > around it. These indicate here a non-equilibrium average, of course; in thermal equilibrium, $<M(\vec{r},t)>_{eq}$ is independent of $\vec{r},t$, and $<\vec{j}^M(\vec{r},t)>_{eq}$ vanishes.

Inserting (2.6) in (2.3) we get the familiar diffusion equation,

$$\partial_t <M(\vec{r},t)> - D\,\nabla^2 <M(\vec{r},t)> = 0 \,, \tag{2.7}$$

which is now complete and can be solved. Note that this equation is only valid if all the properties of the system vary slowly in space and time. This assumption is clearly implicit in (2.6), and will be analyzed a little further below.

We are only interested here in an infinitely extended system. This eliminates boundary conditions so that (2.7) is trivially solved by performing a Fourier transformation in space.

$$<M(\vec{k},t)> = \int d\vec{r}\, e^{-i\vec{k}\cdot\vec{r}} <M(\vec{r},t)>, \tag{2.8a}$$

and a Laplace transformation in time,

$$<M(\vec{k}z)> = \int_0^\infty dt\, e^{izt} <M(\vec{k},t)> . \tag{2.8b}$$

$\vec{k}$ is the wave vector of the fluctuation. The complex frequency z must lie in the upper half of the complex plane for the integral in (2.8b) to converge. From eq. (2.7) we then obtain

$$<M(\vec{k}z)> = \frac{i}{z + iDk^2} <M(\vec{k},t=0)>, \tag{2.9}$$

which solves the initial value problem.

The diffusion process is reflected in a pole on the negative imaginary axis, at $z = -iDk^2$. To get a little more familiar with diffusion poles, note that (2.9) says the same as

$$<M(\vec{k},t)> = e^{-Dk^2t} <M(\vec{k},t=0)> . \tag{2.10}$$

This equation displays the characteristic property of a "hydrodynamic" mode: it is a spatially sinusoidal collective fluctuation which for large wavelength $\lambda = 2\pi/k$ is exponentially damped, with a lifetime

$$\tau(k) = 1/Dk^2 \tag{2.11}$$

which becomes infinite as $k \to 0$.

It is well to appreciate that this behavior is very unusual in as chaotic a many-body system as a liquid. There is an enormous number of channels available into which an arbitrary degree of freedom can decay after the initial excitation. Most degrees of freedom will relax within a short time $\tau_c$ which is determined by the microscopic interactions. For a system of classical particles of mass m, interacting with a pair potential of strength $\epsilon$ and range $a$, dimensional arguments suggest that $\tau_c$ is of the order

$$\tau_c \sim a(m/\epsilon)^{1/2}, \tag{2.12}$$

which for Helium would give $\tau_c \approx 10^{-12}$ sec. Even though, in a quantum liquid like $He^3$ at low temperatures, the Pauli principle severely restricts the number of decay channels--dimensionally, the small thermal energy $k_B T$ and $\hbar$ become available to correct (2.12)--the microscopic decay times at all but the very lowest temperatures are still very small on a macroscopic scale.

What is special about the degree of freedom described by $M(\vec{k},t)$ is that the magnetization is a conserved quantity. A local excess of this quantity cannot disappear locally (which could happen rapidly) but can only relax by spreading slowly over the entire system. A sinusoidal fluctuation as depicted in fig. 2.1

Fig. 2.1

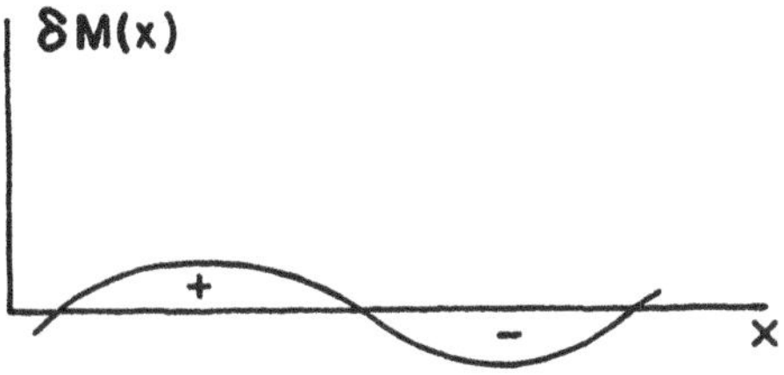


can only relax by the magnetization being physically transported from the excess to the deficiency regions over a distance of order $\lambda/2$, a process that requires an infinitely long time as $\lambda \to \infty$. Indeed, if the transport process occurs via a random walk, then $(\Delta x)^2 \approx D\tau$ or $\tau \approx \lambda^2/D$ which is what we found in (2.11).

Returning to eq. (2.10), let us assume that initially, the magnetization is non-zero only at $\vec{r} = 0$, i.e., that $\langle M(\vec{r},t=0)\rangle = M\delta(\vec{r})$. Then

$$\langle M(\vec{r},t)\rangle = M(4\pi Dt)^{-3/2} \exp-(r^2/Dt) , \tag{2.13}$$

which displays the characteristic Gaussian spreading of a random walk process.

## 2.2 Spin Correlation Function (Roughly)

The last few equations solve the problem of spin diffusion in an infinitely extended system. Now, since this is a book about correlation functions, let us try to extract some information about the spin correlation function. We begin by defining the magnetization correlation function by

$$S(\vec{r},t) = \langle M(\vec{r},t)\, M(\vec{0},0)\rangle_{eq} , \tag{2.14}$$

where $M(\vec{r},t)$ is the magnetization operator employed above. The average in (2.14) is a thermal equilibrium average, by contrast to the average in (2.6) which is meant to describe a system not yet in full equilibrium. Of course, even though $\langle M(\vec{r},t)\rangle_{eq} = 0$, there will be spontaneous, usually small, fluctuations on a local scale. $S(\vec{r},t)$ describes these fluctuations. Because of the magnetic interaction of neutrons with local magnetization fluctuations, the function $S(\vec{r},t)$ can be measured by magnetic neutron scattering.

$S(\vec{r},t)$ presumably vanishes rapidly when r and/or t are very large since then, $M(\vec{r},t)$ and $M(\vec{0},0)$ are statistically independent so that $\langle M(\vec{r},t)M(\vec{0},0)\rangle = \langle M(\vec{r},t)\rangle\langle M(\vec{0},0)\rangle = 0$. Therefore, it can be Fourier transformed,

$$S(k\omega) = \int_{-\infty}^{\infty} dt \int d\vec{r}\, e^{i\omega t - i\vec{k}\cdot\vec{r}}\, S(\vec{r},t) . \tag{2.15}$$

This function represents the spectral density of magnetization fluctuations, and is real and positive. Because of the rotational invariance of the system, $S(k\omega)$ depends only on the magnitude of $\vec{k}$. Of use is also the one-sided (Laplace) transform

$$\tilde{S}(kz) = \int_{0}^{\infty} dt\, e^{izt}\, S(k,t), \quad \text{for } \operatorname{Im} z > 0 . \tag{2.16}$$

It is an easy exercise to show that

$$\tilde{S}(kz) = \int \frac{d\omega}{2\pi i} \frac{S(k\omega)}{\omega - z} , \tag{2.17}$$

which last equation has a meaning for z in both the upper and lower halves of the complex plane. $\tilde{S}(kz)$ is a complex function, analytic in z except for a branch cut along the real axis. In fact, using the identity

$$\frac{1}{x \mp i\epsilon} = P\frac{1}{x} \pm i\pi\delta(x) \tag{2.18}$$

where $\epsilon$ is, here and throughout this book, positive and infinitesimal, and P indicates the Cauchy principal value, we find that $S(k\omega)$ is the discontinuity across the branch line,

$$S(k\omega) = \lim_{\epsilon \to 0} [\tilde{S}(k,\omega+i\epsilon) - \tilde{S}(k,\omega-i\epsilon)] . \tag{2.19}$$

More useful is the equation which follows from the reality of $S(k\omega)$, namely

$$S(k\omega) = 2 \,\mathrm{Re}\, \tilde{S}(k,\omega+i\epsilon) . \tag{2.20}$$

Now let us first use a hit-and-run technique to obtain the correlation function from our hydrodynamic analysis. Why would we want to do that? First and importantly because the correlation function is of immediate experimental interest since it gives the intensity distribution measured by inelastic neutron scattering. Second, because $S(\vec{r},t)$ is a mathematically and operationally well-defined object; we know, in principle at least, how to perform thermal equilibrium averages as in eq. (2.14). The phenomenological fluctuation $\langle M(\vec{r},t)\rangle_{\text{non-eq}}$ of section 2.1 is a little more hazy an object since it is harder to give precise meaning to the non-equilibrium average.

What we shall assume is that the constitutive equation (2.6) and therefore the diffusion equation (2.7) are valid, in some sense, even if we omit the average signs $<>_{non-eq}$, i.e., that they can be understood as operator equations. If so, all we have to do is to multiply eq. (2.7) (without the brackets) from the right by $M(\vec{0},0)$ and then do an equilibrium average, to obtain

$$[\partial_t - D\nabla^2]\, S(\vec{r},t) = 0. \tag{2.21}$$

Our rough assumption therefore says that spontaneous equilibrium fluctuations--described by S--relax according to the same diffusion equation as do induced non-equilibrium fluctuations--described by $<M>_{non-eq}$. This entirely reasonable hypothesis was first proposed by Onsager (1931), and it is quite correct.

Eq. (2.21) is solved just like (2.7) was, and the result is

$$\tilde{S}(kz) = \frac{i}{z+iDk^2}\; S(k,t=0). \tag{2.22}$$

Note, however, that the initial condition is now not arbitrary but is perfectly well defined by eq.. (2.14). In fact, in section 2.4 we will show that $S(k\to 0,\, t=0)$ is $k_B T$ times the spin magnetic susceptibility $\chi$,

$$\lim_{k\to 0}\; S(k,t=0) = \beta^{-1}\chi \tag{2.23}$$

and therefore, for small k,

$$\tilde{S}(kz) = \frac{i\beta^{-1}\chi}{z+iDk^2}\;. \tag{2.24}$$

To extract the spectral density, we cannot use (2.19) since (2.24) holds, by derivation, only for $\text{Im}\, z > 0$. However, we can use (2.20) and obtain the final result

$$S(k\omega) = \frac{Dk^2}{\omega^2 + (Dk^2)^2}\; \frac{2}{\beta}\chi, \tag{2.25}$$

whose experimental significance we shall discuss below.

Of course, eq. (2.25) is an approximation to the real world. It is valid only for small k and $\omega$ which means, in the loose language that is common, it is valid when $k^{-1}$ is much larger than all "natural lengths in the system," and when $\omega^{-1}$ is much larger than all "natural times in the system". The important length is the mean free path, which in a liquid is of the order of the interparticle distance.

Note that the behavior of the correlation function at small k and $\omega$ is by no means simple. For example, $\lim_{\omega\to 0}[\lim_{k\to 0} S(k\omega)] = 0$ but $\lim_{k\to 0}[\lim_{\omega\to 0} S(k\omega)] = \infty$. The order of the limits matters, and one must be extremely careful if he wants to make expansions of correlation functions in terms of k and $\omega$. However, the inverse of the complex function $\tilde{S}(kz)$,

$$\tilde{S}^{-1}(kz) = (i\beta^{-1}\chi)^{-1}\,[z + iDk^2]\,, \tag{2.26}$$

looks smooth, like the beginning of a Taylor series in k and z. We shall see that the rational thing to do is usually to make approximations for $\tilde{S}^{-1}$ rather than S.

Equation (2.23) is equivalent to

$$\lim_{k\to 0} \beta \int \frac{d\omega}{2\pi} S(k\omega) = \chi\,. \tag{2.27}$$

This is called a thermodynamic sum rule since it gives a thermodynamic derivative, $\chi = (dM/dH)$ which is the spin magnetic susceptibility, from a frequency integral over the correlation function. This sum rule is exact and our hydrodynamic approximation (2.25) exhausts it.

Furthermore, from (2.25) we see that

$$\frac{\beta}{2} \lim_{\omega\to 0} [\lim_{k\to 0} \frac{\omega^2}{k^2} S(k\omega)] = D\chi\,, \tag{2.28}$$

which gives the spin diffusion coefficient D in terms of the correlation function. This is a Kubo relation though it doesn't quite look like Kubo's original expression

(Kubo 1957). However, it is a simple and useful exercise to manipulate eq. (2.28) into the form

$$D\chi = (\beta/2)\frac{1}{3}\int d\vec{r} \int_{-\infty}^{\infty} dt \langle \vec{j}^M(\vec{r},t)\, \vec{j}^M(0,0)\rangle , \qquad (2.29a)$$

where use is made of translational and rotational invariance. Or finally, defining the total magnetization current operator by

$$\vec{J}^M(t) = \int d\vec{r}\, \vec{j}^M(\vec{r},t) ,$$

we get what Kubo got,

$$D\chi = \lim_{\epsilon\to 0} (1/3\, V k_B T) \int_0^{\infty} dt \langle \frac{1}{2}\left\{\vec{J}^M(t), \vec{J}^M(0)\right\}\rangle e^{-\epsilon t} , \qquad (2.29b)$$

where V is the volume of the system, and we have put in a convergence factor $e^{-\epsilon t}$, to be safe just in case convergence at large t should be subtle.

## 2.3 Magnetic Neutron Scattering

So far, we have been doing pretty well. Using simple arguments we have obtained an experimentally relevant correlation function. $S(k\omega)$ can, as we said, be measured by neutron scattering. Neutrons possess a magnetic moment which interacts with the magnetization of the medium by the magnetic dipole interaction, and leads to scattering. What one does is shoot into the liquid neutrons which have initial energy $\epsilon_i$ and momentum $\vec{p}_i$. One then looks for scattered neutrons with energy $\epsilon_f = \epsilon_i - \hbar\omega$ and momentum $\vec{p}_f = \vec{p}_i - \hbar\vec{k}$. Obviously, the neutrons have lost (or picked up, depending on the sign of $\omega$) energy and momentum to (from) excitations in the system, namely the collective fluctuations of the magnetization; see fig. 2.2.

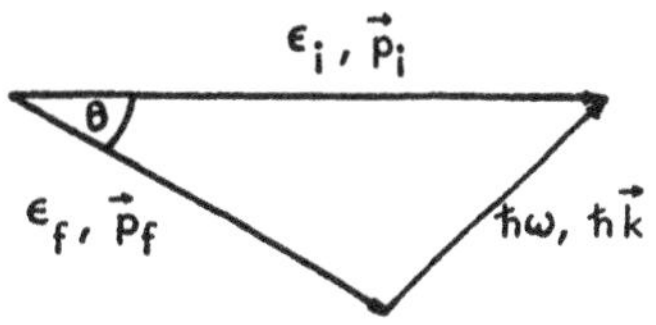


Fig. 2.2

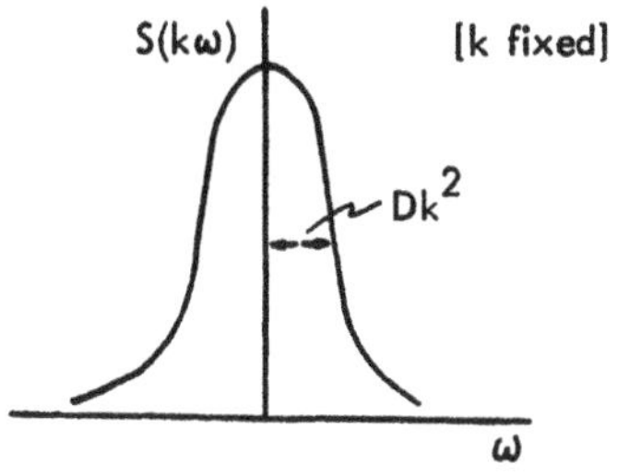


Fig. 2.3

The spectrum of these fluctuations will therefore determine the inelastic scattering cross section. This is shown in the Appendix. For the moment, it will suffice to note that the intensity of the scattered beam, $I_{scatt}$, is given by

$$I_{scatt}\begin{pmatrix}\vec{p}_i \to \vec{p}_i - \hbar\vec{k} \\ \\ \epsilon_i \to \epsilon_i - \hbar\omega\end{pmatrix} = [\text{ factors times }]\, S_{MM}(\vec{k}\omega)\,. \qquad (2.30)$$

Eq. (2.25) tells us to expect a Lorentzian line shape for this process, see fig. 2.3. The width of the Lorentzian, at half maximum, is given by $Dk^2$. Thus, one can measure the spin diffusion coefficient since the [factors] do not involve the frequency shift $\omega$. The total area under the Lorentzian, $\int d\omega\, S(k\omega)$, is given by $2\pi k_B T\chi$, and one could also measure the spin susceptibility if the [factors] were accurately known. Unfortunately, absolute intensity measurements are difficult, and the [factors] involve, moreover, magnetic form factors which are often not well known. We also note, finally, that eq. (2.25) holds only for small k and $\omega$ which is a region difficult to resolve by neutron scattering. Nevertheless, we meant here to demonstrate the principle of measuring correlation functions by scattering and for this purpose, our example will suffice.

## 2.4 The Static Susceptibility

In this section, eq. (2.23) will be demonstrated, to make good on at least one promise made above. The proof is simple. What we have to show is that

$$\beta^{-1}\chi = \lim_{k\to 0} \int d\vec{r}\, e^{-i\vec{k}.\vec{r}} \langle M(\vec{r})\, M(0)\rangle_{eq} . \tag{2.31}$$

No time argument is needed here; both operators $M(\vec{r})$ and $M(0)$ are taken at $t = 0$, i.e., in Schrödinger representation. We can, instead of the dipole moment per unit volume $M(\vec{r})$, introduce the total dipole moment operator by

$$M^{tot} = \int d\vec{r}\; M(\vec{r}) \equiv MV \tag{2.32}$$

so that we get, using translational invariance,

$$\chi = (\beta/V) \langle M^{tot}\, M^{tot}\rangle . \tag{2.33}$$

Now what do we mean by "magnetic susceptibility $\chi$"? Phenomenologically, we mean that if we measure the average magnetization (per unit volume) $\langle M\rangle^h$ in the presence of a constant magnetic field h, we find $\langle M\rangle^h = \chi h$ if the external field h is sufficiently small. (We use h here rather than the customary H to avoid confusion with the Hamiltonian.) More precisely, $\chi$ is experimentally defined by the equation

$$\chi = \frac{\partial \langle M\rangle^h}{\partial h}\Big|_{h=0} , \tag{2.34}$$

the derivative to be taken at constant volume and temperature. So we have to compute this derivative from statistical mechanics. Now in the presence of a magnetic field h, the total Hamiltonian is given by

$$\mathcal{H} = H - \int d\vec{r}\, M(\vec{r})\, h = H - M^{tot} h , \tag{2.35}$$

where H is the full many-body Hamiltonian of the isolated system. Therefore, the average magnetization is

$$<M>^h = \frac{1}{V}<M^{tot}>^h = \frac{1}{V}\frac{\mathrm{tr}\, e^{-\beta\mathcal{H}}M^{tot}}{\mathrm{tr}\, e^{-\beta\mathcal{H}}} = \frac{1}{V}\frac{\mathrm{tr}\, e^{-\beta[H-M^{tot}h]}M^{tot}}{\mathrm{tr}\, e^{-\beta[H-M^{tot}h]}} \qquad (2.36)$$

in a canonical ensemble. Take the derivative with respect to h (this can be done painlessly since the operator, $M^{tot}$, is conserved, and thus commutes with H) and then set h = 0. We get

$$\frac{\partial<M>^h}{\partial h}\Big|_{h=0} = (\beta/V)\Big[\frac{\mathrm{tr}\, e^{-\beta H}M^{tot}M^{tot}}{\mathrm{Tr}\, e^{-\beta H}} - \Big(\frac{\mathrm{tr}\, e^{-\beta H}M^{tot}}{\mathrm{tr}\, e^{-\beta H}}\Big)^2\Big]$$

or

$$\chi = (\beta/V)<(M^{tot}-<M^{tot}>_{eq})(M^{tot}-<M^{tot}>_{eq})>_{eq} \quad . \qquad (2.37)$$

Of course, $<M^{tot}> = 0$ in the absence of the field h so that we have derived eq. (2.23) or the equivalent eq. (2.27).

We have written $\chi$ in the form (2.37) to indicate clearly that the susceptibility is given by the fluctuations of the magnetization from its equilibrium value; static fluctuations, nota bene. This connection should not be too surprising. $\chi$ measures how easy it is to change the average magnetization by means of an external magnetic field. Clearly, this should be the easier the larger, or more probable, are spontaneous fluctuations of the magnetization from its average value. Similarly, we will see later that the response of the particle density to an increase of the pressure, i.e., the compressibility, is given by the spontaneous fluctuations of the total particle number. And the specific heat, describing the change of the energy with changing temperature, is given by the fluctuations of the energy operator, i.e., by $<H^2-<H>^2>$. And so on.

Sometimes things are not quite as simple as we have made them. We have calculated the limit $\lim_{k\to 0} S(k,t=0)$ by simply setting it equal to $S(k=0, t=0)$. Sometimes this is not allowed. For example, for the particle density correlation function at $t=0$, $S_{nn}(k)$, the value $S_{nn}(k=0)$ depends on whether a canonical or grand canonical ensemble is used but the limit $\lim_{k\to 0} S_{nn}(k)$ does not. Things also have to be reconsidered when the forces between particles are of long range. For dipole-dipole forces, the measured susceptibility depends on the shape of the sample, for instance. And finally, even if there are only short-ranged forces, there might be long-ranged correlations in some cases which makes the limit as $k\to 0$ subtle. We will worry about that when we get to it. In $He^3$ where we can neglect dipole-dipole forces, and there is no long range order, there are no problems.

## 2.5 Linear Dynamical Response

What we have just calculated can be called the linear thermodynamic or static response to a constant magnetic field. We will now analyze how the system responds dynamically to an external magnetic field $\delta h^{ext}(\vec{r},t)$ which varies in space and time in some prescribed fashion. Our reason for doing so is, of course, that this is the way most experiments are performed: You apply an external force of some sort to the system, look what happens, and infer from that the properties of the system itself. Some experiments do not fit this Procrustean bed, of course, but many do; you can't win them all. In particular, in discussing spin diffusion in section 2.1 we assumed that, at $t=0$, the system started out from some non-equilibrium state so that $\langle M(\vec{r},t)\rangle \neq 0$ initially. Such a non-equilibrium state can be produced, in the laboratory or on paper, by slowly turning on a magnetic field sometime in the distant past, and following the development of the system until $t = 0$ when a $\langle M(\vec{r},0)\rangle \neq 0$ will have resulted. If we then switch off the field, the further time dependence of $\langle M(\vec{r},t)\rangle_{\text{non-eq}}$ should coincide with our simple spin diffusion theory if that theory is correct.

Now in an external magnetic field $\delta h^{\text{ext}}(\vec{r},t)$ the Hamiltonian is explicitly time-dependent, and given by

$$\mathcal{H}(t) = H + \delta H^{\text{ext}}(t) = H - \int d\vec{r}\; M(\vec{r})\; \delta h^{\text{ext}}(\vec{r},t) \tag{2.38}$$

in Schrödinger representation where the operator $M(\vec{r})$ is time-independent. The time dependence is carried by the density matrix, or ensemble operator, $\rho(t)$ which describes the state of the system such that the average of $M(\vec{r})$, or any other operator, at time $t$ is given by

$$\langle M(\vec{r},t)\rangle = \operatorname{tr}\,\rho(t)\, M(\vec{r})\,, \qquad \text{with } \operatorname{tr}\,\rho(t) = 1\,, \tag{2.39}$$

where tr is the quantum-mechanical trace.

What follows is entirely parallel to the traditional derivation of the Heisenberg-Kramers formula for the dielectric constant given in elementary quantum mechanics texts. We have to solve the Schrödinger equation for the density matrix,

$$i\hbar\,\partial_t\,\rho(t) = [\mathcal{H}(t), \rho(t)] = [H, \rho(t)] + [\delta\mathcal{H}^{\text{ext}}(t), \rho(t)]\,, \tag{2.40a}$$

subject to the initial condition

$$\rho(t=-\infty) = \rho^0, \qquad \text{with } [H, \rho^0] = 0\,. \tag{2.40b}$$

The initial condition expresses the fact that the system is stationary before $\delta h^{\text{ext}}$ is turned on; we require, of course, that $\delta h^{\text{ext}} \to 0$ sufficiently rapidly as $t \to -\infty$. For the manipulations, it does not matter what $\rho^0$ is in detail but it is presumed known. Since the system usually starts out from thermal equilibrium, an appropriate choice for $\rho^0$ will be a canonical ensemble $\rho^0 = e^{-\beta H}/\operatorname{tr} e^{-\beta H}$, with

N fixed, or a grand canonical ensemble $\rho^0 \sim e^{-\beta(H-\mu N)}$, or some other stationary state.

Now all we want is the linear (in $\delta h^{ext}$) response. But to first order, (2.40) is easily "solved". Namely, $\rho(t) = \rho^0 + \delta\rho(t)$ with

$$\delta\rho(t) = \frac{1}{i\hbar} \int_{-\infty}^{t} d\tau\, e^{-iH(t-\tau)/\hbar} [\delta\mathcal{H}^{ext}(\tau), \rho^0] e^{iH(t-\tau)/\hbar} . \qquad (2.41)$$

From here, a few simple manipulations will convince you that the induced change in the average magnetization, $\delta\langle M(\vec{r},t)\rangle = \mathrm{tr}\, \rho(t)\, M(\vec{r}) - \mathrm{tr}\, \rho^0 M(\vec{r})$, can be written in the form

$$\delta\langle M(\vec{r},t)\rangle = \int_{-\infty}^{t} dt' \int d\vec{r}\,' \langle \frac{i}{\hbar} [M(\vec{r},t), M(\vec{r}\,',t')]\rangle_{eq}\, \delta h^{ext}(\vec{r}\,',t') \qquad (2.42)$$

where $[A,B] \equiv AB-BA$ is the commutator, and $\langle\rangle_{eq}$ indicates an equilibrium average, $\langle A\rangle_{eq} = \mathrm{tr}\, \rho^0 A$. Henceforth, we will drop the subscript "eq". In (2.42), $M(\vec{r},t)$ are the Heisenberg operators for the unperturbed system,

$$M(\vec{r},t) = e^{iHt/\hbar} M(\vec{r})\, e^{-iHt/\hbar} . \qquad (2.42a)$$

Eq. (2.42) is the fundamental result of linear response theory. It shows that the response function is the averaged commutator, rather than the correlation function $S(\vec{r},t)$ as one might have expected. Small matter; the two functions turn out to be essentially equivalent.

It is customary to define the response function by

$$\chi''_{MM}(\vec{r}t, \vec{r}\,'t') = \langle \frac{1}{2\hbar} [M(\vec{r},t), M(\vec{r}\,',t')]\rangle . \qquad (2.43)$$

Since a liquid in equilibrium is translationally invariant in space and time, $\chi''(\vec{r}t, \vec{r}\,'t') = \chi''(\vec{r}-\vec{r}\,', t-t')$, and we can define a Fourier transform by

$$\chi''(\vec{r}-\vec{r}\,', t-t') = \int \frac{d\omega}{2\pi} \int \frac{d\vec{k}}{(2\pi)^3}\, e^{-i\omega(t-t') + i\vec{k}(\vec{r}-\vec{r}\,')} \chi''(\vec{k}\omega) . \qquad (2.44)$$

Here and henceforth, all frequency integrals extend over the whole real axis: $\int d\omega \equiv \int_{-\infty}^{\infty} d\omega$. It is easy to show that $\chi''(\vec{k}\omega)$ is real, an odd function of $\omega$, and that it only depends on $k = |\vec{k}|$, since $\chi''$ is a commutator of hermitian operators, and the equilibrium state is invariant under time reversal, parity, and is spatially isotropic. We will also show that $\omega\chi''(k\omega) \geq 0$ in a stable system.

A useful function is the complex response function $\chi(kz)$, defined by

$$\chi(kz) = \int \frac{d\omega}{\pi} \frac{\chi''(k\omega)}{\omega - z} . \tag{2.45}$$

This is an analytic function of the complex frequency variable z as long as $\text{Im} z \neq 0$. On the real axis it has a branch cut. Of course, if z is in the upper half plane, $\chi(kz)$ is identical with the Laplace transform

$$\chi(kz) = 2i \int_0^{\infty} dt\, e^{izt} \chi''(k,t) , \qquad \text{for } \text{Im} z > 0 . \tag{2.46a}$$

On the other hand, if z is in the lower half plane, $\chi(kz)$ is determined from negative times,

$$\chi(kz) = (-2i) \int_{-\infty}^{0} dt\, e^{izt} \chi''(k,t), \text{ for } \text{Im} z < 0 . \tag{2.46b}$$

The physical response is given by the limit of $\chi(kz)$ as we approach the real frequency axis from above (from the "physical sheet"):

$$\chi(k\omega) \equiv \lim_{\epsilon \to 0} \chi(k, \omega + i\epsilon) = \chi'(k\omega) + i\chi''(k\omega) , \tag{2.47}$$

where

$$\chi'(k\omega) = P \int \frac{d\omega'}{\pi} \frac{\chi''(k\omega')}{\omega' - \omega} . \tag{2.47a}$$

P indicates the Cauchy principal value, and $\omega$ is real.

To demonstrate this statement, and make physical sense out of these definitions, let us go back to eq. (2.42). The response is a convolution in space and time. We decompose $\delta h^{ext}$ and similarly $\delta\langle M\rangle$ in Fourier amplitudes,

$$\delta h^{ext}(\vec{r},t) = \int \frac{d\omega}{2\pi} \int \frac{d\vec{k}}{(2\pi)^3} e^{-i\omega t + i\vec{k}\vec{r}} \delta h^{ext}(\vec{k},\omega), \tag{2.48}$$

and we obtain

$$\delta \langle M\rangle (\vec{k}\omega) = \chi(k\omega)\ \delta h^{ext}(\vec{k}\omega)\ . \tag{2.49}$$

$\chi(k\omega)$ is therefore the complex dynamical magnetic susceptibility as it is usually defined in electromagnetism. Its imaginary part, $\chi''(k\omega)$, must describe absorption, its real part, $\chi'(k\omega)$, dispersion just as the standard texts show. Eq. (2.47a), which connects the two, is a Kramers-Kronig relation, expressing causality which is implicit in (2.42). And our as yet unproven positivity statement, $\omega\chi''(k\omega) \geq 0$, expresses the fact that a dissipative many-body system takes more energy out of the external field than it gives back.

## 2.6 Hydrodynamics and Correlation Function

We are now in a position to give better than hit-and-run arguments for what we attempted to do in section 2.2--to establish the connection between the hydrodynamic diffusion equation and correlation functions. We approach the matter as we would in the laboratory. First a spatially varying magnetic field $\delta h(\vec{r})$ is slowly, adiabatically, turned on, to mechanically produce a state with non-zero magnetization. At $t = 0$, the field is switched off, and we can follow the relaxation of the induced magnetization as the system returns to equilibrium. Into the general relation (2.42) between force and response, we insert the external field

$$\delta h^{ext}(\vec{r},t) = \begin{cases} \delta h(\vec{r})\, e^{\epsilon t} & \text{for } t < 0 \\ 0 & \text{for } t > 0 \end{cases} \quad . \tag{2.50}$$

At $t = 0$, this force has induced the magnetization

$$\delta M(\vec{r},t=0) = 2i \int_0^{\infty} d\tau \int d\vec{r}\,' \chi''(\vec{r}-\vec{r}\,',\tau)\, e^{-\epsilon\tau}\, \delta h(\vec{r}\,'), \tag{2.51a}$$

or
$$\delta\langle M(\vec{k},t=0)\rangle = \chi(k)\, \delta h(\vec{k}) \tag{2.51b}$$

by spatial Fourier transformation, where

$$\chi(k) = \lim_{\epsilon\to 0} \chi(k,z)\big|_{z=i\epsilon} = \int \frac{d\omega}{\pi}\, \chi''(k\omega)/\omega\,. \tag{2.52}$$

For positive times, $t > 0$, (2.42) and (2.50) give

$$\delta\langle M(\vec{r},t)\rangle = 2i \int_{-\infty}^{0} d\tau \int d\vec{r}\,' \chi''(\vec{r}-\vec{r}\,',t-\tau)\, e^{\epsilon\tau}\, \delta h(\vec{r}\,'), \tag{2.53a}$$

whose Laplace transform, defined as in (2.8b), is

$$\delta\langle M(\vec{k}z)\rangle = \int \frac{d\omega}{\pi i}\, \frac{\chi''(k\omega)}{\omega(\omega-z)}\, \delta h(\vec{k})\,. \tag{2.53b}$$

And if we use (2.51b) to eliminate the external field, we obtain

$$\delta\langle M(\vec{k}z)\rangle = (1/iz)[\chi(kz)\, \chi^{-1}(k) - 1]\, \delta\langle M(\vec{k},t=0)\rangle\,. \tag{2.54}$$

We have inserted the response function (2.45) in an obvious way.

The fundamental result (2.54) is an exact expression. It is, moreover, of the same general form as the hydrodynamic result (2.9), and it describes the same process. Therefore, for small k and z where hydrodynamics is correct, we can compare (2.54) and (2.9), and obtain

$$\chi(kz) = \frac{iDk^2}{z + iDk^2}\, \chi(k) \tag{2.55a}$$

and in particular, by setting $z = \omega + i\epsilon$ and taking the imaginary part,

$$\chi''(k\omega) = \frac{Dk^2\omega}{\omega^2 + (Dk^2)^2}\, \chi(k)\,. \tag{2.55b}$$

What we have obtained is the correct limiting expression for the magnetic response function. If everything is consistent, $\chi(k)$ must be the static susceptibility at small k since then, (2.51b) states that $[\chi \equiv \chi(k\to 0)]\ \delta\langle M(\vec{k})\rangle = \chi\ \delta h(\vec{k})$ or $\delta\langle M(\vec{r})\rangle = \chi\,\delta h(\vec{r})$ which is how the susceptibility is defined. From (2.52) we can therefore write the thermodynamic sum rule as

$$\chi = \frac{\partial M}{\partial h}\Big|_{h=0} = \lim_{k\to 0} \int \frac{d\omega}{\pi}\, \chi''(k\omega)/\omega\ . \tag{2.56}$$

And from (2.55b) we find

$$D\chi = \lim_{\omega\to 0}\ [\lim_{k\to 0} \frac{\omega}{k^2}\, \chi''(k\omega)], \tag{2.57}$$

which is, again, a Kubo-type relation for the transport coefficient.

Don't miss the point of these results: $\chi''(k\omega)$ is, by eqs. (2.43, 44), a mathematically well-defined object. It may not be easy to calculate from a microscopic theory but at least one knows exactly what one should, in principle and in approximation, calculate. And indeed, there are several powerful techniques available for such a calculation. Eqs. (2.56, 57) show how from $\chi''(k\omega)$ to obtain the macroscopic parameters $\chi$ and D. Undoubtedly, these expressions are now on a much firmer basis than what we have, haphazardly, derived in section 2.2. We shall now show that the respective results are completely equivalent.

## 2.7 The Fluctuation-Dissipation Theorem

In order to see whether or not we have goofed in section 2.2, i.e., whether or not (2.57) and (2.28) are consistent, for example, we have to establish a connection between the functions $S(k\omega)$ and $\chi''(k\omega)$. This connection is given by the celebrated fluctuation-dissipation theorem, discovered by H. Nyquist in 1928 as a relation between noise and dissipation in electric resistors. For our case, this theorem (Callen and Welton 1951) states that

$$\chi''(k\omega) = (1/2\hbar)(1-e^{-\hbar\omega\beta})\, S(k\omega) \,. \tag{2.58}$$

Accepting the theorem for the moment, we see that the Kubo expressions (2.57) and (2.28) are indeed equivalent. With the thermodynamic sum rules (2.56) and (2.27), things are just a little more subtle. If we use (2.58) in (2.56) we get

$$\lim_{k\to 0} \int \frac{d\omega}{2\pi} \frac{1}{\hbar\omega}(1-e^{-\hbar\omega\beta})S(k\omega) \overset{?}{=} \chi = \lim_{k\to 0} \beta \int \frac{d\omega}{2\pi} S(k\omega) \,. \tag{2.59}$$

After our derivations in sections 2.4-6, we cannot very well doubt either (2.27) or (2.56). Now notice that if it were true that $\lim_{k\to 0} S(k\omega) \sim \delta(\omega)$ we would be all set since $\omega^n \delta(\omega) = 0$ for $n > 0$, and therefore $(1/\hbar\omega)(1-e^{-\beta\hbar\omega})\,\delta(\omega) = \beta\,\delta(\omega)$ so that the two sides in (2.59) would indeed be identical. Now is $S(k=0,\omega) \sim \delta(\omega)$? Well,

$$S(k=0,t) = \int d\vec{r} \langle M(\vec{r},t)M(0,0)\rangle = \langle M^{tot}(t)\; M(0,0)\rangle \tag{2.60}$$

as in section 2.4. But $M^{tot}(t)$, the total dipole moment, is conserved, vide eq. (2.5). Therefore, $S(k=0,t)$ is in fact independent of time, and its Fourier transform

$$S(k=0,\omega) = \text{const} \int_{-\infty}^{\infty} dt\, e^{i\omega t} = \text{const}\;\; 2\pi\,\delta(\omega) \,. \tag{2.61}$$

So indeed, the ? in (2.59) is unnecessary, and the two expressions (2.56) and (2.27) for the susceptibility are completely equivalent. We might note, parenthetically, that if $M^{tot}$ were not conserved we would have to revisit section 2.4 in fact since we assumed there that $[M^{tot}, H] = 0$. No such assumption was implicit in the derivation of (2.56), and in this case, (2.56) is the correct equation as one can also show from (2.36) and (2.58).

Since our derivation of $\chi''$ in section 2.6 is much more convincing than that of S in section 2.2, let us use the former to correct the latter. Instead of (2.25), we should have

$$S(k\omega) = \frac{\hbar\omega\beta}{1-e^{-\hbar\omega\beta}} \times \frac{Dk^2}{\omega^2+(Dk^2)^2}\, \frac{2}{\beta}\chi \ . \tag{2.62}$$

Thus $S(k\omega)$ is not quite symmetric in $\omega$. At positive frequency $\omega > 0$ it is a little stronger than at negative frequency $\omega < 0$. Indeed, since $\omega\, \chi''(k\omega)$ is always even in $\omega$, it is generally true that

$$S(k,-\omega) = e^{-\hbar\omega\beta}\, S(k,\omega) \ . \tag{2.63}$$

This result makes eminent sense in the light of what we said above about neutron scattering. According to (2.30), $\epsilon_f = \epsilon_i - \hbar\omega$. Positive frequency means the neutron has lost energy to the system (by creating an excitation of energy $\hbar\omega$) while negative frequency describes a process in which the neutron has picked up energy from the system (by destroying an excitation). Of course, to destroy an excitation you must first have one, and their relative abundance is given by $e^{-\hbar\omega\beta}$. "Die Nürnberger hängen keinen, sie hätten ihn denn." (Schinderhannes)

The dissymmetry of the scattering intensity, $\sim S(k\omega)$, is only pronounced at low temperatures, $k_B T \lesssim \hbar\omega$. It is absent classically. It is an important effect in Raman spectra in solids which probe optical phonons of relatively large energy. For hydrodynamic modes, the frequency is so small that the prefactor in Eq. (2.62) is, effectively, unity.

And now we had better prove (2.58). Since we will use this theorem often, and since its proof is simple, let us consider correlations between the arbitrary observables $A_i(t)$ and $A_j(t)$, defining

$$S_{ij}(t) = <A_i(t)A_j(0)> - <A_i(t)><A_j(0)>, \tag{2.64a}$$

$$\chi''_{ij}(t) = <\frac{1}{2\hbar}[A_i(t), A_j(0)]> . \tag{2.64b}$$

We have subtracted the equilibrium averages in (2.64a) so that $S_{ij}(t) \to 0$ as $t \to \infty$, and thus its Fourier transform $S_{ij}(\omega)$ presumably exists. Of course, $<A_i(t)>$ is independent of time.

Let us perform the average over a canonical ensemble $\rho = e^{-\beta H}/\mathrm{tr}\, e^{-\beta H}$. Because the operator, $e^{-\beta H}$, effects a time translation by the imaginary time $\tau = i\hbar\beta$, see (2.42a),

$$\begin{aligned} \mathrm{tr}\, e^{-\beta H} A_i(t) A_j(0) &= \mathrm{tr}\, A_i(t+i\hbar\beta)\, e^{-\beta H} A_j(0) \\ &= \mathrm{tr}\, e^{-\beta H} A_j(0) A_i(t+i\hbar\beta) \ , \end{aligned} \tag{2.65}$$

where we have used the cyclic invariance of the trace, tr AB = tr BA. Because of time translation invariance, $<A_i(t)A_j(0)> = <A_i(0)A_j(-t)>$, we therefore obtain

$$S_{ji}(-t) = S_{ij}(t - i\hbar\beta) = e^{-i\beta\hbar\partial_t} S_{ij}(t) . \tag{2.66}$$

And thus easily from (2.64)

$$2\hbar\, \chi''_{ij}(t) = S_{ij}(t) - S_{ji}(-t) = [1 - e^{-i\beta\hbar\partial_t}] S_{ij}(t) , \tag{2.67a}$$

whose Fourier transform, $\partial_t \to -i\omega$, is

$$2\hbar\, \chi''_{ij}(\omega) = (1 - e^{-\hbar\omega\beta})\, S_{ij}(\omega) . \tag{2.67b}$$

This is essentially eq. (2.58). We only have to recognize that $\vec{r}$, the argument of $M(\vec{r})$, is an operator label like i above. For the magnetization correlation functions we get therefore the equation

$$2\hbar\, \chi''_{MM}(\vec{r}-\vec{r}\,',\omega) = (1-e^{-\hbar\omega\beta})\, S_{MM}(\vec{r}-\vec{r}\,',\omega)\,, \tag{2.68}$$

whose spatial Fourier transform, eq. (2.58), we set out to prove. We will have more comments about this theorem in the next chapter.

## 2.8 Positivity of $\omega\chi''(k\omega)$

We pointed out after eq. (2.44) that

$$\omega\chi''(k\omega) \geq 0 \quad \text{for all } k \text{ and } \omega\ . \tag{2.69}$$

The significance of this property is already clear from the foregoing. From (2.56) we see that it implies that the spin magnetic susceptibility $\chi$ is positive which is a necessary condition for the thermodynamic stability of the system. From (2.57) we see that it also implies that the spin diffusion coefficient D is positive which is necessary for the system to be dynamically stable, see eq. (2.10). And finally, since (2.69) is tantamount, because of the fluctuation dissipation theorem, to the assertion that $S(k\omega) \geq 0$, it is necessary for the interpretation of the function $S(k\omega)$ as a spectral density of fluctuations.

(2.69) is quickly proved. We take again an arbitrary set $\{A_i(t)\}$ of observables, and consider $S_{ij}(t,t')$ in the form

$$S_{ij}(t,t') = \langle (A_i(t) - \langle A_i(t)\rangle)(A_j(t') - \langle A_j(t')\rangle)\rangle\ . \tag{2.70}$$

Multiplying with any set of functions $a_i(t)$ and integrating over some large time T, we obtain

$$\sum_{i,j} (2T)^{-1} \int_{-T}^{T} dt \int_{-T}^{T} dt'\, a_i{}^*(t)\, S_{ij}(t,t')\, a_j(t') = \langle A^* A\rangle \geq 0\,, \tag{2.71}$$

where

$$A = \sum_i (2T)^{-1/2} \int_{-T}^{T} dt\, [A_i(t) - \langle A_i(t) \rangle] a_i(t) . \tag{2.71a}$$

If we choose, in particular, $a_i(t) = a_i\, e^{-i\omega t}$ and remember that $S_{ij}(t,t') = S_{ij}(t-t')$, then in the limit of large T we find

$$\langle A^* A \rangle = \sum_{i,j} a_i^* \, S_{ij}(\omega)\, a_j \geq 0 \tag{2.72a}$$

which is equivalent, because of (2.67b), to

$$\sum_{i,j} a_i^* \, \omega \chi''_{ij}(\omega)\, a_j \geq 0 . \tag{2.72b}$$

Again, the operator label i will, in cases considered here, include the continuous variable $\vec{r}$, and $\sum_i$ is replaced by $\int d\vec{r}$. The choice $a_i \to a(\vec{r}) = e^{i\vec{k}\cdot\vec{r}}$ and translational invariance in space yields

$$S_{MM}(k\omega) \geq 0 \quad \text{or} \quad \omega \chi''_{MM}(k\omega) \geq 0 . \tag{2.73}$$

This demonstration is, of course, a bit relaxed. It can be fancied up considerably, but too much mathematical rigor may not be in place when one deals with $10^{23}$ particles. A more physical dynamical proof,which elucidates the fundamental connection between (2.73) and the dissipative property of many-particle systems, is given in section 3.3.

## 2.9 Sum Rules

From (2.43) one sees that

$$\left(i\frac{\partial}{\partial t}\right)^n \chi''(\vec{r}t, \vec{r}'t') = \langle \frac{1}{2\hbar} [\left(i\frac{\partial}{\partial t}\right)^n M(\vec{r},t), M(\vec{r}',t')] \rangle . \tag{2.74}$$

Taken at equal times $t = t'$, this means that

$$\int \frac{d\omega}{\pi} \omega^n \chi''(k\omega) = \int d(\vec{r}-\vec{r}')\, e^{-i\vec{k}(\vec{r}-\vec{r}')} \langle \frac{1}{\hbar} [i^n M^{(n)}(\vec{r},t), M(\vec{r}',t)] \rangle , \tag{2.75}$$

where

$$i^n M^{(n)}(\vec{r},t) \equiv (i\frac{\partial}{\partial t})^n M(\vec{r},t) = (\frac{1}{\hbar})^n [\ldots[M(\vec{r},t),H]\ldots,H] . \quad (2.75a)$$

Thus the right hand side of (2.75) contains a sequence of equal time commutators which can in principle, and in some cases in fact, be exactly calculated. The simplest of these is the sum rule for $n = 1$. (Clearly, since $\chi''(k\omega)$ is an odd function of $\omega$ all sum rules for even n vanish.) Namely,

$$i\frac{\partial}{\partial t}\chi''(\vec{r}-\vec{r}', t-t') = \frac{-i}{2\hbar}\vec{\nabla}\cdot\langle[\vec{j}^M(\vec{r}t), M(\vec{r}'t')]\rangle \quad (2.76)$$

because of the conservation law (2.3). At equal times, the commutator is easily evaluated. From the explicit expressions for the magnetization operator and its current, Eqs. (2.2,4), one finds

$$[\vec{j}^M(\vec{r}), M(\vec{r}')] = \frac{\mu^2}{m} i\hbar\vec{\nabla}'[n(\vec{r})\,\delta(\vec{r}-\vec{r}')], \quad (2.77)$$

which is "a very disguised version of the fundamental statement that the commutator of the position and the momentum is $i\hbar$". $n(\vec{r}) = \sum_a \delta(\vec{r}-\vec{r}^a)$ is the particle density (operator). Thus we find the sum rule

$$\int\frac{d\omega}{\pi}\,\omega\,\chi''(k\omega) = \frac{n}{m}\mu^2k^2 . \quad (2.78)$$

This is the spin analog of the famous f-sum rule. By contrast to the thermodynamic sum rule which we found earlier, eq. (2.56), and which holds only as $k \to 0$, (2.78) is exact for all k. Further sum rules can be derived, but with rapidly increasing labor.

The sum rules provide the coefficients for an expansion of $\chi(kz)$ for large z. From its definition in (2.45), we see that for large z

$$\chi(kz) = -\sum_{n=1}^{\infty}\frac{\langle\omega^{(n)}(k)\rangle}{z^n}\chi(k) , \quad (2.79a)$$

where

$$\langle\omega^{(n)}(k)\rangle = \frac{\int \frac{d\omega}{\pi}\, \omega^n \chi''(k\omega)/\omega}{\int \frac{d\omega}{\pi}\, \chi''(k\omega)/\omega} \; . \tag{2.79b}$$

From its derivation which expands $[1-\omega/z]^{-1} = 1 + (\omega/z) + (\omega/z)^2 + \ldots$ it is clear that this expansion can only be asymptotic. It is valid when $|z|$ is "large compared to all frequencies in the system" which means, all frequencies $\omega$ for which $\chi''(k\omega)$ is not effectively zero.

We can also relate the sum rules to a Taylor expansion in time $t$ as should be apparent from (2.74,75). (2.79) is equivalent to

$$\chi''(k,t) = \frac{1}{2} \sum_{n=0}^{\infty} \frac{(-it)^n}{n!} \langle\omega^{(n+1)}(k)\rangle\, \chi(k) \, , \tag{2.80}$$

which makes it clear that the high-frequency expansion is equivalent to a short time expansion.

## 2.10 Relaxation Time Approximation

An interesting feature of the sum rules is their very existence. There is no reason why the thermodynamic average of the multiple commutators in (2.75) should not exist, for all n, and in many cases this can in fact be rigorously proved. This means, then, that $\chi''(k\omega)$ has to fall off sufficiently rapidly at large $\omega$ so that all of its moments are finite.

Evidently, the hydrodynamic approximation for $\chi''(k\omega)$, eq. (2.55b), does not have this property. Indeed, while it exhausts the "thermodynamic" sum rule (2.56), it fails to satisfy even the first high-frequency sum rule (2.78). The Lorentzian dies off too slowly in the wings.

This situation can be remedied in a simple fashion. Since, so far, our explicit expression (2.55b) for $\chi''$ is not much more than a fancy way of writing the diffusion equation, let us now try to improve the theory by improving the phenomenological equation which it represents.

The constitutive equation

$$\langle \vec{j}^M(\vec{r},t) \rangle = -D \vec{\nabla} \langle M(\vec{r},t) \rangle \tag{2.6}$$

from which we started, implies that the current follows changes in the magnetization instantaneously. This is true when $\langle M \rangle$ varies extremely slowly in time, i.e., for small frequencies. However, when $\langle M(\vec{r},t) \rangle$ varies more rapidly, i.e., for larger frequencies, the current cannot keep up, and there must be a time lag between M and $\vec{j}^M$. Instead of (2.6), let us therefore try

$$\langle \vec{j}^M(\vec{r},t) \rangle = -\int_0^t dt' \, D(t-t') \, \vec{\nabla} \langle M(\vec{r},t') \rangle . \tag{2.81}$$

This equation incorporates causality: the current must follow the magnetization fluctuation which causes it. We have fixed the lower limit at $t = 0$ since for negative times, when the initial disturbance is adiabatically created, the current surely vanishes. Indeed, it is a good exercise to prove that the f-sum rule cannot be satisfied as long as $\langle j^M \rangle \neq 0$ at $t = 0$. That is what is wrong with eq. (2.6).

The function $D(t-t')$, appropriately called a memory function, incorporates all the complicated rapid processes which are set in motion by the initial disturbance. Let us assume, for simplicity, that all these processes can be described by a single relaxation time $\tau$, i.e., that

$$D(t-t') = (D/\tau) \, e^{-(t-t')/\tau} . \tag{2.82}$$

Putting it all into the continuity equation (2.3), we solve as before:

$$\langle M(\vec{k}z) \rangle = \frac{i}{z + ik^2 D/(1-iz\tau)} \langle M(\vec{k},t=0) \rangle . \tag{2.83a}$$

And just as before, we obtain from this the response function

$$\chi(kz) = \frac{ik^2 D/(1-iz\tau)}{z + ik^2 D/(1-iz\tau)}\ \chi \qquad (2.83b)$$

and its absorptive part

$$\chi''(k\omega) = \frac{\omega k^2 D}{\omega^2 + D^2(k^2 - \omega^2 \tau/D)^2}\ \chi\ . \qquad (2.83c)$$

Note that with $D(t-t')$ given by (2.82), the current can be obtained from

$$(\tau\partial_t + 1)\ \langle \vec{j}^M(\vec{r},t)\rangle = -D\ \vec{\nabla}\langle M(\vec{r},t)\rangle\ , \qquad (2.84a)$$

so that eqs. (2.83) are equivalent to the phenomenological equation of motion

$$[\partial_t^2 + (1/\tau)(\partial_t - D\nabla^2)]\ \langle M(\vec{r},t)\rangle = 0\ . \qquad (2.84b)$$

The last few equations are the direct analog of the description employed by Drude and Maxwell for the dielectric response.

What have we gained? For small frequencies, $(\omega/Dk^2)\omega\tau << 1$, (2.83c) is our old Lorentzian of the hydrodynamic theory. However, $\chi''(k\omega)$ now falls off faster in the wings, and sufficiently fast that the first moment is finite:

$$\langle \omega^{(2)}(k)\rangle = \chi^{-1} \int \frac{d\omega}{\pi}\ \omega\ \chi''(k\omega) = k^2\ D/\tau\ . \qquad (2.85)$$

Eqs. (2.83) will therefore give an interpolation formula which is correct at both long and short times, or small and large frequencies. To this end, we require that the exact f-sum rule (2.78) be fulfilled, and we obtain an expression for the spin diffusion coefficient, namely

$$D = \frac{n\mu^2}{m\chi}\ \tau\ . \qquad (2.86)$$

Thus we have done our first sum rule calculation. (For similar ideas, see de Gennes 1959, Mori and Kawasaki 1962, Bennett and Martin 1965.) As a calculation of D, it seems barely worth the effort at first since it just replaces one parameter, D, by another, $\tau$. However, $\tau$ is much more closely related to the microscopic dynamics. It is clearly of the nature of a microscopic collision time, and its numerical value can be estimated from atomic collision cross sections in a gas, for example, or even more crudely, in a classical liquid, assessed to be roughly given by $\tau_c$ of eq. (2.11). Semiquantitatively, (2.86) is a very useful result.

In $He^3$ which is a degenerate Fermi liquid at low temperatures, scattering is sharply reduced because of the Pauli principle, and $\tau \sim 1/T^2$. The spin susceptibility reduces to the well-known Pauli susceptibility, and is independent of temperature. We therefore expect from eq. (2.86) that $D \sim T^{-2}$ at low temperatures (Martin 1968), and this is experimentally verified.

Eq. (2.86) also makes an interesting prediction about the behavior of the diffusion constant near a critical point, say in a Heisenberg paramagnet to which our considerations apply with no essential change. As the ferromagnetic transition is approached, the spin susceptibility $\chi(T)$ increases without bound while there is no reason to expect that the microscopic decay time $\tau$ is much affected. We therefore expect that the diffusion constant goes to zero. This critical slowing down (see eq. (2.10)) is a general phenomenon. It is coupled, as eq. (2.83c) shows, with a tremendous increase in the amplitude of spontaneous fluctuations which leads to strong scattering. Near the liquid-gas transition in normal fluids, this accounts for the entirely analogous phenomenon of critical opalescence (for critical phenomena, see Stanley 1971).

While the result for $\chi''(k\omega)$ is somewhat improved from eqs. (2.55) to (2.83), all higher frequency moments $\langle \omega^{(n)}(k) \rangle$, $n \geq 4$, still diverge. This can now be easily remedied. Remember that the relaxation time ansatz (2.82) was quite ad hoc. For arbitrary memory D(t), we would obtain

$$\chi(kz) = \frac{ik^2 D(z)}{z + ik^2 D(z)} \chi , \tag{2.87a}$$

which makes it plain that what the "constitutive equation with memory" (2.81) does is introduce frequency dependence into the transport coefficient. Here

$$D(z) = \int_0^\infty e^{izt} D(t) = \int \frac{d\omega}{2\pi i} \frac{D'(\omega)}{\omega - z} \qquad (\text{Im}\, z > 0), \tag{2.87b}$$

where $D(t)$ and $D'(\omega) = \int_{-\infty}^{\infty} dt\, e^{i\omega t} D(t)$ can always be chosen as real and even functions of their argument. Also

$$\chi''(k\omega) = \frac{\omega k^2 D'(\omega)/2}{[\omega + k^2 P \int \frac{d\omega'}{2\pi} \frac{D'(\omega')}{\omega' - \omega}]^2 + [k^2 D'(\omega)/2]^2} . \tag{2.87c}$$

The function $D'(\omega)$ must therefore be positive.

From (2.87c) or more easily from (2.87a) it is easy to see that the first n moments of $\chi''(k\omega)$ will be finite if we choose a $D'(\omega)$ whose first n moments are finite. In particular, the f-sum rule is fulfilled if

$$\chi^{-1} \int \frac{d\omega}{\pi} \omega\, \chi''(k\omega) = k^2 \int \frac{d\omega}{\pi} D'(\omega) \tag{2.88a}$$

or

$$D(t=0) = \int \frac{d\omega}{2\pi} D'(\omega) = \frac{n\mu^2}{m\chi} . \tag{2.88b}$$

All this sum rule does, therefore, is to prescribe the value of the memory function $D(t)$ at $t = 0$. This may be a helpful reminder for the occasional reader who should be overly impressed by these general results. The complexity of a many-body problem does not disappear quite so easily. Since it leads to finite moments of any order, one often prefers a Gaussian memory

$$D(t) = \frac{n\mu^2}{m\chi} e^{-\pi(t/2\tau)^2} \quad \text{or} \quad D'(\omega) = \frac{n\mu^2}{m\chi} (2\tau)\, e^{-(\omega\tau)^2/\pi} , \tag{2.89}$$

but many other functions are equally good candidates. (The Gaussian has also some appeal for stochastic reasons.)

For hydrodynamically long times, one probes only the integrated effect of the rapid processes contained in D(t). For any D(t), the hydrodynamic diffusion coefficient is given by

$$D = \lim_{z \to 0} D(z) = \int_0^\infty dt\, D(t) = \frac{1}{2} \lim_{\omega \to 0} D'(\omega) \tag{2.90}$$

which follows, for example, from (2.87c) and the Kubo relation (2.57). This also gives a more precise meaning to the relaxation time of eq. (2.86). $\tau$ is the average "memory time",

$$\tau = [1/D(t=0)] \int_0^\infty dt\, D(t) \quad . \tag{2.91}$$

There is one simple fact about the time dependence of D(t) which we know for certain, namely that

$$|D(t)| \leq D(t=0) , \tag{2.92}$$

which is a simple consequence of $D'(\omega) \geq 0$. Loss of memory as time passes, is a general phenomenon in nature.

## 2.11 Dispersion Relation Representation

We have said almost everything that can be said without doing more detailed microscopic calculations. However, the linear constitutive relation (2.81) can be generalized a little further. We have first relaxed (2.6) to allow for memory effects. If we in addition allow for a spatially non-local connection between current and magnetization (or gradient of the latter, rather; a spatially constant magnetization will evidently not give rise to a current) we obtain

$$\langle \vec{j}^M(\vec{r},t)\rangle = -\int_0^t dt' \int d\vec{r}' D(\vec{r}-\vec{r}', t-t')\, \vec{\nabla}' \langle M(\vec{r}',t')\rangle . \tag{2.93}$$

This leads to

$$\chi(kz) = \frac{ik^2 D(kz)}{z+ik^2 D(kz)} \chi(k) , \tag{2.94}$$

where

$$D'(k\omega) = \int_{-\infty}^{\infty} dt \int d\vec{r}\, e^{i\omega t - i\vec{k}\cdot\vec{r}} D(\vec{r},t), \tag{2.95a}$$

$$D(kz) = \int \frac{d\omega}{2\pi i} \frac{D'(k\omega)}{\omega - z} . \tag{2.95b}$$

$D'(\vec{k}\omega)$ can again be chosen real, even in $\omega$, and a function of $|\vec{k}|$ only. Its Hilbert transform, $D(kz)$, is closely related to what, in field theory, would be called the self-energy. From (2.94), $\chi''(k\omega)$ is given by

$$\chi''(k\omega) = \frac{\omega k^2 D'(k\omega)/2}{[\omega + k^2 \omega P \int \frac{d\omega}{2\pi} \frac{D'(k\omega)}{\omega'^2 - \omega^2}]^2 + [k^2 D'(k\omega)/2]^2} . \tag{2.96}$$

We will now show that these formulae, which replace the hydrodynamic transport coefficient by a k- and z-dependent object, are in fact completely general. Of course, since (2.94) simply defines the function $D(kz)$ in terms of $\chi(kz)$ by

$$D(kz) = \frac{iz}{k^2} [1 - \frac{\chi(k)}{\chi(kz)}]^{-1} = \frac{\chi(kz)/k^2}{(1/iz)[\chi(kz) - \chi(k)]} , \tag{2.97}$$

the only thing that has to be proved is that this expression is indeed an analytic function of $z$, for $\text{Im} z \neq 0$, as implied by the representation (2.95b). Now we know that $\chi(kz)$ is analytic everywhere, except on the real axis, of course. All that could go wrong, therefore, is that the denominator,

$$\frac{1}{iz}\,[\chi(kz) - \chi(k)] = \int \frac{d\omega}{\pi i}\,\frac{\chi''(k\omega)}{\omega(\omega - z)} \overset{?}{=} 0\,, \tag{2.98}$$

would vanish for some z off the real axis. However, for $z = x + iy$

$$\mathrm{Re}\int \frac{d\omega}{\pi i}\,\frac{\chi''(k\omega)}{\omega(\omega - z)} = \int \frac{d\omega}{\pi}\,\frac{y}{(\omega - x)^2 + y^2}\,\frac{\chi''(k\omega)}{\omega} \neq 0 \quad \text{if } y \neq 0 \tag{2.99}$$

cannot vanish anywhere ($y \neq 0$) since $\chi''(k\omega)/\omega$ is non-negative, and thus (2.98) cannot vanish for $\mathrm{Im}\,z \neq 0$. Hence, $D(kz)$ as defined by (2.94, 97) is analytic, and (2.95b) is a proper representation.

There is a point to being suspicious of exact results of complete generality when they pertain to a complicated many-body system. If they are so general, how can they be practically useful? What we have proved is an exact dispersion relation (2.96) for $\chi''(k\omega)$. That is not awfully much since it just introduces another unknown function $D'(k\omega)$ to describe $\chi''(k\omega)$. And indeed, many different kinds of dispersion relations can easily be proved (Kadanoff and Martin 1963). What makes the present one nonetheless valuable is that in the important region of small k and ω where $\chi''(k\omega)$ has a complicated analytic structure, $D'(k\omega)$ is presumably well-behaved, with its value at $k = 0$ and $\omega = 0$ given by

$$\frac{1}{2}D'(0,0) = D\,, \tag{2.100}$$

the spin diffusion coefficient. Thus, for small k and ω, $D'(k\omega)$ is a simpler object than $\chi''(k\omega)$, and approximations to $D'(k\omega)$ have a better chance of success. All of our previous results, such as (2.55b) or (2.83c), represent such approximations.

It is instructive to recover the hydrodynamic result (2.55) from the general representation (2.94). This equation, or equivalently

$$\frac{1}{iz\beta}[\chi(kz)-\chi(k)] \equiv C(kz) = \frac{1}{z+ik^2 D(kz)}\, i\beta^{-1}\chi(k), \tag{2.101}$$

does incorporate one important feature of the dynamics: the conservation law (2.3) is the origin of the factor $k^2$ in (2.101), and thus at $k=0$, $C(kz)$ has a pole at $z=0$. As k becomes finite, the pole migrates into the lower half z-plane, to a point $z=z^0(k)$ which is the solution of

$$z^0 + ik^2 D(k,z^0) = 0 . \tag{2.102}$$

Note that in (2.102) $D(kz)$ is not the function (2.95b) for $\mathrm{Im}\,z < 0$; instead, it is that function, analytically continued from the upper half z-plane through the branch cut on the real axis, onto a second Riemann sheet. For z not far from the solution $z^0(k)$ of (2.102), we can expand the slowly varying function $D(kz)$, and obtain to first order

$$z + ik^2 D(kz) = (z-z^0(k))\, Z^{-1}(k) ,$$

where

$$Z^{-1}(k) = 1 + ik^2 \frac{\partial D(kz^0)}{\partial z^0} . \tag{2.103}$$

Near $z^0(k)$ therefore

$$C(kz) = \frac{Z(k)}{z-z^0(k)}\, i\beta^{-1}\chi(k) . \tag{2.104}$$

The constant $Z(k)$ is the pole strength, akin to the wave function renormalization constant in field theory. And to order $k^2$,

$$z^0(k) = ik^2 D(0,0) \text{ and } Z(k) = 1 , \tag{2.105}$$

which is the hydrodynamic result, and gives precise meaning to the claim that that result, eq. (2.55), is a rigorous asymptotic expression for the correlation function.

Let us, finally, mention one more approximation which is, by contrast, a high-frequency or short-time approximation. It is obtained by replacing D(k,t) by its value at t = 0, D(k). Alternatively, we might say that we replace D(kz) by the first term in its high-frequency expansion, $D(kz) = (i/z)D(k) + D(z^{-3})$. The constant D(k) is again determined by the f-sum rule, and given by

$$D(k) = \int \frac{d\omega}{2\pi} D'(k\omega) = \chi^{-1}(k)k^{-2} \int \frac{d\omega}{\pi} \omega \chi''(k\omega) = \frac{n\mu^2}{m\,\chi(k)}\,, \qquad (2.106)$$

so that D(k) is real and positive. If we put this into (2.94) we obtain

$$\chi(kz) = \frac{-k^2 D(k)}{z^2 - k^2 D(k)}\,\chi(k). \qquad (2.107)$$

At high frequency therefore where this approximation is valid, we find reactive, rather than diffusive, behavior. There is a sound-like mode, with a propagation velocity at small k which is given by

$$\omega = ck; \; c^2 = \lim_{k\to 0} D(k) = \lim_{k\to 0} \frac{\langle \omega^{(2)}(k) \rangle}{k^2} = \frac{n\mu^2}{m\,\chi}\,. \qquad (2.108)$$

This feature, reactive behavior at high frequencies, is a very general phenomenon. For a second example, a normal liquid like water is viscous at low frequency; it exerts no resistance to slowly varying shear forces. If you dive into the ocean softly, it will do you no harm. At high frequencies, however, the liquid becomes elastic: if you fall into the sea from a high-flying plane your last impression might be that the water must have been frozen.

This concludes, for the moment, what we have to say about our simple example--spin diffusion. Clearly, most of what we said applies to other diffusion processes as well. At least for the spectrum at low frequency and wave vector, the main input, really, was the fact that the magnetization is conserved. Our considerations apply literally to the isotropic Heisenberg paramagnet (Bennett and Martin 1965) which can also be described by the simple diffusion equation.

(For large wavelength, the lattice structure does not matter.) Of course, while in the liquid case treated here, diffusion comes about because the spins are carried along by the atoms in the liquid, in the Heisenberg paramagnet the atoms stay put on fixed lattice sites. The diffusion of magnetization takes place via spin-flip interaction between neighbors.

An occasional reader might be concerned about the fact that we have treated the magnetization as a scalar quantity. $\vec{M}(\vec{r},t)$ is a (pseudo-) vector of course, and $j^M_{ij}$ a tensor. However, the vector character of $\vec{M}$ is quite irrelevant. The point is that the Hamiltonian H is separately invariant under rotations in spin space. Because of this property, which is intimately related to the existence of the conservation law (2.5), of course, one can treat $\vec{M}(\vec{r},t)$ which is a vector in spin space, as a scalar in real space. If you wish, you can take $M_x$ and $j^M_{xj}$ everywhere; the y- and z-components obey precisely the same equations, with the same diffusion constant D etc.

Many other processes fit our description as well. An example is the Brownian motion of a heavy particle immersed in a fluid of light particles. In chapter 6, we will consider this process in a slightly different formulation. In this case, the mass ratio m/M plays the role of the wave vector k above.

Transverse momentum transport in a normal fluid also follows a hydrodynamic diffusion law, and everything we said above applies, mutatis mutandis, to this process, too. However, the longitudinal behavior which has to do with compressions and temperature fluctuations, is complicated by the fact that several hydrodynamic modes are coupled. Another, and entirely new, aspect has to be considered in systems such as the Heisenberg ferromagnet, superfluid helium, or liquid crystals. These are ordered systems, and we will have to discuss how the presence of order affects the mode structure.

# CHAPTER 3

# FORMAL PROPERTIES OF CORRELATION FUNCTIONS

The purpose of this chapter is chiefly to summarize the formal results obtained so far, and extend them to the important case where we are interested in the correlations among several physical quantities. In the simple example of spin diffusion, we were only concerned with the "autocorrelation function" $S_{MM}(\vec{r}t, \vec{r}'t')$ of the magnetization, the probability essentially of finding the magnetization at the space-time point $\vec{r},t$ if you know its value at the point $\vec{r}',t'$. Now in a liquid, there are several quantities of interest: the particle density $n(\vec{r},t)$, the momentum density $\vec{g}(\vec{r},t)$, the energy density $\varepsilon(\vec{r},t)$, and maybe others. And these are dynamically coupled. A local imbalance in the energy density (i.e., a temperature inhomogeneity) will result in a spatially varying particle density as well, for example. We are therefore led to consider such correlation functions as $S_{n\varepsilon}(\vec{r}t, \vec{r}'t') = [\langle n(\vec{r}t)\, \varepsilon(\vec{r}'t')\rangle - \langle n(\vec{r}t)\rangle\langle\varepsilon(\vec{r}'t')\rangle]$. Or else, since we found that the averaged commutator was a little closer to the action, such response functions as $\chi''_{n\varepsilon}(\vec{r}t, \vec{r}'t') = \langle(1/2\hbar)[n(\vec{r}t), \varepsilon(\vec{r}'t')]\rangle$. We will therefore consider the general properties of multivariate correlation functions, most of which are obtained by a perfectly straightforward extension from the case of a single variable. We shall treat the general, quantum-mechanical, case which is formally a little easier to handle in fact, and indicate classical limits where appropriate. This chapter follows in much detail Martin 1968; see also Berne and Harp 1970.

## 3.1 Linear Dynamical Response

We are now interested in the dynamics of several observables $\{A_i(\vec{r},t)\}$. If each of those couples to a small external field $\delta a_i^{ext}(\vec{r},t)$, the Hamiltonian in the presence of these fields is given by

$$\mathcal{H}(t) = H - \sum_i \int d\vec{r}\, A_i(\vec{r})\, \delta a_i^{ext}(\vec{r},t) \tag{3.1}$$

in Schrödinger representation. As in section 2.5 we calculate by time-dependent perturbation theory the linear response. That is, we want to know the induced value of $\langle B(\vec{r},t)\rangle_{non-eq.} = \langle B(\vec{r},t)\rangle_{eq.} + \delta\langle B(\vec{r},t)\rangle$ for some variable $B(\vec{r},t)$ if we start out from equilibrium at $t = -\infty$, and let the system evolve under $\mathcal{H}(t)$. Proceeding exactly as in section 2.5, we obtain

$$\delta\langle B(\vec{r},t)\rangle = \sum_i \int_{-\infty}^{t} dt' \int d\vec{r}\,' \langle \frac{i}{\hbar}[B(\vec{r},t), A_i(\vec{r}\,',t')]\rangle\, \delta a_i^{ext}(\vec{r}\,',t'). \tag{3.2}$$

In particular, the induced change in the variable $A_i$ from its equilibrium value is

$$\delta\langle A_i(\vec{r},t)\rangle = \sum_j \int_{-\infty}^{t} dt' \int d\vec{r}\,' 2i\, \chi''_{ij}(\vec{r}t, \vec{r}\,'t')\, \delta a_j^{ext}(\vec{r}\,',t'), \tag{3.3}$$

where the response function is now a matrix, and given by

$$\chi''_{ij}(\vec{r}t, \vec{r}\,'t') = \chi''_{ij}(\vec{r}-\vec{r}\,', t-t') = \langle \frac{1}{2\hbar}[A_i(\vec{r},t), A_j(\vec{r}\,',t')]\rangle, \tag{3.4}$$

in a translationally invariant system. If we define its Fourier transform as usual by

$$\chi''_{ij}(\vec{k}\omega) = \int_{-\infty}^{\infty} d(t-t') \int d(\vec{r}-\vec{r}\,')\, e^{i\omega(t-t')-i\vec{k}\cdot(\vec{r}-\vec{r}\,')} \chi''_{ij}(\vec{r}-\vec{r}\,', t-t') \tag{3.5}$$

and the matrix of response functions by

$$\chi_{ij}(\vec{k}z) = \int \frac{d\omega}{\pi} \frac{\chi''_{ij}(\vec{k}\omega)}{\omega - z} \qquad (\mathrm{Im}\, z \neq 0), \tag{3.6}$$

then we can write (3.3) in the form

$$\delta\langle A_i\rangle(\vec{k}\omega) = \chi_{ij}(\vec{k}\omega)\, \delta a_j^{ext}(\vec{k}\omega), \tag{3.7}$$

where

$$\chi_{ij}(\vec{k}\omega) \equiv \lim_{\epsilon\to 0} \chi_{ij}(\vec{k}z)\Big|_{z=\omega+i\epsilon}$$

$$= P\int\frac{d\omega'}{\pi}\,\frac{\chi''_{ij}(\vec{k}\omega')}{\omega'-\omega} + i\,\chi''_{ij}(\vec{k}\omega)\ , \qquad (3.7a)$$

and where, here and henceforth, summation over repeated indices is implied. These equations are perfectly analogous to those in section 2.5. Note one minor difference: $\chi''_{ij}(\vec{k}\omega)$ need not necessarily be real, for $i \neq j$. However, $\chi''_{ii}(\vec{k}\omega)$ is always real.

Eq. (3.7) is valid for arbitrary external fields; it identifies $\chi_{ij}(\vec{k}z)$ as a matrix of dynamical susceptibilities. Now in particular, we again turn on the external fields slowly, (adiabatically), and we switch them off for positive times:

$$\delta a_i^{ext}(\vec{r},t) = \begin{cases} \delta a_i(\vec{r})\, e^{\epsilon t} & \text{for } t \leq 0\ , \\ 0 & \text{for } t > 0\ . \end{cases} \qquad (3.8)$$

<u>At $t = 0$,</u> such a field (rather, set of fields) will have produced spatially varying values $\delta\langle A_i(\vec{r},t=0)\rangle$ whose spatial Fourier transforms are, from (3.3), given by

$$\delta\langle A_i(\vec{k},t=0)\rangle = \chi_{ij}(\vec{k})\ \delta a_j(\vec{k})\ , \qquad (\textstyle\sum_j \text{ implied}) \qquad (3.9)$$

where

$$\chi_{ij}(\vec{k}) \equiv \int\frac{d\omega}{\pi}\ \frac{\chi''_{ij}(\vec{k}\omega)}{\omega}\ . \qquad (3.10)$$

In the limit in which $\delta a(\vec{r})$ varies slowly in space, i.e., at small $\vec{k}$, $\chi_{ij}(\vec{k}\to 0)$ will again reduce to a set of static susceptibilities, or thermodynamic derivatives. For example, we will find that

$$\lim_{k\to 0} \int\frac{d\omega}{\pi}\ \frac{\chi''_{An}(\vec{k}\omega)}{\omega} = \frac{\partial A}{\partial \mu}\Big|_T = n\,\frac{\partial A}{\partial p}\Big|_T\ . \qquad (3.11)$$

But that is hurrying matters a bit.

For $t > 0$, the system starts out from the non-equilibrium values (3.9), and left to itself it will relax back towards equilibrium. If we define

$$\delta\langle A_i\rangle(\vec{k}z) = \int_0^\infty dt\, e^{izt}\, \delta\langle A_i(k,t)\rangle \ , \tag{3.12}$$

then from (3.3) and (3.8) we get

$$\delta\langle A_i\rangle(\vec{k}z) = \int \frac{d\omega}{\pi i} \frac{\chi''_{ij}(\vec{k}\omega)}{\omega(\omega-z)} \cdot \delta a_i(\vec{k}) \ , \tag{3.13}$$

or, using the definitions above and the static result (3.9), we get the fundamental result

$$\delta\langle A_i\rangle(\vec{k}z) = \frac{1}{iz}\, [\chi(\vec{k}z)\, \chi^{-1}(\vec{k}) - 1]_{ij}\, \delta\langle A_i(\vec{k},t=0)\rangle \ , \tag{3.14}$$

which is to be read as a matrix equation. We will see that the matrix $\chi(\vec{k})$ is positive so that its inverse always exists.

Note the advantage of (3.14) over (3.13). The external fields $\delta a_i$, whose physical realization may sometimes be hard to assess, have been eliminated, and the relaxation process appears now as an initial value problem. Eqs. (3.13) and (3.14) are, of course, the generalization of (2.53a) and (2.54). Indeed, this present section has just rewritten equations from sections 2.5 and 2.6 as matrix equations. Life is that simple. Sometimes.

We have given these considerations in an operational form which has a clear and simple classical limit. Indeed, since $\hbar$ occurs explicitly only in the definition (3.4) of the response function, we obtain the corresponding classical expressions by replacing (3.4) by

$$\chi''_{ij}(\vec{r}t, \vec{r}'t') = \langle \tfrac{i}{2}[A_i(\vec{r},t), A_i(\vec{r}'t')]_{P.B.} \rangle \text{ classically.} \tag{3.15}$$

Otherwise, the whole development above goes through unchanged. The Poisson bracket appears thus as usual as the classical equivalent of a commutator divided by $i\hbar$, in case the observables in question have, themselves, a clear classical meaning (Dirac, 1958). The definition of the Poisson bracket is the standard one. Classically, the dynamical variables $A_i(\vec{r},t)$ are functions of the canonical positions $\vec{r}^{\alpha}$ and momenta $\vec{p}^{\alpha}$ of all particles, at some common time, say $t = 0$. Then

$$[A_i(\vec{r},t), A_j(\vec{r}\,',t')]_{P.B.} \equiv \sum_{\alpha} \left( \frac{\partial A_i(\vec{r},t)}{\partial \vec{r}^{\alpha}} \frac{\partial A_j(\vec{r}\,',t')}{\partial \vec{p}^{\alpha}} - \frac{\partial A_i(\vec{r},t)}{\partial \vec{p}^{\alpha}} \frac{\partial A_j(\vec{r}\,'t')}{\partial \vec{r}^{\alpha}} \right), \tag{3.16}$$

where $A_i(\vec{r},t) \equiv A_i(\vec{r},t;\ldots \vec{r}^{\alpha}\,\vec{p}^{\alpha}\ldots)$. To demonstrate these statements, one starts from the Liouville equation

$$\partial_t \rho(t) = [\mathcal{H}(t), \rho(t)]_{P.B.} \text{ with } \rho(t=-\infty) = \rho^0, \tag{3.17}$$

where classically the density matrix $\rho(t)$ becomes the N-particle phase space distribution function, $\rho(t) \equiv \rho(t; \therefore \vec{r}^{\alpha}\vec{p}^{\alpha}..)$, in terms of which averages are defined as usual,

$$\langle A(\vec{r},t)\rangle = \mathrm{Tr}_{cl}\ \rho(t)A(\vec{r},t)/\mathrm{Tr}_{cl}\ \rho(t)\ . \tag{3.18}$$

Here, $\mathrm{Tr}_{cl}$ is the classical phase space integral. The reader might find it a useful exercise to solve (3.17) to first order in the external fields, and so recover the equations of this section.

## 3.2. Symmetry Properties

The matrix of response functions has a number of symmetry properties which can be directly obtained from its microscopic definition (3.4). The derivation of these properties is quite straightforward, and we leave it as an exercise. The weary will find some help in the paper by Kadanoff and Martin (1963).

### 3.2.1 General Properties

An equilibrium system is time translation invariant, so that

$$\chi''_{ij}(\vec{r},t;\vec{r}',t') = \chi''_{ij}(\vec{r},\vec{r}';t-t') . \qquad (3.19)$$

Therefore, a Fourier transform $\chi''_{ij}(\vec{r},\vec{r}';\omega)$ with respect to the time difference variable can be defined.

Three additional properties reflect the fact that $\chi''_{ij}$ is a commutator of hermitian operators $A_i(\vec{r},t)$ which have, by assumption, a definite signature $\epsilon_i^T$ under time reversal. They are

$$\chi''_{ij}(\vec{r},t;\vec{r}',t') = -\chi''_{ji}(\vec{r}',t';\vec{r},t) \quad \text{(commutator)} \qquad (3.20a)$$

$$= -[\chi''_{ij}(\vec{r},t;\vec{r}',t')]^* \quad \text{(hermiticity)} \qquad (3.20b)$$

$$= -\epsilon_i^T \epsilon_j^T \chi''_{ij}(\vec{r},-t;\vec{r}',-t') \quad \text{(time reversal)}, \qquad (3.20c)$$

or for the Fourier transform

$$\chi''_{ij}(\vec{r},\vec{r}';\omega) = -\chi''_{ji}(\vec{r}', \vec{r};-\omega) \qquad (3.21a)$$

$$= -[\chi''_{ij}(\vec{r},\vec{r}';-\omega)]^* \qquad (3.21b)$$

$$= -\epsilon_i^T \epsilon_j^T \chi''_{ij}(\vec{r},\vec{r};-\omega) . \qquad (3.21c)$$

$\epsilon_i^T$ is + 1 for the mass or energy density, e.g., and -1 for the momentum or spin density. (3.20c) and (3.21c) hold forsystems for which both the Hamiltonian and the ensemble are time reversal invariant, the situation most frequently encountered. Of course, if the system is in a magnetic field $\vec{B}$, this changes sign under time reversal so that (3.20c), (3.21c) more generally read

$$\chi''_{ij}(\vec{r},t;\vec{r}',t';\vec{B}) = -\epsilon_i^T\epsilon_j^T\chi''_{ij}(\vec{r},-t;\vec{r}',-t';-\vec{B}), \tag{3.22a}$$

$$\chi''_{ij}(\vec{r},\vec{r}';\omega;\vec{B}) = -\epsilon_i^T\epsilon_j^T\chi''_{ij}(\vec{r},\vec{r}';-\omega;-\vec{B}). \tag{3.22b}$$

$\vec{B}$ need not be an external field. In a ferromagnet, it might be the spontaneously produced internal field. Namely, while in a ferromagnet, in the absence of external fields, the Hamiltonian is invariant under time reversal, the state is not. If you turn all spins backwards, the axis of the spontaneous magnetization will flip over.

### 3.2.2 Isotropic Systems

We will be mostly concerned with isotropic systems which are, in addition, invariant under spatial translations, rotations, and reflections (parity). Then

$$\chi''_{ij}(\vec{r},t;\vec{r}',t') = \chi''_{ij}(\vec{r}-\vec{r}',t-t') \tag{3.23}$$

$$= \epsilon_i^P\epsilon_j^P\chi''_{ij}(-\vec{r},t;-\vec{r}',t') \quad \text{(parity)}, \tag{3.24}$$

where $\epsilon_i^P$ is the signature of $A_i(\vec{r},t)$ under parity; i.e., $\epsilon^P = 1$ for the mass and energy density, spin etc., and $\epsilon^P = -1$ for the momentum density. Corresponding relations hold for the Fourier transforms, $\chi''_{ij}(\vec{k},\omega)$.

It is useful to separate the various symmetry properties into their effects on $\vec{k},\omega$, and the variable indices $i,j$. We get

$$\chi''_{ij}(\vec{k},\omega) = \epsilon_i^T\epsilon_j^T\epsilon_i^P\epsilon_j^P[\chi''_{ij}(\vec{k},\omega)]^* = \epsilon_i^T\epsilon_j^T\epsilon_i^P\epsilon_j^P\chi''_{ji}(\vec{k},\omega)$$

$$= -\epsilon_i^T\epsilon_j^T\chi''_{ij}(\vec{k},-\omega) = \epsilon_i^P\epsilon_j^P\chi''_{ij}(-\vec{k},\omega). \tag{3.25}$$

Consequently, $\chi''_{ij}(\vec{k}\omega)$ is either real and symmetric in $i \leftrightarrow j$, or imaginary and antisymmetric. Usually the former; one of the few correlation functions $\chi''(\vec{k}\omega)$ of some practical use which is imaginary is that which links the momentum density $\vec{g}$ with the spin density $\vec{S}$.

We have not yet used rotational invariance. In this case, little is to be gained from general formulae. Yet, the matter is simple enough. For example, the density-density correlation function $\chi''_{nn}(\vec{k}\omega)$ in a liquid must be a scalar. But there is only one scalar that one can form from $\vec{k}$: $k^2$. Thus, $\chi''_{nn}(\vec{k}\omega)$ must be a function of $k = |\vec{k}|$ only. Or consider the momentum density correlation function $\chi''_{g_i g_j}(\vec{k}\omega)$. It must be a tensor. Now from $\vec{k}$, one can form three tensors: $\delta_{ij}$, $k_i k_j$, and $\epsilon_{ijk} k_k$ where $\delta_{ij}$ is the Kronecker symbol, and $\epsilon_{ijk}$ the Levi-Civita symbol. Thus

$$\chi''_{g_i g_j}(\vec{k}\omega) = \delta_{ij}\, A(k\omega) + k_i k_j\, B(k\omega) + \epsilon_{ijk}\, k_k\, C(k\omega)\,, \tag{3.26}$$

where the three functions A, B, and C depend only on $k = |k|$. Actually $C(k\omega) = 0$ since $\chi''_{g_i g_j}$ is even under parity. So this is how symmetry is used.

3.2.3. Crystals

An ideal crystal is, of course, not invariant under arbitrary translations and rotations but only under those which bring every lattice point into an equivalent position within the lattice. This means, in particular, that (3.23) is no longer true. A Fourier transform must now contain two wavevectors of which one is discrete, reflecting the periodicity of the lattice. The matter is essentially trivial. One normally gets by with

$$\chi''_{ij}(\vec{k}\omega) = V^{-1} \int d\vec{r} \int d\vec{r}\,'\, e^{-i\vec{k}(\vec{r}-\vec{r}\,')}\, \chi''_{ij}(\vec{r},\vec{r}\,';\omega), \tag{3.27}$$

if one uses experimental probes whose wavelength $2\pi/k$ is much larger than the lattice constant.

## 3.3 Positivity of $\omega\chi''(;\omega)$ and Dissipation

In section 2.8, we proved an important positivity property, namely that $\omega\chi''(\vec{k}\omega)$ is non-negative, in a fasion that easily applies to multivariate functions. The general statement is that, for any stable system, $\omega\chi''_{ij}(\vec{r},\vec{r}\,';\omega)$ is positive (semi-) definite if considered as a supermatrix with indices $i,\vec{r}$ and $j,\vec{r}\,'$. Or in translationally invariant systems,

$$\sum_{i,j} a_i^* \, \omega\chi''_{ij}(\vec{k}\omega)\, a_j \geq 0 \quad \text{for all } \vec{k} \text{ and } \omega\,, \tag{3.28}$$

for arbitrary $a_i$ .

Our previous proof was statistical, involving the fluctuation-dissipation theorem which continues to hold true here. It is very instructive, however, to consider the matter by an alternative, dynamical route which is taken from Kadanoff and Martin (1963), and inquires into the dynamical significance of (3.28). We consider the total energy, $W(t) = \mathrm{tr}\ \rho(t)\,\mathcal{H}(t)$, and its rate of change

$$\frac{dW}{dt} = \mathrm{tr}\ \rho(t)\,\frac{\partial \mathcal{H}(t)}{\partial t} = -\sum_i \int d\vec{r}\,[\mathrm{tr}\ \rho(t)\,A_i(\vec{r})]\frac{\partial}{\partial t}\delta a_i^{ext}(\vec{r},t)$$

$$= -\sum_i \int d\vec{r}\,[\langle A_i(\vec{r},t)\rangle_{eq.} + \delta\langle A_i(\vec{r},t)\rangle]\frac{\partial}{\partial t}\delta a_i^{ext}(\vec{r},t) \tag{3.29}$$

Only the external change, $\sim \partial_t\, \delta a^{ext}(t)$, contributes. The internal change, $\sim \mathrm{tr}\ \dot{\rho}(t)\,\mathcal{H}(t) = (i\hbar)^{-1}\,\mathrm{tr}[\rho(t),\mathcal{H}(t)]\,\mathcal{H}(t) = (i\hbar)^{-1}\,\mathrm{tr}\ \rho(t)[\mathcal{H}(t),\mathcal{H}(t)]$, vanishes. The equilibrium term in (3.29) does not contribute, on the average. The total work done on the system must be positive, and it is given by

$$\Delta W = \int_{-T}^{T} dt\,\frac{dW}{dt} = \sum_i \int_{-T}^{T} dt \int d\vec{r}\,\delta a_i^{ext}(\vec{r},t)\,\frac{\partial}{\partial t}\,\delta\langle A_i(\vec{r},t)\rangle\,, \tag{3.30}$$

where T is so large that the external fields vanish before -T and after T. Using the

Wiener-Khinchin theorem (which is simply the statement that $\int dt\, A^*(t)B(t) = \int \frac{d\omega}{2\pi} A^*(\omega)B(\omega)$) and the general result (3.7), we obtain, to order $(\delta a)^2$,

$$\Delta W = \int \frac{d\omega}{2\pi} \int \frac{d\vec{k}}{(2\pi)^3} (-i) \sum_{i,j} \delta a_i^*(\vec{k}\omega)\, \omega\chi_{ij}(\vec{k},\omega)\delta a_j(\vec{k}\omega) , \tag{3.31}$$

where $\chi_{ij}(\vec{k}\omega)$ is as given in (3.7a). The fields $\delta a_i(\vec{r},t)$ are real so that $\delta a_i^*(\vec{k},\omega) = \delta a_i(-\vec{k},-\omega)$. Hence the principal value part of $\omega\chi_{ij}(\vec{k},\omega)$, eq. (3.7a), which is odd under $\vec{k},\omega;ij \rightarrow -\vec{k},-\omega;ji$, does not contribute to (3.31). We therefore obtain

$$\Delta W = \int \frac{d\omega}{2\pi} \int \frac{d\vec{k}}{(2\pi)^3} \sum_{i,j} \delta a_i^*(\vec{k}\omega)\, \omega\chi''_{ij}(\vec{k}\omega)\, \delta a_j(\vec{k}\omega) \geq 0 , \tag{3.32}$$

which must be positive in a dissipative system. Since the external fields can be arbitrarily chosen, we obtain eq. (3.28).

Of course, our conclusion holds only in a stable system. In network theory, one would say it holds in a "passive network". A laser is not dissipative after being pumped, and (3.28) will fail at the lasing frequency. However, the equation (3.28) will certainly be true for any system which is in thermal equilibrium. Undercooled liquids? Yes, for those it holds, too. (3.28) indicates stability with respect to infinitesimal disturbances. In this sense, an undercooled liquid is quite stable (metastable). You have to shake it, softly but with some determination, to make it crystallize.

## 3.4 Sum Rules

Exactly as in section 2.9, one obtains from (3.4)

$$\int \frac{d\omega}{\pi} \omega\chi''_{ij}(\vec{r},\vec{r}\,';\omega) = \langle \frac{i^n}{\hbar} [\partial_t^n A_i(\vec{r},t), A_j(\vec{r}\,',t)] \rangle$$

$$= \langle \frac{1}{\hbar} [[\ldots [A_i(\vec{r}t), \frac{H}{\hbar}] \ldots, \frac{H}{\hbar}], A_j(\vec{r}\,',t)] \rangle . \tag{3.33}$$

These expressions, which can be exactly evaluated in some cases, furnish coefficients for a large z expansion of the response functions $\chi_{ij}(;z)$. Their practical importance is in their use for interpolation schemes as discussed in sections 2.10,11, since the sum rules represent essentially the only results of numerical use which can be rigorously obtained from first principles. The most important sum rule, similar to eq. (2.78), is the f-sum rule for the particle density correlation function,

$$\int \frac{d\omega}{\pi}\, \omega \chi''_{nn}(\vec{r},\vec{r}';\omega) = \frac{1}{m}(\vec{\nabla}\cdot\vec{\nabla}')\langle n(\vec{r})\rangle\,\delta(\vec{r}-\vec{r}') \tag{3.34a}$$

or

$$\int \frac{d\omega}{\pi}\, \omega \chi''_{nn}(\vec{k}\omega) = \frac{n}{m}k^2 , \tag{3.34b}$$

for a translationally invariant system. Its proof is obvious from the analogous proof of eq. (2.78).

## 3.5 The Fluctuation-Dissipation Theorem

Here again, we have done all the work in section 2.7. If we define fluctuation functions in thermal equilibrium by

$$S_{ij}(\vec{r},\vec{r}';\omega) = \int_{-\infty}^{\infty} d(t-t')e^{i\omega(t-t')}\langle(A_i(\vec{r},t)-\langle A_i(\vec{r},t)\rangle)(A_j(\vec{r}',t')-\langle A_j(\vec{r}',t')\rangle)\rangle, \tag{3.35}$$

we obtain from eq. (2.67b)

$$\chi''_{ij}(\vec{r},\vec{r}';\omega) = (2\hbar)^{-1}(1-e^{-\hbar\omega\beta})\, S_{ij}(\vec{r},\vec{r}';\omega), \tag{3.36a}$$

or

$$\chi''_{ij}(\vec{k}\omega) = (2\hbar)^{-1}(1-e^{-\hbar\omega\beta})\, S_{ij}(\vec{k}\omega), \tag{3.36b}$$

in a translationally invariant system. Another useful function is the symmetrized fluctuation function

$$\varphi_{ij}(\vec{r},\vec{r}\,';t-t') = \langle \tfrac{1}{2}\{A_i(\vec{r},t), A_j(\vec{r}\,')\}\rangle - \langle A_i(\vec{r},t)\rangle\langle A_j(\vec{r}\,',t')\rangle, \qquad (3.37)$$

whose Fourier transform $\varphi_{ij}(;\omega)$ is given by

$$\varphi_{ij}(\vec{r},\vec{r}\,';\omega) = \hbar \coth(\hbar\omega\beta/2)\, \chi''_{ij}(\vec{r},\vec{r}\,';\omega), \qquad (3.38)$$

and so its symmetry properties can be inferred from those of $\chi''_{ij}(;\omega)$.

Classically, the functions S and $\varphi$ are identical, and the fluctuation-dissipation theorem becomes

$$\chi''_{ij}(\vec{r},\vec{r}\,';\omega) = (\beta/2)\omega\, S_{ij}(\vec{r},\vec{r}\,';\omega) \qquad \text{classically} \qquad (3.39a)$$

or

$$\chi''_{ij}(\vec{r},\vec{r}\,';t-t') = (\beta/2)\, i\partial_t\, S_{ij}(\vec{r},\vec{r}\,';t-t'). \qquad (3.39b)$$

It is a useful exercise to derive eqs. (3.39) classically, without taking the detour via quantum mechanics. For some help, see Martin (1968).

The name of the celebrated theorem is now clear. It relates, for any system in thermal equilibrium, two physically distinct quantities of fundamental experimental significance: the spontaneous fluctuations on the one hand which arise, even in the absence of external forces, from the thermal motion of the constituent particles. Described by S, these fluctuations give rise to the scattering of neutrons or light. And the dissipative behavior of many-body systems, on the other hand, describing the fact that all or part of the work done by external stirring forces is irreversibly disseminated into the infinitely many degrees of freedom of thermal systems. This characteristic property is described by $\chi''$ as we saw in the last section. The former property is an essentially statistical one (albeit ultimately of mechanical origin), the latter property is an essentially mechanical one (albeit not without statistical implications).

Our proof has been based, specifically, on the canonical ensemble $\sim e^{-\beta H}$; it goes through unchanged in the grand canonical ensemble $\sim e^{-\beta(H-\mu N)}$ as well. In fact, as our discussion of neutron scattering after (2.63) indicates, the theorem is more generally valid. Its fundamental connection with detailed balancing can be exploited in a dynamical proof. Indeed, in a sense the theorem can be used to introduce temperature into a correlation function description. This is its fundamental significance, most clearly recognized in the Green's function techniques developed by Martin and Schwinger (1959) and others. See also Kadanoff and Baym (1962).

This remark is of practical import for systems such as ferromagnets or superfluids in which there is spontaneous order present. If such systems are described by the canonical or grand canonical ensemble, the fluctuation-dissipation theorem fails for certain variables. What has to be done, then, is to restrict the ensemble (by specifying order parameters) in such a fashion that the theorem is reinstituted again. We shall discuss examples of this procedure.

We end this section with another but related remark. There are two features which enter any theory of many-particle systems: one which is mechanical, summarized in the Schrödinger equation or Newton's equations of motion, and one which is statistical, summarized in the averaging recipe over a properly chosen ensemble. The former is straightforward in principle though formidable in practice, the latter is much more subtle. These two aspects can, of course, never be completely disentangled.

However, as we have pointed out above, the response function $\chi''(\vec{r},\vec{r}\,';t)$ is largely a mechanical quantity. This is indicated by our derivation of linear response which is characterized by $\chi''$, a purely mechanical derivation which makes but implicit reference to statistical questions. It is very clearly indicated by the f-sum rule (2.78) which holds in any stationary ensemble, expression nothing but the conservation law (2.3) and the commutator (2.77). The fluctuation functions $S(\vec{r},\vec{r}\,';t)$ or $\omega(\vec{r},\vec{r}\,';t)$ are much more statistical. Even in their definitions, eqs. (3.35) or (3.37), we had to subtract constant terms to avoid contributions $\sim\delta(\omega)$ in their spectrum, and we had to argue that because of the presumed statistical independence of $A(\vec{r},t)$ from $A(\vec{r}\,',t')$ for large $|t-t'|$, these terms were given by $\langle A\rangle^2$. This assumption is one of ergodicity or more precisely, "mixing" which implies ergodicity (Lebowitz 1972), and it is a subtle affair. It is best, therefore, to base theoretical considerations upon the more mechanical and straightforward response functions $\chi''(\vec{r},\vec{r}\,';t)$, inferring fluctuation properties later from the Nyquist theorem.

The response functions $\chi''$ have an additional property which is very welcome in systems with long-ranged order (superfluids etc.). We will always be interested in local observables $A(\vec{r})$, i.e., variables which depend only on properties (position, momentum, spin) of particles in a small neighborhood of $\vec{r}$. Even for quantum systems in which particles are indistinguishable, this is a valid concept: $A(\vec{r})$ depends only on creation and annihilation fields $\psi^+(\vec{r}\,')$, $\psi(\vec{r}\,')$ with $|\vec{r}-\vec{r}\,'|$ small, say of the order of the force range. This means that the commutators, $[A_i(\vec{r}),A_j(\vec{r}\,')]$, vanish if $\vec{r}\,'$ is far from $\vec{r}$, and therefore the functions $\chi''_{ij}(\vec{r},\vec{r}\,';t=0)$ and their time derivatives at $t=0$, or else the sum rules

$$\int\frac{d\omega}{\pi}\,\omega^n\,\chi''_{ij}(\vec{r},\vec{r}\,';\omega)\,,\qquad\qquad n\geq 0\,,\tag{3.40}$$

fall off very rapidly as $|\vec{r}-\vec{r}\,'|\to\infty$. They are of short range. One might argue, quantum-mechanically, that the commutator function $\chi''_{ij}(\vec{r}t,\vec{r}\,'t')$ ought to vanish if $\vec{r}\,'$ is far from $\vec{r}$ since then, measurements of $A_i(\vec{r},t)$ and $A_j(\vec{r}\,'t')$ do not interfere.

This is not always true of the statistical functions, e.g.,

$$S_{ij}(\vec{r},\vec{r}\,';t=0) = \int\frac{d\omega}{2\pi}\,S_{ij}(\vec{r},\vec{r}\,';\omega)\,.\tag{3.41}$$

It is a fundamental property of ordered systems that certain functions of this type fall off very slowly as $|\vec{r}-\vec{r}'| \to \infty$, as $|\vec{r}-\vec{r}'|^{-3}$ (vide, the momentum density correlation function in superfluids, chapter 10) or even as $|\vec{r}-\vec{r}'|^{-1}$ (vide, the transverse spin correlation function in ferromagnets, chapter 7). This appearance of long-ranged order can, of course, not be avoided; it is real and of great experimental significance. However, the static correlation function $S(\vec{r},\vec{r}';t=0)$ is essentially a time integral over the function $\chi''(\vec{r},\vec{r}';t)$, as seen from the formal expression (2.67a) or more clearly the classical equation (3.39b). In a formulation based on $\chi''$, the long-ranged nature of correlations appears therefore, properly, as a cumulative effect.

It is helpful in this respect to introduce yet another correlation function C(t) defined so that (3.39b) is true quantum-mechanically, i.e., so that

$$i\partial_t C_{ij}(\vec{r},\vec{r}';t) = (2/\beta)\, \chi''_{ij}(\vec{r},\vec{r}';t). \qquad (3.42)$$

This is the Kubo function (Kubo 1959), and it can be written in the form

$$C_{ij}(\vec{r},\vec{r}';t-t') = \beta^{-1}\int_0^\beta d\beta' [\langle A_i(\vec{r},t)A_j(\vec{r}',t'+i\hbar\beta')\rangle - \langle A_i\rangle\langle A_j\rangle], \qquad (3.43)$$

as one shows easily from the fluctuation-dissipation theorem. We have already encountered its Laplace transform

$$C_{ij}(\vec{r},\vec{r}';z) = \int_0^\infty dt\, e^{izt}\, C_{ij}(\vec{r},\vec{r}';t) \qquad \text{(for Im} z > 0) \qquad (3.44a)$$

$$= \beta^{-1}\int \frac{d\omega}{\pi i} \frac{\chi''_{ij}(\vec{r},\vec{r}'\ \omega)}{\omega(\omega-z)} \qquad \text{(for Im} z \neq 0) \qquad (3.44b)$$

$$= (1/iz\beta)[\chi_{ij}(\vec{r},\vec{r}';z) - \chi_{ij}(\vec{r},\vec{r}';i0)], \qquad (3.44c)$$

which describes relaxation in the most direct fashion.

With the last few remarks, we are a few steps ahead of the development since the system treated in the next chapter, the normal fluid, does not exhibit any long-ranged order. Who, though, would not like a little advance notice of future complications?

# CHAPTER 4

# THE NORMAL FLUID

In this chapter we will use the methods developed above, to analyze fluctuations in a normal (isotropic, not superfluid) liquid or gas. Of interest are fluctuations of several physical quantities: the particle number, momentum, energy and entropy densities, and a few others. The most important of these are density fluctuations whose spectral function is given by

$$S_{nn}(k\omega) = \int_{-\infty}^{\infty} dt \int d\vec{r}\, e^{i\omega t - i\vec{k}\cdot\vec{r}} \langle (n(\vec{r},t)-n)(n(\vec{0},0)-n) \rangle , \qquad (4.1)$$

where $n(\vec{r},t)$ is the particle density operator, $n = \langle n(\vec{r},t)\rangle_{eq.}$ the equilibrium value of the density.

$S_{nn}(k\omega)$ is called the dynamical structure factor, and it is one of the most important quantities in the theory of many-particle systems. It is the density fluctuation spectrum that is measured in inelastic light, X-ray, and neutron scattering experiments, to name only a few applications. In a typical light scattering experiment, for example, one sends a laser beam of frequency $\omega_i$ and wave vector $\vec{k}_i$ into a liquid cell. One then looks for scattered light with frequency $\omega_f = \omega_i - \omega$ and wave vector $\vec{k}_f = \vec{k}_i - \vec{k}$, see fig. 4.1.

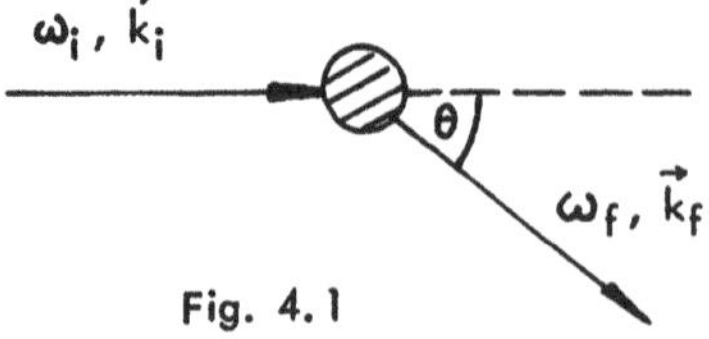


Fig. 4.1

$$\omega_f = \omega_i - \omega$$

$$\vec{k}_f = \vec{k}_i - \vec{k}$$

$$|k| \simeq 2\,|k_i|\,\sin(\theta/2) \qquad (4.2)$$

In analogy to magnetic neutron scattering as discussed before, the scattered intensity is given by

$$I_{scatt.}\begin{pmatrix}\vec{k}_i \to \vec{k}_i - \vec{k} \\ \\ \omega_i \to \omega_i - \omega\end{pmatrix} = [\text{ factors times }]\ S_{nn}(\vec{k}\omega) \tag{4.3}$$

as shown in the Appendix. The bracket contains kinematical factors which are not important for our purpose.

Light scattering experiments have been performed on many liquids and gases (see, for example, the recent book by Berne and Pecora 1973), and at low frequencies the spectrum found looks as shown in fig. 4.2.

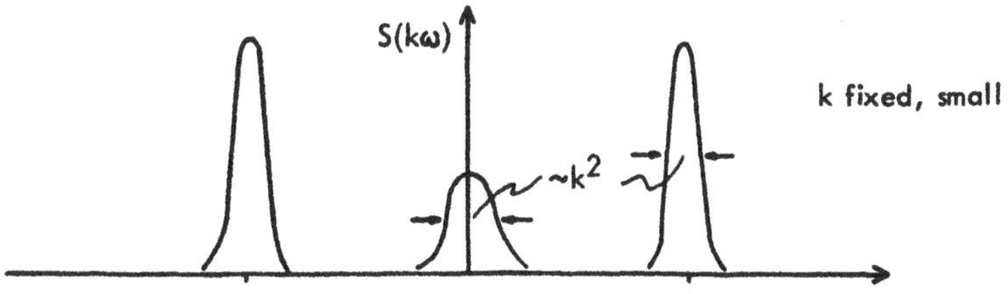


Fig. 4.2

There are three Lorentzian peaks, the central "Rayleigh peak" (caused by heat diffusion) and two symmetrically displaced "Brillouin peaks" (caused by sound waves). Much of the effort in this chapter will go into deriving this spectrum.

Since the wavelength of light is so long compared to interatomic distances in a liquid, we expect that a continuum theory will be sufficient to explain this spectrum. Indeed, we will derive it in all detail from the phenomenological, linearized Navier-Stokes equations of fluid dynamics, just as we obtained, in chapter 2, a simple Lorentzian spectrum for magnetization fluctuations from the spin diffusion equation (2.7). (In fact, as will be more apparent later, our presentation,which follows the paper by Kadanoff and Martin 1963, is tantamount to a derivation of the equations of hydrodynamics, a fact which does not seem to be generally appreciated.)

Fig. 4.2 contains an experimental clue that shows that a hydrodynamic theory should be sufficient: the widths of the three peaks are measured to be proportional to $k^2$; they become extremely narrow in forward scattering as $k \to 0$. Consequently, each of the peaks must reflect a collective process whose lifetime $\tau(k) \sim k^{-2}$ becomes infinite as $k \to 0$: the trademark of a hydrodynamic process.

Now we saw in section 2.1 that there is a hydrodynamic process associated with each conserved local variable. And so we begin with the five continuity equations for the particle density $n(\vec{r},t)$, the momentum density $\vec{g}(\vec{r},t)$, and the energy density $\epsilon(\vec{r},t)$:

$$\partial_t n(\vec{r},t) + \vec{\nabla}\cdot\vec{g}(\vec{r},t)/m = 0 \quad \text{number conservation}, \tag{4.4a}$$

$$\partial_t g_i(\vec{r},t) + \nabla_j \tau_{ij}(\vec{r},t) = 0 \quad \text{momentum conservation}, \tag{4.4b}$$

$$\partial_t \epsilon(\vec{r},t) + \vec{\nabla}\cdot\vec{j}^{\epsilon}(\vec{r},t) = 0 \quad \text{energy conservation}. \tag{4.4c}$$

Here, $\vec{j}^{\epsilon}$ is the microscopic energy current density (operator), and the microscopic stress tensor $\tau_{ij}$ functions as the current for the momentum.

For definiteness, let us write down microscopic expressions for these quantities even though we will make little explicit use of them. We take the case of a "simple" liquid, i.e., an assembly of identical point particles of mass m which interact through a central pair potential $v(r) = v(|\vec{r}|)$ which is of short range. For the noble gases like argon, this is adequate. The Hamiltonian is then

$$H = \sum_\alpha \frac{\vec{p}^{\alpha 2}}{2m} + \frac{1}{2}\sum_{\alpha,\beta}^{(\alpha\neq\beta)} v(|\vec{r}^\alpha - \vec{r}^\beta|). \tag{4.5}$$

And the microscopic expressions for a classical system (quantum-mechanically, they simply have to be symmetrized in $\vec{r}^{\alpha}$ and $\vec{p}^{\alpha}$, as in (2.4)) are the following:

$$n(r,t) = \sum_{\alpha} \delta(r-r^{\alpha}), \tag{4.6}$$

$$g_i(r,t) = \sum_{\alpha} p_i^{\alpha}\delta(r-r^{\alpha}),$$

$$\epsilon(r,t) = \sum_{\alpha}[\frac{p^{\alpha 2}}{2m} + \frac{1}{2}\sum_{\beta\neq\alpha} v(r^{\alpha\beta})]\delta(r-r^{\alpha}),$$

$$\tau_{ij}(r,t) = \sum_{\alpha}\frac{p_i^{\alpha}p_j^{\alpha}}{m}\delta(r-r^{\alpha}) - \frac{1}{2}\sum_{\alpha\neq\beta} r_i^{\alpha\beta} v_j(r^{\alpha\beta})\int_0^1 d\lambda\, \delta(r-\frac{r^{\alpha}+r^{\beta}}{2} - \frac{\lambda}{2}r^{\alpha\beta}),$$

$$j_i^{\epsilon}(r,t) = \sum_{\alpha}[\frac{p^{\alpha 2}}{2m} + \frac{1}{2}\sum_{\beta\neq\alpha} v(r^{\alpha\beta})]\frac{p_i^{\alpha}}{m}\delta(r-r^{\alpha})$$
$$-\sum_{\alpha\neq\beta}\frac{1}{4m} r_i^{\alpha\beta} v_j(r^{\alpha\beta})(p_j^{\alpha}+p_j^{\beta})\int_0^1 d\lambda\, \delta(r-\frac{r^{\alpha}+r^{\beta}}{2} - \frac{\lambda}{2}r^{\alpha\beta}),$$

where $\vec{r}^{\alpha\beta} = \vec{r}^{\alpha} - \vec{r}^{\beta}$ and $v_i(r) = \nabla_i v(r)$, and where the positions $\vec{r}^{\alpha}$ and momenta $\vec{p}^{\alpha}$ of all particles are to be taken at time t. For clarity, all vector arrows have been omitted. Similar if a little more complicated expressions can be written down for a molecular liquid. In fact, all the hydrodynamic results obtained in this chapter are rigorously valid for isotropic molecular liquids, too.

The method used in this chapter is the one presented in chapter 2. A different approach which we will develop in later chapters, is applied to superfluids in chapter 10 and can easily be used for the description of normal fluids as well.

## 4.1 The Equations of Fluid Dynamics

The continuity equations are microscopically rigorous but they do not yet make a complete theory. As in the spin diffusion case, they have to be supplemented by macroscopic constitutive relations. In order to have all the equations together which constitute the linearized hydrodynamic theory, we shall write them all down first, and then explain.

Conservation Laws:

$$\partial_t\, n(\vec{r},t) + \vec{\nabla}\cdot \vec{g}(\vec{r},t)/m = 0, \tag{4.7a}$$

$$\partial_t\, g_i(\vec{r},t) + \nabla_j \tau_{ij}(\vec{r},t) = 0, \tag{4.7b}$$

$$\partial_t\, \varepsilon(\vec{r},t) + \vec{\nabla}\, \vec{j}^{\,\varepsilon}(\vec{r},t) = 0. \tag{4.7c}$$

Constitutive Relations:

$$\langle \vec{g}(\vec{r},t)\rangle = mn\, \vec{v}(\vec{r},t)\,, \tag{4.8a}$$

$$\langle \tau_{ij}(\vec{r},t)\rangle = p(\vec{r},t)\delta_{ij} - \eta[\nabla_i v_j(\vec{r},t) + \nabla_j v_i(\vec{r},t) - \tfrac{2}{3}\vec{\nabla}\cdot\vec{v}(\vec{r},t)\delta_{ij}] \tag{4.8b}$$

$$- \zeta\, \vec{\nabla}\cdot\vec{v}(\vec{r},t)\,\delta_{ij}\,,$$

$$\langle \vec{j}^{\,\varepsilon}(\vec{r},t)\rangle = (\varepsilon + p)\,\vec{v}(\vec{r},t) - \varkappa\vec{\nabla}\, T(\vec{r},t). \tag{4.8c}$$

Thermodynamic Relations:

$$\vec{\nabla} p(\vec{r},t) = \left(\frac{\partial p}{\partial n}\right)_\varepsilon \vec{\nabla}\langle n(\vec{r},t)\rangle + \left(\frac{\partial p}{\partial \varepsilon}\right)_n \vec{\nabla}\langle \varepsilon(\vec{r},t)\rangle\,, \tag{4.9a}$$

$$\vec{\nabla} T(\vec{r},t) = \left(\frac{\partial T}{\partial n}\right)_\varepsilon \vec{\nabla}\langle n(\vec{r},t)\rangle + \left(\frac{\partial T}{\partial n}\right)_\varepsilon \vec{\nabla}\langle \varepsilon(\vec{r},t)\rangle\,. \tag{4.9b}$$

Here $n, \varepsilon, p$ are the equilibrium values of density, energy density, and pressure, e.g., $n = \langle n(\vec{r},t)\rangle_{eq.}$. $p(\vec{r},t)$, $T(\vec{r},t)$, and $v(\vec{r},t)$ are the local values of pressure, temperature, and (average) velocity, in the non-equilibrium flow state. The coefficients in (4.9) are normal thermodynamic derivatives. And the three transport coefficients in (4.8) are called shear viscosity ($\eta$), bulk viscosity ($\zeta$), and heat conductivity ($\varkappa$).

These are the Navier-Stokes equations of fluid dynamics. (See, for example, Landau and Lifshitz 1959.) They are linear because we have made them that way; more generally, there are terms $\sim v_i v_j$ in $\langle \tau_{ij} \rangle$, for example. However, $\chi''$ gives the linear response, and so to obtain $\chi''$ the linearized phenomenological theory is sufficient. Note that the system of equations is now closed. Inserting eqs. (4.8,9) into (4.7) one obtains 5 coupled equations for the five densities $\langle n \rangle$, $\langle \vec{g} \rangle$, $\langle \varepsilon \rangle$. These equations involve, of course, a number of coefficients which remain undetermined, namely thermodynamic derivatives and transport coefficients, the parameters of the hydrodynamic theory, just as the spin diffusion coefficient was in chapter 2.

And now to the explanation. Hydrodynamics is valid when, after a brief initial period of rapid and complicated motion, the system has reached local equilibrium, a state in which, for example, the pressure at $\vec{r},t$ is at equilibrium with the local values of density and energy density. This state of affairs, $p(\vec{r},t) = p_{eq.}(n(\vec{r},t), \varepsilon(\vec{r},t))$, is expressed in eqs. (4.9).

The non-derivative, <u>reactive terms</u> in (4.8) are of the same nature, and can be inferred from a Galilei transformation. Generally, the relations between quantities in a stationary medium versus one that moves with constant velocity $\vec{v}$ are given by

$$\vec{g}(\vec{r},t) = \vec{g}^0(\vec{r}-\vec{v}t,t) + \vec{v}\, mn^0(\vec{r}-\vec{v}t,t)\,, \tag{4.10a}$$

$$\tau_{ij}(\vec{r},t) = \tau_{ij}^0 + v_i g_j^0 + v_j g_i^0 + v_i v_j\, mn^0\,, \tag{4.10b}$$

$$j_i^{\varepsilon}(\vec{r},t) = j_i^{\varepsilon 0} + v_j(\tau_{ij}^0 + \varepsilon^0 \delta_{ij}) + v_i v_j g_j^0 + \frac{1}{2} v^2 (g_i^0 + v_i mn^0), \tag{4.10c}$$

as can be easily verified from eqs. (4.6). All quantities with superscript 0 refer to the system at rest, and their arguments are $(\vec{r}-\vec{v}t,t)$. But in a normal fluid at rest, $\vec{g}^0 = \vec{j}^{\varepsilon 0} = 0$, and $\tau_{ij}^0 = p\,\delta_{ij}$. In a system which moves with the local velocity $\vec{v}(\vec{r},t)$, and omitting terms of order $v^2$, we obtain the reactive terms of (4.8).

The dissipative terms in (4.8) account for the fact that, for example, a temperature gradient $\vec{\nabla} T(\vec{r},t)$ will produce an energy current even if the average velocity is zero. Similarly, the two terms in (4.8b), which are the only ones compatible with isotropic symmetry, account for frictional stresses due to velocity gradients $\nabla_i v_j$. A term $\delta j_i^{\varepsilon} \sim \lambda (\nabla \times v)_i$ would be possible by rotational symmetry, but is ruled out by parity.

Eqs. (4.8) are thus the result of an expansion of the current densities to first order in gradients of the local conjugate forces $\vec{v}(\vec{r},t)$, $T(\vec{r},t)$, and $p(\vec{r},t)$. If so, we have to justify the absence of dissipative currents driven by a pressure gradient. Why are there, more generally, no terms of the form

$$\delta\vec{g} = -\lambda_{nn} \frac{\vec{\nabla} p}{n} - \lambda_{nq} \frac{\vec{\nabla} T}{T}, \tag{4.11a}$$

$$\delta[\vec{j}^{\varepsilon} - \frac{\varepsilon + n}{mn} \vec{g}] = -\lambda_{qn} \frac{\vec{\nabla} p}{n}, \tag{4.11b}$$

which would be allowed by symmetry? The answer is that these terms would violate momentum conservation. To demonstrate this fact is a good but not trivial exercise; note the Onsager relation $\lambda_{nq} = \lambda_{qn}$, to be proved below. Actually, this question is much more easily answered in terms of correlation functions, and we will therefore keep it in store until section 4.6. (The argument given by Landau-Lifshitz, 1959, p. 187, is not correct on this point. It is true that $\lambda_{nn} \cdot \varkappa T - \lambda_{nq}^2 \geq 0$ so that if $\lambda_{nn}$ vanishes, $\lambda_{nq}$ must vanish, too. But $\lambda_{nn} = 0$ can only be inferred from the fact that the flux of the particle density, i.e., the momentum current, is itself conserved.)

## 4.2 Solution of the Hydrodynamic Equations

The Navier-Stokes equations present, in general, a problem of considerable complexity (see e.g., Chandrasekhar 1961). In (4.7-9), we have already simplified this problem enormously by omitting non-linear terms. We shall furthermore eliminate boundary conditions by considering a simply connected medium of infinite extent. In terms of spatial Fourier transforms, we are then faced with a simple initial value problem.

### 4.2.1. Transverse Fluctuations

Inserting (4.8b) into (4.7b), one obtains

$$\partial_t \langle \vec{g}(\vec{r},t)\rangle + \vec{\nabla} p(\vec{r},t) - \frac{\zeta+\eta/3}{mn}\vec{\nabla}(\vec{\nabla}\cdot\langle\vec{g}(\vec{r},t)\rangle) - \frac{\eta}{mn}\nabla^2\langle\vec{g}(\vec{r},t)\rangle = 0. \quad (4.12)$$

This equation separates into a longitudinal and a transverse part. The momentum density $\vec{g}$, like any vector, can be split up in the form

$$\vec{g}(\vec{r},t) = \vec{g}_\ell(\vec{r},t) + \vec{g}_t(\vec{r},t) , \quad (4.13)$$

where

$$\vec{\nabla}\times\vec{g}_\ell = 0 \quad \text{and} \quad \vec{\nabla}\cdot\vec{g}_t = 0 . \quad (4.13a)$$

For the transverse component, we then obtain

$$(\partial_t - \frac{\eta}{mn}\nabla^2)\langle\vec{g}_t(\vec{r},t)\rangle = 0 , \quad (4.14)$$

which is our old friend, the diffusion equation. And as we did in section 2.1, we solve by performing the Fourier-Laplace transformation (Imz > 0)

$$\langle\vec{g}_t(\vec{k}z)\rangle = \int_0^\infty dt\, e^{izt}\int d\vec{r}\, e^{-i\vec{k}\cdot\vec{r}}\langle\vec{g}_t(\vec{r},t)\rangle . \quad (4.15)$$

The solution of the transverse initial value problem is then

$$\langle \vec{g}_t(\vec{k}z)\rangle = \frac{i}{z+ik^2\eta/mn} \langle \vec{g}_t(\vec{k},t=0)\rangle . \tag{4.16}$$

Witness the diffusion pole at $z=-ik^2\eta/mn$, in the lower half plane, corresponding to a hydrodynamic lifetime $\tau(k) = mn/(\eta k^2) \to \infty$ as $k \to 0$.

4.2.2. Longitudinal Fluctuations

The other equations are only a little more complicated because they couple the longitudinal variables. By using the constitutive relations (4.8) and the conservation laws (4.7), we find

$$\partial_t\langle n(\vec{r},t)\rangle + \vec{\nabla}\cdot\langle \vec{g}_\ell(\vec{r},t)\rangle/m = 0 , \tag{4.17a}$$

$$\partial_t\langle \vec{g}_\ell(\vec{r},t)\rangle + \vec{\nabla}p(\vec{r},t) - (\zeta+\tfrac{4}{3}\eta)/(mn)\,\vec{\nabla}(\vec{\nabla}\cdot\langle \vec{g}_\ell(\vec{r},t)\rangle) = 0 , \tag{4.17b}$$

$$\partial_t\langle \epsilon(\vec{r},t)\rangle + (\epsilon+p)\vec{\nabla}\cdot\langle \vec{g}_\ell(\vec{r},t)\rangle/(mn) - \varkappa\nabla^2T(\vec{r},t) = 0 . \tag{4.17c}$$

The second equation is really a scalar equation since all of its terms are parallel to $\vec{\nabla}$, i.e., $\vec{k}$.

Eqs. (4.17a,c) combine to

$$\partial_t[\langle\epsilon(\vec{r},t)\rangle - \frac{\epsilon+p}{n}\langle n(\vec{r},t)\rangle] - \varkappa\nabla^2T(\vec{r},t) = 0, \tag{4.18}$$

which indicates that it may be convenient to introduce the variable

$$q(\vec{r},t) \equiv \epsilon(\vec{r},t) - \frac{\epsilon+p}{n} n(\vec{r},t) \tag{4.19}$$

in place of the energy density. Its physical significance becomes apparent from the thermodynamic relations (4.9) which, after the usual headache with thermodynamic manipulations, take the form

$$\vec{\nabla}p(\vec{r},t) = \left(\frac{\partial p}{\partial n}\right)_S \vec{\nabla}\langle n(\vec{r},t)\rangle + \frac{V}{T}\left(\frac{\partial p}{\partial S}\right)_n \vec{\nabla}\langle q(\vec{r},t)\rangle , \tag{4.20a}$$

$$\vec{\nabla}T(\vec{r},t) = \left(\frac{\partial T}{\partial n}\right)_S \vec{\nabla}\langle n(\vec{r},t)\rangle + \frac{V}{T}\left(\frac{\partial T}{\partial S}\right)_n \vec{\nabla}\langle q(\vec{r},t)\rangle , \tag{4.20b}$$

where S is the total entropy. These equations identify $q(\vec{r},t)$ as an operator for (T times) an entropy density. The corresponding thermodynamic identity is

$$Tn\, d(S/N) = d\varepsilon - \frac{\varepsilon+p}{n}\, dn , \tag{4.21}$$

where S/N is the entropy per particle.

And now let us, once again, write down the three longitudinal equations (4. 17). With (4. 19,20), they become

$$\partial_t\langle n\rangle + \nabla\langle g_\ell\rangle/m = 0,$$

$$[\partial_t - (\tfrac{4}{3}\eta + \zeta)/(mn)\,\nabla^2]\langle g_\ell\rangle + \left(\frac{\partial p}{\partial n}\right)_S \nabla\langle n\rangle + \frac{V}{T}\left(\frac{\partial p}{\partial S}\right)_n \nabla\langle q\rangle = 0,$$

$$[\partial_t - \kappa\frac{V}{T}\left(\frac{\partial T}{\partial S}\right)_n \nabla^2]\langle q\rangle - \kappa\left(\frac{\partial T}{\partial n}\right)_S \nabla^2\langle n\rangle = 0, \tag{4.22}$$

where we have omitted the arguments $(\vec{r},t)$. These equations are again of the type (2.7) or (4. 14), with the minor difference that there are now three of them, coupled To simplify their appearance a little, we introduce a few names:

$$D_\ell = (\tfrac{4}{3}\eta + \zeta)/mn ,$$

$$mnc_v = \frac{T}{V}\left(\frac{\partial S}{\partial T}\right)_n ; \quad mnc_p = \frac{T}{V}\left(\frac{\partial S}{\partial T}\right)_p ,$$

$$c^2 = \left(\frac{\partial p}{\partial mn}\right)_S = \frac{c_p}{c_v}\left(\frac{\partial p}{\partial mn}\right)_T . \tag{4.23}$$

$D_\ell$ is called the longitudinal diffusion coefficient. $mc_v$ and $mc_p$ are the specific heats per particle, at constant volume and pressure, respectively. And the inverse adiabatic compressibility, $c^2$, will be found to be the (square of the) speed of sound.

Again we do a Fourier-Laplace transformation as in (4.15). This brings (4.22) into the matrix form

$$\begin{bmatrix} z & -k/m & 0 \\ -kmc^2 & z+ik^2D_\ell & -\frac{V}{T}\left(\frac{\partial p}{\partial S}\right)_n k \\ ik^2\kappa\left(\frac{\partial T}{\partial n}\right)_S & 0 & z+ik^2\frac{\kappa}{mnc_v} \end{bmatrix} \begin{bmatrix} \langle n(\vec{k}z)\rangle \\ \langle g_\ell(\vec{k}z)\rangle \\ \langle q(\vec{k}z)\rangle \end{bmatrix} = i \begin{bmatrix} \langle n(\vec{k},t{=}0)\rangle \\ \langle g_\ell(\vec{k},t{=}0)\rangle \\ \langle q(\vec{k},t{=}0)\rangle \end{bmatrix} . \tag{4.24}$$

For the moment, let us consider this a result, leaving the inversion of the 3x3 matrix for homework. However, to get some physics out of what we have, we will inquire into the poles of $\langle n(\vec{k}z)\rangle$ etc. on the complex z-plane. That means to set the determinant of the matrix in (4.24) equal to zero. For small k, the resulting cubic equation for z is solved with little difficulty, and yields

$$z = \pm ck - \frac{i}{2}k^2\Gamma \qquad \text{(sound poles)}, \tag{4.25a}$$

$$z = -ik^2D_T \qquad \text{(heat pole)}, \tag{4.25b}$$

up to terms of order $k^3$. Here, the heat diffusion constant $D_T$ is given by

$$D_T = \kappa/mnc_p \tag{4.25c}$$

and the sound attenuation constant $\Gamma$ by

$$\Gamma = D_\ell + D_T\left(\frac{c_p}{c_v} - 1\right) = D_T\left(\frac{c_p}{c_v} - 1\right) + \left(\frac{4}{3}\eta + \zeta\right)/mn . \tag{4.25d}$$

Note that all three poles lie on the lower half of the complex z-plane which is a relief for the stability-minded, and that they are all, of course, hydrodynamic: $\pi(k) \sim k^{-2}$.

## 4.3 Thermodynamic Sum Rules

We have solved the hydrodynamic initial value problem, obtaining eqs. (4.16) and (4.24) which are of the same form as (3.14), and therefore allow a comparison with the formulation of the same problem in terms of correlation functions. To make that comparison, however, we need the static susceptibilities $\chi_{ij}(\vec{k})$, at least for small k, which are not determined by the equations of motion. These susceptibilities are instead determined by equilibrium statistical mechanics as we saw already in section 2.4. Before listing them, let us introduce a useful notation. We analyzed the symmetry restrictions on the momentum density correlation function in (3.24). Accordingly, we can separate this function into a longitudinal and a transverse part by

$$\chi''_{g_i g_j}(\vec{k}\omega) = \frac{k_i k_j}{k^2}\chi''_{\ell}(k\omega) + (\delta_{ij} - \frac{k_i k_j}{k^2})\chi''_t(k\omega) . \tag{4.26}$$

This splitting is clearly equivalent to that in eq. (4.13). $\chi''_{\ell,t}(k\omega)$ are real functions of $k = |\vec{k}|$, odd in $\omega$.

And now the susceptibilities:

$$\lim_{k\to0} \chi_{nn}(k) = \lim_{k\to0} \int\frac{d\omega}{\pi}\chi''_{nn}(k\omega)/\omega = n(\frac{\partial n}{\partial p})_T , \tag{4.27a}$$

$$\lim_{k\to0} \chi_{qq}(k) = \lim_{k\to0} \int\frac{d\omega}{\pi}\chi''_{qq}(k\omega)/\omega = mnc_pT , \tag{4.27b}$$

$$\lim_{k\to0} \chi_{nq}(k) = \lim_{k\to0} \int\frac{d\omega}{\pi}\chi''_{nq}(k\omega)/\omega = T(\frac{\partial n}{\partial T})_p , \tag{4.27c}$$

$$\chi_\ell(k) = \int\frac{d\omega}{\pi}\,\chi''_\ell(k\omega)/\omega = mn\,, \tag{4.27d}$$

$$\lim_{k\to 0}\chi_t(k) = \int\frac{d\omega}{\pi}\,\chi''_t(k\omega)/\omega = mn\,, \tag{4.27e}$$

$$\chi_{g_i n}(\vec{k}) = \chi_{g_i q}(\vec{k}) = 0\,. \tag{4.27f}$$

The last two equations, (4.27f), are a consequence of time reversal symmetry. By eq. (3.25), $\chi''_{g_i n}(\vec{k}\omega)/\omega$ and $\chi''_{g_i q}(\vec{k}\omega)/\omega$ are odd functions of $\omega$, and their frequency integral vanishes at all $\vec{k}$.

The three equations, (4.27a-c), follow by an analysis parallel to that given in section 2.4. Considering (4.27a), the fluctuation-dissipation theorem (3.36) gives

$$\chi_{nn}(k) = \int\frac{d\omega}{2\pi}\,S_{nn}(k\omega)(1-e^{-\hbar\omega\beta})/\hbar\omega\,. \tag{4.28}$$

Next, we notice that in the limit as $k\to 0$, $S_{nn}(k\omega) \sim \delta(\omega)$ because the total number of particles, $N = \int d\vec{r}\,n(\vec{r},t)$, is conserved. Thus we can write

$$\lim_{k\to 0}\chi_{nn}(k) = \lim_{k\to 0}\beta\int\frac{d\omega}{2\pi}\,S_{nn}(k\omega) = \lim_{k\to 0}\beta\,S_{nn}(k,t=0) \tag{4.29}$$

or

$$\chi_{nn}(k=0) = \beta\langle n(N-\langle N\rangle)\rangle = (\beta/V)\langle(N-\langle N\rangle)^2\rangle. \tag{4.30}$$

Finally, in a grand canonical ensemble,

$$n = \frac{\langle N\rangle}{V} = \mathrm{tr}\,e^{-\beta(H-\mu N)}N/\mathrm{tr}\,e^{-\beta(H-\mu N)} \tag{4.31a}$$

and therefore

$$\left(\frac{\partial n}{\partial\mu}\right)_{\beta,V} = (\beta/V)\langle N^2-\langle N\rangle^2\rangle = n\left(\frac{\partial n}{\partial p}\right)_T. \tag{4.31b}$$

The last equation follows from the thermodynamic identity $dp = n d\mu + (S/V)\, dT$ relating the pressure p to the chemical potential $\mu$. This proves (4.27a). The other two thermodynamic sum rules, (4.27b,c), follow similarly but after a bit of thermodynamic manipulation.

We add two remarks. The first which is minor is to point out that formally, whatever A and B,

$$\frac{\partial}{\partial\lambda} \operatorname{tr} e^{-\beta(H-\lambda B)} A \Big/_{\lambda=0} = \operatorname{tr} e^{-\beta H} \int_0^1 d\beta' \, e^{\beta' H} B \, e^{-\beta' H} A \, , \quad (4.32)$$

from which it follows that, for example,

$$\left(\frac{\partial n}{\partial \mu}\right)_{\beta, V} = \beta \, C_{nn}(k=0, t=0) = \int \frac{d\omega}{\pi} \, \chi''_{nn}(k=0, \omega)/\omega \, , \quad (4.33)$$

where $C_{nn}$ is the Kubo function defined in eq. (3.43). To establish (4.27) therefore, the fluctuation-dissipation theorem is necessary but the conservation laws are not.

The second remark has to do with our choice of ensemble. If we were to interpret (4.30), rashly, in the canonical ensemble which holds the total number of particles fixed, we would find $\chi_{nn}(k=0) = 0$. At finite k, with $k^{-1}$ small compared to the linear dimensions of the container, $\chi_{nn}(k)$ describes local fluctuations of the particle density, and is independent of the choice of ensemble (among the proper ensembles: microcanonical, canonical, etc.) as well as of boundary conditions. To eliminate such dependence, we have to consider an infinite medium, i.e., take the limit $V \to \infty$ first, with $N/V = n$ fixed as usual, and $k \to 0$ second. If, for practical reasons, we want to calculate $\lim_{k \to 0} \chi(k)$ by equating it with $\chi(k=0)$, we can do so only in an ensemble which allows, even for finite V, fluctuations of N as they take place locally, i.e., as if the system enclosed in volume V where a part of a much larger system with which it freely exchanges particles (and energy). This, however, is the characteristic property of the grand canonical ensemble.

Eq. (4.27d) is an immediate consequence of the f-sum rule (3.34) which in turn expresses the equal time commutator

$$[n(\vec{r}), \vec{g}(\vec{r}\,')] = -i\hbar\, \vec{\nabla}\, n(\vec{r})\, \delta(\vec{r}-\vec{r}\,') . \tag{4.34}$$

We only have to use particle conservation,

$$\partial_t n(\vec{r},t) = -\vec{\nabla}\cdot \vec{g}(\vec{r},t)/m = -\nabla g_\ell(\vec{r},t)/m , \tag{4.35a}$$

in the form

$$\omega^2 \chi''_{nn}(k\omega) = k^2 \chi''_\ell(k\omega)/m^2 \tag{4.35b}$$

to obtain eq. (4.27d) from (3.34). Note that this sum rule holds for arbitrary k.

For a classical system, the transverse sum rule (4.27e) is also exact for all k. By the classical fluctuation-dissipation theorem (3.39),

$$\chi_{g_i g_j}(\vec{k}) = \beta \int \frac{d\omega}{2\pi} S_{g_i g_j}(\vec{k}\omega) = \beta \int d(\vec{r}-\vec{r}\,') e^{-i\vec{k}(\vec{r}-\vec{r}\,')} \langle g_i(\vec{r}) g_j(\vec{r}\,')\rangle . \tag{4.36}$$

Now the classical equilibrium average is easy to calculate. In a canonical or grand canonical ensemble the momenta of different particles are not correlated so that $\langle p_i^\alpha p_j^\beta\rangle = \delta_{ij}\delta_{\alpha,\beta} m/\beta$. And since, classically, averages over positions and momenta factorize, we obtain

$$\langle g_i(\vec{r})\, g_j(\vec{r}\,')\rangle = \delta_{ij}\, mn/\beta\, \delta(\vec{r}-\vec{r}\,') \quad \text{(classically)}, \tag{4.37a}$$

where we have used the explicit expression (4.6) for the densities; and therefore for any k

$$\chi_{g_i g_j}(\vec{k}) = \int \frac{d\omega}{\pi} \chi''_{g_i g_j}(\vec{k}\omega)/\omega = mn\delta_{ij} \quad \text{(classically)} . \qquad (4.37b)$$

Quantum-mechanically, the transverse equation (4.27e) holds only in the limit as $k \to 0$. It holds in this limit since for very small k, or very large wavelength, longitudinal and transverse fluctuations are indistinguishable.

Unless something unforeseen happens. Naturally, it does happen: in superfluid helium, as well as in superconductors, there is macroscopic quantum phase coherence, of which one consequence is the fact that longitudinal and transverse fluctuations are distinct even at infinite wavelength, or more precisely,

$$\lim_{k \to 0} \int \frac{d\omega}{\pi} \chi''_t(k\omega)/\omega = \rho_n < mn \quad \text{(superfluid)}. \qquad (4.38)$$

The macroscopic manifestations of this property of superfluids are profound, and chapter 10will be devoted to their discussion. For now, we return to normal fluids where everything is nice and simple.

## 4.4 The Hydrodynamic Correlation Functions

We have two representations for the small amplitude relaxation problem: the first, given by eqs. (4.16) and (4.24), is macroscopic, and valid only for small wave vector and frequency; the second, generally given by eqs. (3.14), is microscopically rigorous, for any k and z. Comparing the two we obtain the limiting, hydrodynamic expressions for the microscopic correlation functions. The simplest of these is the transverse momentum density correlation function $\chi''_t(k\omega)$. From (4.16) and (3.14), and using the Kubo function introduced in (3.42-44), we find

$$C_t(kz) = \beta^{-1}\int \frac{d\omega}{\pi}\, \frac{\chi_t''(k\omega)}{\omega(\omega - z)} = \frac{i\beta^{-1} mn}{z + ik^2\eta/mn}\,, \tag{4.39}$$

for $\text{Im}\, z > 0$. And taking the imaginary part at $z = \omega + i0$, we obtain

$$\frac{1}{\omega}\chi_t''(k\omega) = \frac{k^2\eta}{\omega^2 + (k^2\eta/mn)^2}\,. \tag{4.40}$$

This function has, again, the familiar diffusion structure. Note in passing that the shear viscosity, $\eta$, must of course be positive; eq. (4.40) proves that since $\omega\chi_t''(k\omega) \geq 0$.

The longitudinal correlation functions follow in the same fashion. Of course, we do have to invert the matrix in eq. (4.24) now, but for k so small that

$$(D_T k^2)^2 \ll c^2 k^2\,, \tag{4.41}$$

that is not so fearsome. For the Kubo functions, we obtain:

$$C_{nn}(kz) = i\beta^{-1} n\left(\frac{\partial n}{\partial p}\right)_T \left[\frac{c_v}{c_p}\, \frac{z + ik^2(\Gamma + D_T[c_p/c_v - 1])}{z^2 - c^2k^2 + izk^2\Gamma} + \left(1 - \frac{c_v}{c_p}\right)\frac{1}{z + ik^2 D_T}\right], \tag{4.42a}$$

$$C_{qq}(kz) = \frac{i\beta^{-1}\, mnc_p T}{z + ik^2 D_T}\,, \tag{4.42b}$$

$$C_{nq}(kz) = i\beta^{-1} T\left(\frac{\partial n}{\partial T}\right)_p \left[\frac{ik^2 D_T}{z^2 - c^2k^2 + izk^2\Gamma} + \frac{1}{z + ik^2 D_T}\right]. \tag{4.42c}$$

These expressions are rigorous in the following sense: If the Kubo functions $C_{ij}(kz)$ are written as a sum of individual pole contributions,

$$C_{ij}(kz) = i\beta^{-1}\chi_{ij}(k)\left[\frac{Z_{ij}^{(+)}(k)}{z-ck+\frac{i}{2}k^2\Gamma} + \frac{Z_{ij}^{(-)}(k)}{z+ck-\frac{i}{2}k^2\Gamma} + \frac{Z_{ij}^{(T)}(k)}{z+ik^2 D_T}\right], \tag{4.43}$$

then hydrodynamics correctly determines the positions of the poles to order $k^2$, and the residues $Z_{ij}$ to order k. Note that all the non-vanishing susceptibilities are even in k, so that e.g., $\chi_{nn}(k) = n\left(\frac{\partial n}{\partial p}\right)_T + O(k^2)$.

From eqs. (4.42), we obtain the absorptive parts $\chi''(k\omega)$ easily:

$$\frac{1}{\omega}\chi''_{nn}(k\omega) = n\left(\frac{\partial n}{\partial p}\right)_T \left[\frac{(c_v/c_p)\, c^2 k^4 \Gamma}{(\omega^2-c^2k^2)^2+(\omega k^2\Gamma)^2} + \frac{(1-c_v/c_p)\, k^2 D_T}{\omega^2+(k^2 D_T)^2}\right.$$

$$\left. - \left(1-\frac{c_v}{c_p}\right)\frac{(\omega^2-c^2k^2)\, k^2 D_T}{(\omega^2-c^2k^2)^2+(\omega k^2\Gamma)^2}\right], \tag{4.44a}$$

$$\frac{1}{\omega}\chi''_{qq}(k\omega) = mnc_p T\, \frac{k^2 D_T}{\omega^2+(k^2 D_T)^2}, \tag{4.44b}$$

$$\frac{1}{\omega}\chi''_{nq}(k\omega) = T\left(\frac{\partial n}{\partial T}\right)_p \left[\frac{k^2 D_T}{\omega^2+(k^2 D_T)^2} - \frac{(\omega^2-c^2k^2)\, k^2 D_T}{(\omega^2-c^2k^2)^2+(\omega k^2\Gamma)^2}\right]. \tag{4.44c}$$

After this onslaught, the reader will appreciate that density correlation functions do indeed have a very complicated analytic structure at hydrodynamically small k and $\omega$. We have not written down correlation functions which involve the longitudinal momentum density $g_\ell$ since for any A,

$$\omega\chi''_{nA}(k\omega) = \frac{1}{m}k\,\chi''_{g_\ell A}(k\omega) \tag{4.45}$$

because of particle conservation.

Now let us try to understand what we have obtained. There are two processes, one propagating and one diffusive. To order k, the diffusive mode corresponds to fluctuations of the entropy variable $q(\vec{r},t) = \varepsilon(\vec{r},t) - \frac{\varepsilon+p}{n} n(\vec{r},t)$. As is apparent from eq. (4.44b), a local entropy fluctuation will spread out by a random walk process, with diffusivity $D_T = \varkappa/mnc_p$. Only at very low temperature, when $(\frac{\partial T}{\partial n})_S = 0$, is the diffusive mode a pure temperature fluctuation, see eq. (4.20b).

The other modes are, of course, identical with sound waves, propagating with the Laplacean speed of sound $c = (\frac{\partial p}{\partial mn})_S^{1/2}$. Their damping is viscous, as seen from eq. (4.25d), but contains a thermal admixture $\sim \varkappa$ as well. Only at very low temperature, when $c_p \approx c_v$, is the propagating mode a purely mechanical density fluctuation.

The discrepancy between the Newtonian, isothermal, speed of sound $c_N = (\partial p/\partial mn)_T^{1/2}$ and the observed adiabatic one $c = (\partial p/\partial mn)_S^{1/2}$ is interesting. Historically, Newton's theory, which was purely mechanical, could only be corrected with the later recognition of thermal--as different from mechanical--processes. It is the coupling of density and energy fluctuations which converts $c_N$ to c. In simple microscopic theories, such coupling is often omitted, and the isothermal speed of sound obtained. As an interesting exercise, the reader may amuse himself by repeating the calculation which led us to eq. (2.107), applying the same reasoning to the particle density instead of the magnetization. He will find the dispersion relation $\omega = c_N k + O(k^2)$: high-frequency sound is isothermal.

The contrast in complexity between, say, $\chi''_{nn}$ and $\chi''_t$ is striking. The transverse momentum density is conserved, and not coupled to any other conserved quantity. This fact results in its simple diffusion structure. The density variable and its correlation function are complicated by two features: First, its current, the longitudinal momentum density, is itself conserved; because of this, a density fluctuation can "overshoot" and come to equilibrium in an oscillatory (propagating) rather than a diffusive fashion. And second, the density is coupled to yet another conserved variable, the energy density.

The last term in (4.44a) is of only minor importance as far as light scattering experiments are concerned, and it is often omitted. However, in order for $\chi''_{nn}(k\omega)$ to satisfy the f-sum rule (3.34), and therefore to be in accord with momentum conservation, this term is necessary. In addition, the hydrodynamic expressions (4.44) exhaust all the thermodynamic sum rules (4.27). This is no surprise; our derivation guarantees that these sum rules are obeyed, to order k. However, none of the higher sum rules, such as $\int d\omega\, \omega^3 \chi''$, can be fulfilled. This is in direct parallel to our discussion in section 2.10; the hydrodynamic correlation functions are rigorous at small k and $\omega$, but they fall off too slowly at large frequency.

## 4.5 Light Scattering

Eq. (4.44a) is an old and famous result which has been first derived by Landau and Placzek (1934). Its experimental importance is, of course, the fact that $\chi''_{nn}(k\omega)$, or rather

$$S_{nn}(k\omega) = 2\hbar(1-e^{-\hbar\omega\beta})^{-1} \chi''_{nn}(k\omega) \simeq 2k_BT\, \chi''_{nn}(k\omega)/\omega \; , \qquad (4.46)$$

is what is measured in (many) light scattering experiments, see eq. (4.3). (With $\omega$ of order ck where $c \approx 10^5$ cms$^{-1}$ say and $k \approx 10^4$ cm$^{-1}$, $2\hbar(1-e^{-\hbar\omega\beta})^{-1} = 2/\beta\omega$ for reachable temperatures.) Indeed, since $k = 2k_i \sin(\theta/2)$ is maximally $k_{max} \simeq 2k_{in} \approx 10^{-4}$cm, k is much smaller than the inverse mean free path in liquids and in all but very dilute gases. $\omega$ is also small, of order $10^9$ s$^{-1}$ or less, so that our hydrodynamic theory should be perfectly appropriate. And light scattering experiments produce a wealth of information.

To see this, consider fig. 4.3 which gives the spectrum in some detail.

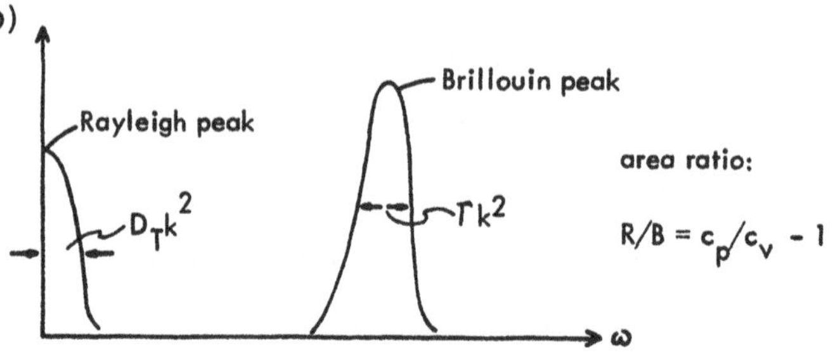


Fig. 4.3

The most immediate quantities to obtain are the speed of sound, from the position of the Brillouin line, the sound attenuation constant $\Gamma$ from its width, and the thermal diffusivity from the width of the central peak. To measure the isothermal compressibility $(\partial n/\partial p)_T$ or the specific heat ratio $c_p/c_v = mc^2(\partial n/\partial p)_T$ seems to require the sum rule (4.27a) and therefore an absolute intensity measurement. Such measurements are difficult, and accurate knowledge of the "factors" in eq. (4.3), in particular the polarizability contained in them, is hard to come by. Fortunately, relative intensity measurements suffice to obtain $c_p/c_v$ since the ratio (called the Landau-Placzek ratio) of the areas under half the Rayleigh peak and one Brillouin peak is $\frac{c_p}{c_v} - 1$. The reader will convince himself of this fact from eq. (4.44a) or, much more easily, by looking at the residues in eq. (4.42a). In summary then, one obtains from Brillouin and Rayleigh scattering these quantities:

$$\frac{c_p}{c_v};\ \left(\frac{\partial p}{\partial mn}\right)_T;\ D_T = \frac{\varkappa}{mnc_p};\ D_\ell = \left(\frac{4}{3}\eta + \zeta\right)/mn\ . \qquad (4.47)$$

Quite a harvest. Note, however, what these experiments involve. The frequency shifts $\omega$, of order ck, are extremely small, $\lesssim 10^8\ s^{-1}$ which is to be compared to $\sim 10^{15}\ s^{-1}$ for the frequency of the incident light. This requires light sources which are extremely monochromatic; lasers, in other words. It also requires very accurate measurements of frequency shifts which is accomplished by self-beating techniques, specially invented to awe the theoretician.

Suppose we could extend these measurements of $\chi''_{nn}(k\omega)$ or $S_{nn}(k\omega)$ to arbitrary k and $\omega$. Through X-ray and neutron scattering, and through the computer studies of molecular dynamics which are a theoretician's experiment (see the brief review of Berne and Forster, 1971), one can go at least part of the way. What else would we get? Well, knowing the particle mass m we could, even from the hydrodynamic spectrum, obtain the mass density n because of the f-sum rule $\int d(\omega/\pi)\,\omega\,\chi''_{nn}(k\omega) = k^2(n/m)$. Moreover, we can obtain the static structure factor $S(k) \equiv S_{nn}(k,t=0)$ since

$$S(k) = k_B T \int \frac{d\omega}{\pi}\, \chi''_{nn}(k\omega)/\omega \;, \tag{4.48}$$

where we have taken a classical system. If we know the interaction potential v(r), this determines the pressure and the energy. Namely, from the eqs. (4.6), we find that

$$\epsilon = \frac{3}{2} n k_B T + \frac{1}{2} n^2 \int d\vec{r}\; v(r)\, g(r) \;, \tag{4.49}$$

$$p = n k_B T - \frac{1}{6} n^2 \int d\vec{r}\; r\, v'(r)\, g(r) \;, \tag{4.50}$$

where g(r) is the pair correlation function,

$$\begin{aligned} n^2 g(|\vec{r}-\vec{r}\,'|) &= \langle \sum_{\alpha\neq\beta} \delta(\vec{r}-\vec{r}^{\,\alpha})\, \delta(\vec{r}\,'-\vec{r}^{\,\beta}) \rangle \\ &= n^2 + \int \frac{d\vec{k}}{(2\pi)^3}\, e^{i\vec{k}(\vec{r}-\vec{r}\,')} [S(k)-n] \;, \end{aligned} \tag{4.51}$$

and $v'(r) = \frac{dv(r)}{dr}$. From $\epsilon$ in particular, we could obtain the specific heat, $mnc_v = (\partial\epsilon/\partial T)_n$. A theoretician should therefore calculate $\chi''_{nn}$, or better the momentum density correlation function $\chi''_{g_i g_j}$ since its transverse part $\chi''_t$ determines the shear viscosity $\eta$. From $\chi''_{g_i g_j}(k\omega)$ he would get all the thermodynamic and hydrodynamic parameters in a classical system, and most of them in a quantum system.

Let us mention, finally, the special features that appear when the liquid is near its critical point where $(\partial p/\partial n)_T \to 0$. There are three effects. First, the total light scattering intensity becomes enormous because of the thermodynamic sum rule (4.27a). This is the phenomenon of critical opalescence. Second, while at low temperatures the Brillouin peaks are much more prominent than the central peak since $c_p \approx c_v$, near the critical point the Landau-Placzek ratio $\frac{c_p}{c_v} - 1$ becomes very large since

$$c_p - c_v = \frac{T}{mn^2} \left(\frac{\partial p}{\partial T}\right)_n^2 \left(\frac{\partial n}{\partial p}\right)_T \to \infty \,. \tag{4.52}$$

Therefore, most of the intensity near the critical point is in the central peak. The entropy fluctuates wildly while the more mechanical sound waves are essentially unaffected which is physically as one should expect. Third, even though the total Rayleigh intensity increases, its spectral width becomes very narrow since $D_T = \varkappa/mnc_p$ and $c_p \to \infty$. This is the phenomenon of critical slowing down already encountered in section 2.10. $10^{-2}$ degrees away from the critical point, you can, after each change of temperature, go to the movies while the system slowly comes to equilibrium.

As explained in section 2.10, the transport coefficient, which is here $\varkappa$, is determined by local, rapid fluctuations and should be insensitive to the large-scale rearrangement that goes on at the critical point. This argument, implicit in the early theory by Van Hove (1954), is qualitatively borne out: $\varkappa/c_p \to \infty$ at $T_c$. More recently, however, it has been found that $\varkappa$ becomes weakly singular as well. The experimental and theoretical situation in critical phenomena is reviewed in the book by Stanley (1971). This has been field of continued intensive activity since Wilson's (1971) renormalization group ideas appeared; another book would already be necessary to describe it. ( It is being written: Ma, 1975).

## 4.6 Kubo Expressions for the Transport Coefficients

From the results we have obtained, eqs. (4.44), we can extract the transport coefficients by performing appropriate limits. The limits involved are those for small k and $\omega$ where the hydrodynamic theory becomes rigorous. The Kubo-type relations which we obtain are therefore exact expressions, relating macroscopic measurements (of $\eta$, $\kappa$ etc.) to the underlying microscopic structure.

From eq. (4.40), we obtain the shear viscosity as

$$\eta = \lim_{\omega\to 0} [\lim_{k\to 0} \frac{\omega}{k^2} \chi''_t(k\omega)] . \tag{4.53a}$$

The longitudinal viscosity follows similarly from (4.44a) as

$$\frac{4}{3}\eta + \zeta = \lim_{\omega\to 0} [\lim_{k\to 0} \frac{m^2\omega^3}{k^4} \chi''_{nn}(k\omega)] = \lim_{\omega\to 0} [\lim_{k\to 0} \frac{\omega}{k^2} \chi''_\ell(k\omega), \tag{4.53b}$$

since $m^2\omega^2\chi''_{nn}(k\omega) = k^2\chi''_\ell(k\omega)$ from particle conservation, see eq. (4.45). And finally, the entropy correlation function (4.44b) gives the heat conductivity,

$$\kappa T = \lim_{\omega\to 0} [\lim_{k\to 0} \frac{\omega}{k^2} \chi''_{qq}(k\omega)] . \tag{4.53c}$$

Thus all three transport coefficients involved in the hydrodynamics of a normal liquid are given by expressions of the same structure we already encountered in eq. (2.57), namely

$$\lambda_{AB} = \lim_{\omega\to 0} [\lim_{k\to 0} \frac{\omega}{k^2} \chi''_{AB}(k\omega)] \tag{4.54}$$

for the transport coefficient $\lambda_{AB}$ associated with the conserved variables A, B. As it happens, the normal liquid has only diagonal $\lambda$'s. $\lambda_{AB}$ would appear in constitutive relations in the form

$$\delta\langle \vec{j}^{A}(\vec{r},t)\rangle = -\lambda_{AB}\,\vec{\nabla}\,\delta b(\vec{r},t)\,, \tag{4.55}$$

where $\delta b$ is the thermodynamic force conjugate to the variable $\delta\langle A\rangle$: $\delta b_n = \delta p/n$, $\delta b_q = \delta T/T$ etc., as we will see more systematically later; see also Martin (1965). The current densities are defined by

$$\partial_t\, A(\vec{r},t) + \nabla_i j_i^A(\vec{r},t) = 0 \quad\text{and}\quad \partial_t B(\vec{r},t) + \nabla_i j_i^B(\vec{r},t) = 0\,. \tag{4.56}$$

Note that $\chi''_{AB}(k\omega) = \chi''_{BA}(k\omega)$ if A and B have the same signature under time reversal and parity. Then

$$\lambda_{AB} = \lambda_{BA}\,, \tag{4.57}$$

which are the famous Onsager relations.

If we use the conservation laws (4.56) and the fluctuation-dissipation theorem (3.38), we can express the transport coefficients in terms of current correlations, as we did in eq. (2.29a). Generally, we obtain from (4.54)

$$\lambda_{AB} = \lim_{\omega\to 0}\lim_{k\to 0}\frac{1}{4k_BT}\int_{-\infty}^{\infty} dt \int d\vec{r}\; e^{i\omega t - i\vec{k}\cdot\vec{r}}\langle\{ j_\ell^A(\vec{r},t), j_\ell^B(\vec{0},0)\}\rangle, \tag{4.58}$$

where $j_\ell = \hat{k}\cdot\vec{j}$ is the longitudinal component of the current density. (There is no need to subtract the constant terms $\langle j^A\rangle\langle j^B\rangle$ as in (3.37) since their Fourier transform is $\sim\delta(\omega)$ and does not contribute to the limit as $\omega\to 0$.) This procedure, and a pinch of rotational symmetry, gives the hydrodynamic transport coefficients in the invariant form

$$\kappa T = \lim_{\omega\to 0}\lim_{k\to 0} \frac{1}{12k_BT}\int_{-\infty}^{\infty} dt \int d\vec{r}\, e^{i\omega t - i\vec{k}\cdot\vec{r}} \langle\{\vec{j}^q(\vec{r},t), \vec{j}^q(\vec{0},0)\}\rangle, \quad (4.59a)$$

$$\eta(\delta_{ij} + \frac{1}{3}\frac{k_ik_j}{k^2}) + \zeta\frac{k_ik_j}{k^2} = \lim_{\omega\to 0}\lim_{k\to 0}\frac{1}{4k_BT}\int_{-\infty}^{\infty} dt \int d\vec{r}\, e^{i\omega t - i\vec{k}\cdot\vec{r}}$$

$$\times \sum_{m,n} \frac{k_mk_n}{k^2} \langle\{\tau_{im}(\vec{r},t), \tau_{jn}(\vec{0},0)\}\rangle. \quad (4.59b)$$

These results make it evident that not only $\kappa$, $\eta$ and $\frac{4}{3}\eta + \zeta$, but also $\zeta$ are positive. Pretty expressions, aren't they? They invite a few comments of a general nature:

Consider eq. (4.58), expressing a transport coefficient in terms of current densities of conserved variables. These currents (i.e., $\tau_{ij}$, $j^q$) are not, in general, themselves conserved. Consequently, fluctuations of the total current $\mathcal{J}^A(t) = \int d\vec{r}\, j^A(\vec{r},t)$ will decay within a finite, microscopic time, as explained in section 2.1. In the absence of long-ranged correlations, the correlation functions in (4.58) will therefore be different from zero, effectively, only over some finite range of $\vec{r}$ and t. There is thus no convergence problem with the integral

$$\lambda_{AB} = \frac{1}{k_BT}\int_0^{\infty} dt \int d\vec{r}\, [\langle\frac{1}{2}\{ j_\ell^A(\vec{r},t), j_\ell^B(\vec{0},0)\}\rangle - \langle j_\ell^A\rangle\langle j_\ell^B\rangle], \quad (4.60)$$

and a finite transport coefficient results. What we have just said is that the limit (4.54) will, in general, be finite if A and B are conserved but $j^A$ and $j^B$ are not.

Now the longitudinal momentum density $g_\ell(\vec{r},t)$ is the current of the conserved mass density $mn(\vec{r},t)$. However, $g_\ell$ obeys itself a continuity equation, with the current $\tau_{zz}$, taking $\vec{k} = k\hat{z}$. According to the argument given above, the zz-component of (4.59b) which gives $\frac{4}{3}\eta + \zeta$, should thus converge, and following the equations backwards to (4.53b) this means that for small but finite $\omega$

$\lim(k\to 0)\, k^{-4}\, \chi''_{nn}(k\omega)$ must be finite. Consequently, $\lim(k\to 0)\, k^{-2}\, \chi''_{nn}(k\omega)$ vanishes, and therefore

$$\lambda_{nn} = \lim_{\omega\to 0} [\lim_{k\to 0} \frac{\omega}{k^2} \chi''_{nn}(k\omega)] = 0 . \tag{4.61}$$

By the same token,

$$\lambda_{nq} = \lambda_{qn} = \lim_{\omega\to 0} [\lim_{k\to 0} \frac{\omega}{k^2} \chi''_{nq}(k\omega)] = 0 . \tag{4.62}$$

These coefficients would appear in constitutive relations in the form already given in eqs. (4.11). We have now given the answer promised there. Since we omitted the coefficients $\lambda_{nn}$ and $\lambda_{nq}$ from the outset, it is hardly startling that our expressions (4.44) are in agreement with eqs. (4.61) and (4.62).

It is apparent from (4.61) and (4.62) that

$$\lim_{k\to 0} \frac{1}{k^2} \chi''_{qq}(k\omega) = \lim_{k\to 0} \frac{1}{k^2} \chi''_{q+\lambda n, q+\lambda n}(k\omega) , \tag{4.63}$$

whatever $\lambda$. If we choose $\lambda = (\epsilon+p)/n$ and remember that $\delta q + \frac{\epsilon+p}{n}\delta n = \delta\epsilon$, eq. (4.19), we find that the Kubo expression for the heat conductivity can also be written in terms of the energy-energy correlation function,

$$\varkappa T = \lim_{\omega\to 0} [\lim_{k\to 0} \frac{\omega}{k^2} \chi''_{\epsilon\epsilon}(k\omega)] \tag{4.64}$$

instead of eq. (4.53c). This fact, a consequence as we saw of momentum conservation, is given a thorough analysis by Kadanoff and Martin (1963, p. 458) which we urge our reader to look up, even if it means that he might become aware of just how much we have pilferaged from this fundamental paper.

## 4.7 Free Particle Behavior

The expressions for correlation functions which we have obtained in the last few sections, are asymptotically rigorous. They hold in the hydrodynamic regime when the system, as a result of many collisions, has reached local equilibrium. In terms of frequency and wave vector, they hold when $k\ell \ll 1$ and $\omega\tau \ll 1$ where $\ell$ is the mean free path of a particle, and $\tau$ the mean collision time. In a dense liquid, this region is large: the mean free path is of the order of the atomic force range, a few Angstroms maybe, and the collision time, by the dimensional argument that led to (2.12), is also very small except at the lowest temperatures where most liquids freeze. In a gas, the range of validity of the hydrodynamic expressions shrinks as $\ell$ and $\tau$ are much larger than in a liquid.

At the extreme other end is the simple gas of non-interacting point particles. It may be worthwhile to write down a few expressions for this simple system for comparison. The free classical gas is particularly simple. Since the equations of motion are $\vec{r}^{\alpha}(t) = \vec{r}^{\alpha}(0) + (\vec{p}^{\alpha}/m)t$ and $\vec{p}^{\alpha}(t) = \vec{p}^{\alpha}(0)$ for the $\alpha$-th particle, one shows easily that

$$\chi''_{nn}(k\omega) = \sqrt{\pi/2}\,\frac{n}{mv_o^2}\left(\frac{\omega}{v_o k}\right) e^{-\frac{1}{2}\left(\frac{\omega}{v_o k}\right)^2} \tag{4.65a}$$

and

$$\chi''_{t}(k\omega) = \sqrt{\pi/2}\;mn\left(\frac{\omega}{v_o k}\right) e^{-\frac{1}{2}\left(\frac{\omega}{v_o k}\right)^2}, \tag{4.65b}$$

where $v_o = (m\beta)^{-\frac{1}{2}}$ is the thermal velocity. The corresponding expressions for the non-interacting Fermi gas are somewhat more complicated, and shall not be given here (see, for example, Nozières 1964).

It is interesting, though not too significant, that the transport coefficients vanish. For example, the Kubo expressions (4.58) and eqs. (4.65) give

$\eta = 0$ and $\zeta = 0$ since $\lim_{k \to 0} k^{-3} e^{-a/k^2} = 0$ for $a > 0$; similarly, the heat conductivity $\kappa$ vanishes. Operationally, of course, transport coefficients cannot even be defined for a gas of non-interacting particles. A measurement of the heat conductivity, for example, is only possible if we can apply, quasi-statically, a temperature gradient and maintain it while we measure the heat current. However, only for a system with a finite mean free path can a temperature gradient be maintained quasi-statically. A free gas would "run away", and the standard measurements of transport coefficients cannot be performed. Still, it may be satisfying for some that in this case the Kubo expressions give the most sensible result: zero.

## 4.8 Sum Rule Calculations

In this section, we will attempt to extend our results, eqs. (4.40) and (4.44), from the collision-dominated regime of hydrodynamics to larger frequency and wavenumber. The purpose of this enterprise is twofold. Those rapid microscopic processes which decay over short times, of the order of the collision time $\tau$, are contained in eqs. (4.40) and (4.44) in a summary fashion; they determine the numerical values of the transport coefficients. If one wants to calculate these coefficients, one has to obtain more information about high frequency processes. Moreover, their spectrum can be obtained directly in some cases, for example by neutron scattering.

Now we can obtain at least some information about the short time behavior of correlation functions from their frequency moments. In sections 2.9 and 3.4 we showed how sum rules can be calculated, in principle and sometimes in practice. Let us consider the transverse momentum density correlation function, $\chi''_t(k\omega)$. According to eq. (3.33), its moments,

$$\langle \omega_t^{\nu}(k) \rangle \equiv \frac{1}{mn} \int \frac{d\omega}{\pi} \omega^{\nu} \chi_t''(k\omega)/\omega , \qquad (4.66)$$

are determined in terms of multiple commutators (Poisson brackets classically) with the Hamiltonian. For a classical system with the Hamiltonian (4.5), one obtains (Forster, Martin, and Yip 1968)

$$\beta m \langle \omega_t^2(k) \rangle = k^2 + n\beta \int d\vec{r}\, g(r)(1-\cos \vec{k}\cdot\vec{r}) \frac{1}{2} [\nabla^2 - (\hat{k}\cdot\nabla)^2]\, v(r) , \qquad (4.67)$$

$$\beta^2 m^2 \langle \omega_t^4(k) \rangle =$$

$$3k^4 + 3n\beta \int d\vec{r}\, g(r) \Big\{ k^2 \nabla^2 v - \frac{2}{3} (\vec{k}\cdot\vec{\nabla})^2 v$$

$$+ \sin \vec{k}\cdot\vec{r} [\nabla^2 - (\hat{k}\cdot\vec{\nabla})^2](\vec{k}\cdot\vec{\nabla})v + \frac{1}{3}\beta(1-\cos \vec{k}\cdot\vec{r})[(\vec{\nabla}\vec{\nabla}v)^2 - (\hat{k}\cdot\vec{\nabla}\vec{\nabla}v)^2] \Big\}$$

$$+ \frac{1}{2} n^2\beta^2 \int d\vec{r} \int d\vec{r}\,'\, g_3(\vec{r},\vec{r}\,') \Big\{ 1 + \cos[\vec{k}\cdot(\vec{r}-\vec{r}\,')] - 2\cos \vec{k}\cdot\vec{r} \Big\}$$

$$\times [(\vec{\nabla}\cdot\vec{\nabla}')^2 - (\hat{k}\cdot\vec{\nabla})(\hat{k}\cdot\vec{\nabla}')]\, v(r)\, v(r') , \qquad (4.68)$$

where $\hat{k} = \vec{k}/k$. These expressions are frightening, and higher order moments would become more horrible yet. They have been given here partly to demonstrate that fact. Still, the two sum rules given are integrals that can be evaluated numerically if the interparticle potential $v(r)$ is known, and in addition two static correlation functions: the famous pair correlation function $g(r)$, already introduced in (4.51), which gives the probability of finding a particle at point $\vec{r}$ if we know that there is one at the origin; and the triplet correlation function which is defined by

$$n^3 g_3(\vec{r},\vec{r}\,') = \sum_{\alpha\neq\beta\neq\gamma} \langle \delta(\vec{r}-\vec{r}^{\,\alpha})\, \delta(\vec{r}\,'-\vec{r}^{\,\beta})\delta(\vec{r}^{\,\gamma}) \rangle . \qquad (4.69)$$

Information about these functions can be obtained from the numerical computer calculations of molecular dynamics. (For a recent review, see e.g. Berne and

Forster 1971). In these studies, the most frequently used potential is the Lennard-Jones (6-12) potential

$$v(r) = 4\epsilon[(\frac{\sigma}{r})^{12} - (\frac{\sigma}{r})^6] , \qquad (4.70)$$

which is appropriate for the noble gases. For argon, for example, $\epsilon = 120°$K and $\sigma = 3.4$ Å. For this system therefore, it is feasible to compute the sum rules (4.67) and (4.68). Numerical data for $\langle\omega_t^2\rangle$ and $\langle\omega_t^4\rangle$, near k = 0 are given in Forster et al. (1968), for a range of densities and temperatures. So let us now consider the moments $\langle\omega_t^2\rangle$ and $\langle\omega_t^4\rangle$ as known. What do we do with them?

### 4.8.1. High Frequency Shear Waves

The answer is contained in the work we did in sections 2.10 and 2.11. The hydrodynamic expression for $\chi_t''(k\omega)$, eq. (4.40), is not compatible with the sum rules; it has to be generalized. Now the proof of eq. (2.94), given earlier for the magnetic correlation function, applies to $\chi_t''(k\omega)$ as well. In other words, in terms of a real function $D_t'(k\omega)$ which is positive, even in $\omega$, and has the complex Hilbert transform

$$D_t(kz) = \int\frac{d\omega}{2\pi i} \frac{D_t'(k\omega)}{\omega - z} , \qquad (4.71)$$

we can always write the rigorous dispersion relation

$$C_t(kz) = \frac{i\beta^{-1}mn}{z + ik^2 D_t(kz)} , \qquad (4.72)$$

which is the complete analogue of (2.94). Comparison with eq. (4.39) shows that (4.72) can be understood as a genealization of the hydrodynamic result, to account for a frequency and wave number dependent viscosity

$$\eta(kz) = mn\, D_t(kz) \cdot \tag{4.73}$$

Eq. (4.73) has therefore been called the basic equation of "Generalized Hydrodynamics", for the transverse case.

Let us first jump from extremely low to asymptotically high frequencies. We noted in section 2.10 that to accommodate finite sum rules for $\chi''_t(k\omega)$, the function $D'_t(k\omega)$ has to have finite frequency moments. And using our earlier method, expansion in terms of $1/z$, we find that its zeroth moment is given by

$$c^2_{t\infty}(k) \equiv \int \frac{d\omega}{2\pi} D'_t(k\omega) = \langle \omega^2_t(k)\rangle / k^2 \,. \tag{4.74}$$

For very high frequencies therefore where $D_t(z) = \frac{i}{z} c^2_{t\infty}$, we obtain from (4.72) the asymptotic result

$$C_t(kz) = \frac{iz\beta^{-1} mn}{z^2 - k^2\, c^2_{t\infty}(k)} \,. \tag{4.75}$$

Again, as in (2.107), we find reactive behavior at high frequency. There are shear waves, with a speed which at small k is obtained from (4.67) in the form

$$\beta m c^2_{t\infty}(0) = 1 + \frac{2\pi}{15}\, n\beta \int_0^\infty dr\; g(r)\, \frac{d}{dr}\left[ r^4 \frac{dv(r)}{dr} \right] . \tag{4.76}$$

This result has first been obtained by Zwanzig and Mountain (1965) who identified the quantity $G_\infty = mnc^2_{t\infty}(0)$ as the high-frequency shear modulus. It describes the initial elastic response of the fluid to a suddenly applied shear force. Indeed, since the microscopic expressions for the energy density $\varepsilon$ and the pressure p, namely eqs. (4.49,50), are similar to (4.76), one can, for the case of the (6-12) potential (4.70), express $G_\infty$ in terms of $\varepsilon$ and p. The result,

$$G_\infty = \frac{26}{5} nk_B T + 3p - \frac{24}{5}\varepsilon \;, \tag{4.77}$$

has been discussed by Zwanzig and Mountain in numerical detail. A similar and instructive quantum-mechanical calculation which is applicable to liquid Helium, has been given by Puff (1965).

#### 4.8.2 Interpolation Ansatz

We now have a low-frequency result in (4.39) and a high-frequency result in (4.75). Can we close the gap? Naturally, one must not be immodest: only a detailed microscopic calculation could accomplish this goal. In a rare gas, such a calculation might be feasible if laborious since the Boltzmann equation gives a rather detailed picture of the dynamics. (For related work, see e.g., Mazenko, Wei, and Yip 1972, and references given there.) For a liquid, the problem is infinitely more difficult (see Rice and Gray 1965). On the other hand, the gap between "low" and "high" frequencies in a simple liquid is not all that large since the mean free path and the collision time are small. We can therefore hope for good luck with a relatively simple, and more or less ad hoc, interpolation model.

Again, we utilize the general representation (4.72), in the spirit of our work in section 2.10. We do not know the function $D'(k\omega)$, of course, but we can hope to get by with a simple ansatz. The ansatz function $D'(k\omega)$ should be even and positive, of course. It also should fall off rapidly at large $\omega$ so that its frequency moments exist. Indeed, if we choose a $D'(k\omega)$ which fulfils the sum rule (4.74) (and therefore (4.67)), we will guarantee the correct behavior (4.75) at high frequencies. At small $\omega$ and k, $D_t'(k\omega)$ should be a smooth function of these variables, essentially a constant. The representation (4.72) would then reproduce the correct hydrodynamic behavior in this region, with the shear viscosity given by

$$\eta = mn\, D_t(0,0) = \frac{1}{2} mn\, D_t'(0,0). \tag{4.78}$$

And so let us be courageous, and try a Gaussian ansatz function

$$D_t'(k\omega) = 2\, c_{t\infty}^2(k)\, \tau(k)\, e^{-(\omega\tau(k))^2/\pi}\ , \tag{4.79}$$

which is in accord with all the restrictions given above. If we use (4.78) to determine the parameter $\tau(k)$ at small k in terms of the experimental viscosity, i.e., use

$$\eta = mn\, c_{t\infty}^2(0)\, \tau(0) = G_\infty\, \tau(0)\ , \tag{4.80}$$

(4.79) provides an interpolation model for small k that has at least a good chance of success.

Of course, one must not take the detailed frequency dependence of $D_t'(k\omega)$ too seriously. A much more modest, and very successful, use of this procedure is to obtain a semi-microscopic, parameter-free calculation of the viscosity itself. We have at our avail another bit of knowledge, namely the as yet unused sum rule (4.68). It translates into a second sum rule for $D_t'(k\omega)$, namely

$$\int \frac{d\omega}{2\pi}\, \omega^2\, D_t'(k\omega) = \langle\omega_t^4(k)\rangle\, k^{-2} - k^2 c_{t\infty}^4(k)\,. \tag{4.81}$$

Computing this integral from (4.79), we can express $\tau(k)$ in terms of $\langle\omega_t^4(k)\rangle k^{-2}$ and $c_{t\infty}^2(k)$ both of which have finite limits as $k \to 0$. Eq. (4.80) therefore gives the final result

$$\eta/mn = (\tfrac{1}{2}\pi)^{\frac{1}{2}} \lim_{k\to 0} c_{t\infty}^3(k)\, [\langle\omega_t^4(k)\rangle\, k^{-2}]^{-\frac{1}{2}}\ . \tag{4.82}$$

This simple calculation is remarkably successful. Numerical results, and a comparison with experimental viscosity data, are given in Forster et al. (1968).

Methods similar to those used in this section can be applied to a wide variety of fluctuation phenomena. Closely related techniques have been used, for example, by Chung and Yip (1969) to discuss density fluctuations in the neutron scattering region, as well as several other fluctuations in simple liquids.

## CHAPTER 5

## THE MEMORY FUNCTION FORMALISM

One of the most fundamental notions in the theory of interacting many-particle systems is that (almost) nothing that is physically interesting, can be calculated rigorously from first principles. Once this has been recognized -- sometime within the first year of graduate study -- it becomes of crucial importance for those not inclined to give up to hide their partial ignorance of complicated details away in places where it is least likely to cause harm. Abstract and general results or formulations are useful if they indicate how this should be done, if they suggest which objects of the theory, while unknown in detail, are likely to be insensitive to many of the complicated and uncalculable features of the dynamics; if they indicate, in other words, where one may approximate, parametrize, or fudge with relative impunity.

A prime example of this kind of general result is the dispersion relation (2.101) for the spin density correlation function, or its fluid mechanics equivalent (4.72). By itself it is general, rigorous -- and almost empty. If however as we believe, and as we will explain in section 5.4 below, there is an indication that the memory function D(kz) contains, in its spectrum, neither very low nor extremely high frequencies, the dispersion relation becomes extremely useful: even though the function D(kz) contains nearly the full horrendous complexity of the many-body problem, at low frequency and wave vector it varies slowly, and can be replaced by a constant: the simple and experimentally interesting hydrodynamic expressions result. Corresponding considerations have led us to practical high-frequency results.

In this chapter, which is largely formal, we shall therefore analyze the memory functions further. The tool is the projection operator technique introduced by R. Zwanzig (1961) and extended by H. Mori (1965); see also Berne, Boon and Rice (1966) and the review by Berne and Harp (1970). With the help of this technique, the dispersion relations can be made into a powerful and versatile instrument as our later applications, particularly to systems of broken symmetry, should amply demonstrate.

## 5.1 Projectors and Memory Functions

In order not to be encumbered by incidentals, let us first consider the formally simplest case: fluctuations of a single classical dynamical variable A(t). A physically interesting example of this sort is the momentum $\vec{p}(t)$ of a tagged diffusing particle. We shall be interested in the equilibrium-averaged autocorrelation function

$$C_{AA}(t,t') \equiv C(t-t') = \langle A(t)\, A(t') \rangle . \tag{5.1}$$

Without loss of generality, we can assume that $\langle A(t) \rangle = 0$ so that $C(t) \to 0$ as $t \to \infty$. If $\langle A(t) \rangle \neq 0$ originally, we would simply use $\tilde{A}(t) = A(t) - \langle A(t) \rangle$ in (5.1). As always, we work with the Laplace transform

$$C(z) = \int_0^\infty dt\, e^{izt}\, C(t) = \beta^{-1} \int \frac{d\omega}{\pi i} \frac{\chi''(\omega)}{\omega(\omega - z)} , \tag{5.2}$$

where the first equation holds only for $\mathrm{Im}\, z > 0$ while the second equation, as in (3.44b), defines C(z) in addition for $\mathrm{Im}\, z < 0$, in terms of the absorptive response function

$$\chi''(\omega) = \int_{-\infty}^{\infty} d(t-t')\, e^{i\omega(t-t')} \langle \frac{i}{2} \left[ A(t), A(t') \right]_{P.B.} \rangle . \tag{5.3}$$

$\omega\chi''(\omega)$ is, as always, non-negative. We can therefore repeat the considerations of section 2.11 to show that there always exists a function D(z) which is analytic for Im z $\neq 0$, in terms of which C(z) can be represented as

$$C(z) = \frac{i\beta^{-1}}{z + i\,D(z)}\,\chi\,, \tag{5.4}$$

where

$$\chi = \beta C(t=0) = \int \frac{d\omega}{\pi}\,\chi''(\omega)/\omega\;. \tag{5.5}$$

It is the representation (5.4) which we want to analyze.

Now according to classical mechanics, A(t) obeys the formal equation of motion

$$\partial_t A(t) = \left[A(t),\ H\right]_{P.B.} \equiv iLA(t)\,, \tag{5.6}$$

where L is the Liouville operator -- a differential operator which, for classical point particles with the Hamiltonian (4.5), for example, is given by

$$iL = \sum_{\alpha} \frac{\vec{p}^{\,\alpha}}{m}\frac{\partial}{\partial \vec{r}^{\,\alpha}} - \frac{1}{2}\sum_{\alpha\neq\beta}\left[\vec{\nabla} v(r^{\alpha}-r^{\beta})\right]\left(\frac{\partial}{\partial \vec{p}^{\,\alpha}} - \frac{\partial}{\partial \vec{p}^{\,\beta}}\right). \tag{5.7}$$

It acts, of course, on the canonical position and momentum variables of which A(t) is a function. The formal solution of eq. (5.6), $A(t) = e^{iLt}A(0)$, can be inserted into (5.1) so that

$$C(t) = \langle A(t)\,A(0)\rangle = \langle A(0)\;A(-t)\rangle = \langle A(0)\,e^{-iLt}A(0)\rangle\;. \tag{5.8}$$

The second equation (5.8) uses time translation invariance which always holds for an equilibrium average. Henceforth, we will drop the argument 0, understanding $A \equiv A(t=0)$. If we perform the Laplace transformation on (5.8), we have the formal expression

$$C(z) = \langle A\;\frac{i}{z-L}\,A\rangle \tag{5.9}$$

for the complex relaxation function.

It is tempting to read (5. 9), as in quantum theory, as the "matrix element" of the operator $i(z-L)^{-1}$, sandwiched between "vectors" $\langle A|$ and $|A\rangle$. And there is nothing to prevent us from doing that. One can define a Hilbert space of observables $|A\rangle$, classical dynamical variables, such that the scalar product between any two "vectors" $\langle A|$ and $|B\rangle$ is given the equilibrium-averaged product of the dynamical variables,

$$\langle A|B\rangle \equiv \langle A^* B\rangle_{eq.} \quad . \tag{5. 10}$$

Complex conjugation* is added for variables that are complex. (5. 10) is then in accord with the defining properties of a Hilbert space, namely

$$\langle A|A\rangle \geq 0 ,$$

$$\langle A|\alpha B + \beta C\rangle = \alpha\langle A|B\rangle + \beta\langle A|C\rangle ,$$

$$\langle A|B\rangle^* = \langle B|A\rangle . \tag{5. 11}$$

A mathematician would, in addition, question whether this space is separable and complete; a physicist will hope that it is, for all practical purposes. Note that, with the definition (5. 10), the Liouville operator is hermitian, in the sense that

$$\langle A|LB\rangle = \langle LA|B\rangle, \tag{5. 12}$$

which follows from time translation invariance, $\langle A^*(t)B(0)\rangle = \langle A^*(0)B(-t)\rangle$.

Now let us turn back to (5. 8). The operator $e^{-iLt}$ "rotates" the vector $|A\rangle$. Except for a normalization constant, the correlation function $C(t)$ is the component of the rotated vector parallel to the original one. It is therefore suggestive to define an operator which projects onto the vector $|A\rangle$, namely the projector

$$P \equiv |A\rangle\langle A|A\rangle^{-1}\langle A| \equiv 1 - Q , \tag{5. 13}$$

which we handle as one does with similar objects in quantum theory. It may take some getting-used-to-but it is useful. (5.13) is, of course, equivalent to the definition

$$\langle B|P|C\rangle = \langle B^* A\rangle \frac{1}{\langle A^* A\rangle} \langle A^* C\rangle \tag{5.14}$$

for any two variables B, C. It is clear that P is indeed a projector since

$$P^2 = P \text{ so that } PQ = QP = 0\,, \tag{5.15}$$

and P is obviously hermitian.

Now we want to extract from the Liouville operator that part which changes the vector $|A\rangle$ only parallel to itself. Let us therefore write $L = LP + LQ$ since $P + Q = 1$, and see what we get. It is most convenient to work with eq. (5.9). If we use the algebraic operator identity

$$\frac{1}{X+Y} = \frac{1}{X} - \frac{1}{X} Y \frac{1}{X+Y}\,, \tag{5.16}$$

we obtain

$$\begin{aligned} C(z) &= \langle A| \frac{i}{z-L} |A\rangle = \langle A| \frac{i}{z-LQ-LP} |A\rangle \\ &= \langle A| \left[\frac{i}{z-LQ} + \frac{1}{z-LQ} LP \frac{i}{z-L}\right] |A\rangle\,. \end{aligned} \tag{5.17}$$

Consider the first term in (5.17). It is given by

$$\langle A| \frac{i}{z-LQ} |A\rangle = \frac{i}{z} \langle A|A\rangle = \frac{i}{z} C(t=0)\,.$$

Why? Well, $P|A\rangle = |A\rangle$ so that $Q|A\rangle = 0$. But if you expand the operator

$$\frac{i}{z-LQ} = \frac{i}{z}\left[1 + \frac{1}{z} LQ + \frac{1}{z^2} LQ\, LQ + \ldots\right]$$

each term except the first has a Q at the end which operates on $|A\rangle$ and contributes nothing.

As to the second term in (5.17), we insert P from (5.13),

$$P\,\frac{i}{z-L}\,|A\rangle = |A\rangle\frac{1}{\langle A|A\rangle}\langle A|\frac{i}{z-L}|A\rangle = |A\rangle C^{-1}(t=0)C(z)\,.$$

Therefore, (5.17) reads

$$C(z) = \frac{i}{z}C(t=0) + \langle A|\frac{1}{z-LQ}L|A\rangle C^{-1}(t=0)\,C(z)\,. \tag{5.18}$$

A similarity to (5.4) begins to emerge. Namely, let us multiply by z and monkey around some more, writing

$$\langle A|\frac{z}{z-LQ}L|A\rangle = \langle A|\left[1 + LQ\frac{1}{z-LQ}\right]L|A\rangle$$

$$\equiv \left[\Omega - i\,\Sigma(z)\right]C(t=0), \tag{5.19}$$

where

$$\Omega = \langle A|L|A\rangle\,C^{-1}(t=0), \tag{5.19a}$$

$$\Sigma = \langle A|LQ\,\frac{i}{z-LQ}\,L|A\rangle\,C^{-1}(t=0)\,. \tag{5.19b}$$

Therefore, noting that $\beta\,C(0) = \chi$, equation (5.18) is of the form

$$C(z) = \frac{i\beta^{-1}}{z-\Omega+i\,\Sigma(z)}\,\chi\,. \tag{5.20}$$

And this, of course, is our old dispersion relation (5.4), with $D(z) = \Sigma(z) + i\,\Omega$. We have chosen the symbol $\Sigma(z)$ since it is an equation like (5.20) through which a "self-energy" in quantum mechanics is usually defined; in field theory contexts, an object analogous to $\Sigma(z) + i\,\Omega$ is known as the mass operator; in non-equilibrium systems, $\Sigma(z)$ has been called the memory function: it is all the same thing. Notice that $\Sigma(z)$ vanishes as $z\to\infty$, via eq. (5.19b). So the frequency $\Omega$ which we have separated off in (5.19) is the limit of the self-energy at infinite frequency $z\to\infty$.

In this sense, $\Omega$ bears a correspondence to the Hartree-Fock part of the self-energy in Green's function theory (Kadanoff and Baym 1962).

The factors $C^{-1}(0)$ in (5.19) could be removed by an appropriate normalization of $|A\rangle$, and they often are. For our later applications to systems of broken symmetry, however, we shall find this factor extremely helpful, and thus we will keep it in place. With the help of little cosmetic operations whose proof is left to the reader, we can finally write (5.19) in the form

$$\Omega \equiv \tilde{\omega}\,\chi^{-1} \quad \text{and} \quad \Sigma(z) \equiv \sigma(z)\,\chi^{-1}, \tag{5.20a}$$

where

$$\tilde{\omega} = i\beta\,\langle \dot{A} \,|\, A\rangle = \int \frac{d\omega}{\pi}\,\chi''(\omega), \tag{5.20b}$$

$$\sigma(z) = \beta\,\langle \dot{A}|\,Q\,\frac{i}{z - QLQ}\,Q\,|\,\dot{A}\rangle\,. \tag{5.20c}$$

Eqs. (5.20) are the fundamental results of this chapter. They have been first given, in this form, by Mori (1965).

We need not, at present, pause to discuss the frequency $\Omega$; for the single variable case, it usually vanishes. This happens whenever A has a definite time reversal signature since then $\langle \dot{A} \,|\, A\rangle = 0$. The essential object in (5.20) is the memory function $\Sigma(z)$. As a comparison of (5.9) and the "explicit" result (5.20c) shows $\Sigma(z)$ is itself of the structure of a time correlation function, of the "force" $\dot{A}$. There are two differences, however: first, only the part $Q\,|\,\dot{A}\rangle$ which is orthogonal to $|A\rangle$ enters the determination of $\Sigma(z)$; and second, the dynamical operator QLQ which determines the frequency spectrum of $\Sigma(z)$, is not the full Liouville operator L but has projected from it the part which determines the intrinsic fluctuations of the variable A. More intuitive illumination of these features can be better given in our later applications to specific examples. Suffice it here to note that if the dynamics of the property A is of interest, and we call all the many other degrees of freedom of the system the "bath", then the part $Q|\dot{A}\rangle$ determines the coupling of A to the bath, and QLQ describes the internal dynamics of the bath which feeds back to impose its behavior on A itself.

Additional understanding can be gained, even in this formal analysis, from considering the time equivalent of eq. (5.20). This equation is simply the Laplace transform of the "equation of motion"

$$\left[\partial_t + i\Omega\right] C(t) + \int_0^t d\tau\, \Gamma(t-\tau)\, C(\tau) = 0 \qquad (t > 0), \tag{5.21}$$

where $\Sigma(z)$ is the Laplace transform of the function $\Gamma(t)$. Equation (5.21) explains why $\Gamma(t)$ is called the "memory function": it describes the extent to which past values of $C(t)$ influence its time rate of change at present. The function $\Gamma(t)$ is given by

$$\Gamma(t)\chi \equiv \gamma(t) = \beta\langle \dot{A}|\, Q\, e^{-iQLQt}\, Q|\dot{A}\rangle, \tag{5.21a}$$

which is to be compared with eq. (5.8). In terms of the Fourier transform,

$$\gamma(\omega) = \int_{-\infty}^{\infty} dt\, e^{i\omega t}\, \gamma(t),$$

the complex memory function is given by

$$\sigma(z) = \int \frac{d\omega}{2\pi i}\, \frac{\gamma(\omega)}{\omega - z}.$$

Let us finally assume that A is a real variable. In this case, both $\omega\chi''(\omega)$ and $\gamma(\omega)$ or $\Gamma(\omega) = \gamma(\omega)\chi^{-1}$ are real, and even in $\omega$. Therefore, taking the real part of (5.20) at $z = \omega + i\epsilon$, one obtains the dispersion relation (setting $\Omega = 0$)

$$\frac{1}{\omega}\chi''(\omega) = \frac{\Gamma(\omega)/2}{\left[\omega - P\int\frac{d\omega'}{2\pi}\frac{\Gamma(\omega')}{\omega-\omega'}\right]^2 + \left[\Gamma(\omega)/2\right]^2}\chi , \tag{5.22}$$

which is the equivalent of eq. (2.96). From eq. (5.22) one can infer an important property of the spectral function $\gamma(\omega)$, namely

$$\gamma(\omega) \geq 0 . \tag{5.23}$$

## 5.2 Memory Function Matrices

The equations of the last section are all completely general and rigorous, no matter what the variable A might be. They are useful, however, only if the memory function $\Sigma(z)$ is expected to be a simpler object than $C(z)$ itself. If, for example, $\Sigma(z)$ is essentially a constant, (5.20) gives a Lorentzian line shape for $\chi''(\omega)/\omega$.

Very often one encounters not single lines of this shape but superpositions of Lorentzians, the next simplest case. An example is the light scattering spectrum of fig. 4.2. Phenomenologically, this comes about because several variables are dynamically coupled. We will therefore do well to extend our formal considerations to this case. The procedure is a perfectly straightforward generalization of section 5.1 which just introduces matrix indices and sums here and there. We will hence be brief.

We consider a set of N linearly independent, classical dynamical variables $A_i(t)$, i = 1 to N, and the matrix of time correlation functions

$$C_{ij}(t-t') = \langle A_i(t)A_j(t')\rangle = \langle A_i| e^{-iL(t-t')}|A_j\rangle , \tag{5.24}$$

where we have again chosen to set $\langle A_i\rangle = 0$. The Laplace transforms,

$$C_{ij}(z) = \langle A_i | \frac{i}{z-L} | A_j \rangle = \beta^{-1} \int \frac{d\omega}{\pi i} \frac{\chi''_{ij}(\omega)}{\omega(\omega - z)} , \tag{5.25}$$

can be treated as before. The matrix

$$\chi_{ij} = \int \frac{d\omega}{\pi} \chi''_{ij}(\omega)/\omega = \beta \langle A_i | A_j \rangle \tag{5.26}$$

is positive definite, and thus its inverse is well-defined, by

$$\chi_{ik} \chi^{-1}_{kj} = \delta_{ij} ,$$

where as always a sum over repeated indices is implied. The projector P onto the variables $A_i$, and its complement Q, can then be written in the form

$$P = \sum_{ij} | A_i \rangle \beta \chi^{-1}_{ij} \langle A_j | \equiv 1 - Q . \tag{5.27}$$

With its help, we can play precisely the same game we played before. The final matrix equations, analogous to eqs. (20), are then the following:

$$\left[ z\delta_{ik} - \Omega_{ik} + i \Sigma_{ik}(z) \right] C_{kj}(z) = i\beta^{-1} \chi_{ij} ,$$

where

$$\Omega_{ij} \equiv \omega_{ik} \chi^{-1}_{kj} \text{ and } \Sigma_{ij}(z) \equiv \sigma_{ik}(z) \chi^{-1}_{kj} ,$$

$$\omega_{ij} = i\beta \langle \dot{A}_i | A_j \rangle = \int \frac{d\omega}{\pi} \chi''_{ij}(\omega) ,$$

$$\sigma_{ij}(z) = \beta \langle \dot{A}_i | Q \frac{i}{z - QLQ} Q | \dot{A}_j \rangle . \tag{5.28}$$

In the light of remarks made in section 5.1, these equations require no further comment for now. They will be used in many contexts below.

The "equations of motion" (5.21) are easily generalized to multivariate processes as well; simply read them as matrix equations. The spectral matrix

$$\gamma_{ij}(\omega) = \beta \int_{-\infty}^{\infty} dt \; e^{i\omega t} \langle \dot{A}_i | Q e^{-iQLQt} Q | \dot{A}_j \rangle \,, \tag{5.29}$$

in terms of which $\sigma_{ij}(z)$ is given by

$$\sigma_{ij}(z) = \int \frac{d\omega}{2\pi i} \; \frac{\gamma_{ij}(\omega)}{\omega - z} \,, \tag{5.30}$$

has several noteworthy properties: Assuming that the $A_i$ are real variables, with time reversal signatures $\epsilon_i^T$, its symmetry properties are given by

$$\gamma_{ij}(\omega) = \gamma_{ji}(-\omega) = \gamma_{ji}^*(-\omega) = \epsilon_i^T \epsilon_j^T \gamma_{ij}(-\omega) \,. \tag{5.31}$$

Further, $\gamma_{ij}(\omega)$ is a positive matrix, in the sense that

$$\sum_{i,j} a_i^* \gamma_{ij}(\omega) a_j \geq 0 \text{ for all } \omega \,. \tag{5.32}$$

The reader will notice that these properties are the same as those of the matrix $\omega\chi''_{ij}(\omega)$, see eqs. (3.21) and (3.28), and they are easily proved on this basis. We will find that the matrix $\gamma_{ij}(\omega)$, for small $\omega$, is a matrix of transport coefficients in appropriate systems. Equations (5.31) thus lead to the Onsager symmetry relations (4.57), and eq. (5.32) is the microscopic basis for the phenomenological statement that the entropy production in a thermodynamic system is positive.

## 5.3 Extension to Quantum Mechanics

The formal developments of the last two sections were presented for a classical system because such objects as, say, the Liouville operator are a little more familiar classically. However, the projector formalism can be applied to quantum systems as well. All that has to be done is to clarify what we mean by L, and what by $\langle A|B\rangle$, quantum-mechanically.

Let us first take the Liouville operator (Fano 1957). The quantum-mechanical equation of motion for an observable A(t), in Heisenberg representation, is $\dot{A} = [A, H]/i\hbar$. What matters is that, given the Hamiltonian H, this is a linear relation between $\dot{A}(t)$ and A(t), which we can therefore surely write in the form

$$\partial_t A(t) = \frac{1}{i\hbar}\left[A(t), H\right] \equiv iLA(t) \ . \tag{5.33}$$

in formal analogy to the classical eq. (5.6). If we write this out in the form of matrix elements, taken for a complete set $\varphi_n$ of quantum-mechanical states, the first half of (5.33) reads

$$\partial_t A_{mn}(t) = \frac{1}{i\hbar}\left[A_{m\nu}(t)\, H_{\nu n} - H_{m\nu} A_{\nu n}(t)\right] , \tag{5.34b}$$

so that the Liouville operator has the matrix elements

$$\hbar L_{mn,\mu\nu} = H_{m\mu}\,\delta_{n\nu} - H_{\nu n}\,\delta_{m\mu} \ . \tag{5.34c}$$

Equation (5.34a) treats both A and H as matrices, with row index m and column index n. (5.34b), however, looks on the pairs (mn) or ($\mu\nu$) as one index each; A is like a vector now, while L is a matrix. To say this in different words: one usually considers observables, A or H, as linear operators in a Hilbert space which is spanned by the states of the system, and that is the normal interpretation of the first half of eq. (5.33). Now one can also consider a different, much larger, vector space in which observables A are not operators but vectors, while the Liouvillian is a linear operator on those vectors; this is expressed by the second eq. (5.33). (While reading these words, the reader will have figured out that the super-Hilbert space of operators has $N^2$ dimensions if the original Hilbert space of states had N dimensions.) The formal solution of eqs. (5.33) can now be written in the two alternative forms

$$A(t) = e^{iHt/\hbar} A(0) e^{-iHt/\hbar} = e^{iLt} A(0). \tag{5.35}$$

Now if the property A is to be a vector in a Hilbert space we want to define a scalar product between two "vectors" A and B. For this, there are many possibilities. One could, for example, choose the definition

$$\langle A|B\rangle \equiv \langle A^+ B\rangle_{eq.} = \mathrm{tr}\, \rho^0 A^+ B ,$$

where $\rho^0$ is the equilibrium density matrix as always, and $^+$ indicates hermitian conjugation. This definition would be in line with the basic axioms (5.11) which a scalar product has to fulfil. And the function of time,

$$\langle A(t)|B\rangle = \langle A|\, e^{-iLt} |B\rangle ,$$

would be the equilibrium-averaged time correlation function, earlier denoted by $S_{AB}(t)$.

We shall instead adopt the following more useful definition:

$$\langle A|B\rangle \equiv \beta^{-1} \int_0^\beta d\lambda \left[\langle A^+ B(i\hbar\lambda)\rangle_{eq.} - \langle A^+\rangle_{eq.} \langle B(i\hbar\lambda)\rangle_{eq.}\right], \tag{5.36}$$

where as usual

$$B(i\hbar\lambda) = e^{-\lambda H} B\, e^{\lambda H} . \tag{5.36a}$$

After time out to check that this definition also fulfils the axioms (5.11), which it does, we then define the quantum-mechanical correlation function by

$$C_{AB}(t) = \langle A(t)|B\rangle = \langle A|\, e^{-iLt}|B\rangle$$

$$= \beta^{-1} \int_0^\beta d\lambda \left[ \langle A^+(t)\, B(i\hbar\lambda)\rangle_{eq.} - \langle A^+(t)\rangle_{eq.} \langle B(i\hbar\lambda)\rangle_{eq.} \right]. \tag{5.37}$$

We have encountered eq. (5.37) for the Kubo function before, in eq. (3.43). To see why such a bizarre definition for the scalar product should be sensible, you should convince yourself that, for hermitian operators A and B,

$$\frac{1}{2}\beta\, i\partial_t\, C_{AB}(t) = \chi''_{AB}(t) = \langle \frac{1}{2\hbar} \left[A(t), B\right]\rangle_{eq.} \tag{5.38a}$$

or

$$\tilde{C}_{AB}(\omega) = \int_{-\infty}^{\infty} dt\, e^{i\omega t}\, C_{AB}(t) = 2\chi''_{AB}(\omega)/\beta\omega\ : \tag{5.38b}$$

The proof uses the fluctuation-dissipation theorem, and is easy.

This simple relation of $C_{AB}(t) = \langle A(t)|B\rangle$ to the absorptive response function $\chi''_{AB}(t)$ is very useful. For one thing, the commutator structure of $\chi''_{AB}(t)$ is the whole secret behind the sum rules, and because of (5.38) sum rules for $\chi''(\omega)$ are directly sum rules for $\tilde{C}(\omega)$, unencumbered by $\coth(\hbar\omega\beta/2)$ factors. For another, importantly, the definition (5.36) and eq. (5.37) open the direct route to the dynamical response. Note that the Laplace transform of $C_{AB}(t)$ is

$$C_{AB}(z) = \langle A|\, \frac{i}{z-L}\, |B\rangle = \beta^{-1} \int \frac{d\omega}{\pi i}\, \frac{\chi''_{AB}(\omega)}{\omega(\omega - z)} \tag{5.39}$$

Therefore, by comparison to eq. (3.13), $C_{AB}(z)$ directly describes the relaxation after an external, adiabatically applied field has been switched off. Moreover,

$$\beta \langle A|B\rangle = \int \frac{d\omega}{\pi} \chi''_{AB}(\omega)/\omega = \chi_{AB} \tag{5.40}$$

is the static susceptibility as always defined. Finally, in the classical limit, the definitions (5.36) and (5.10) coincide.

Time translation invariance implies that the quantum-mechanical Liouvillian L is hermitian with the scalar product (5.36), i.e. $\langle A|LB\rangle = \langle LA|B\rangle$. It is thus obvious that all of our considerations and results from sections 5.1 and 5.2 apply to quantum systems as well. We need say no more.

## 5.4 Spin Diffusion Revisited

After these many pages of formal manipulations, let us take a break and look again at the spin dynamics of chapter 2, to see how all of this formalism can be used. We will again discuss the magnetization correlation function,

$$\chi''(\vec{r}-\vec{r}', t-t') = \langle \frac{1}{2\hbar} [M(\vec{r}t), M(\vec{r}'t')] \rangle, \tag{5.41}$$

whose time integral, the Kubo function, we can now write as

$$C(\vec{r}-\vec{r}', t-t') = \langle M(\vec{r}t) | M(\vec{r}'t')\rangle. \tag{5.42}$$

The spatial coordinates are a bit in the way. But because of translational invariance we can write the spatial Fourier transform in this manner:

$$C(k, t) = \frac{1}{V} \int d\vec{r}\, e^{-i\vec{k}\cdot\vec{r}} \int d\vec{r}'\, e^{i\vec{k}\cdot\vec{r}'} C(\vec{r}-\vec{r}', t). \tag{5.43a}$$

In other words, in terms of the collective variable

$$M(\vec{k}, t) = V^{-\frac{1}{2}} \int d\vec{r}\, e^{i\vec{k}\cdot\vec{r}} M(\vec{r}, t) = V^{-\frac{1}{2}} \sum_{\alpha} 2\mu s^{\alpha} e^{i\vec{k}\cdot\vec{r}^{\alpha}(t)} \tag{5.43b}$$

we can write

$$C(k,t) = \langle M(\vec{k}) | e^{-iLt} | M(\vec{k}) \rangle , \tag{5.43c}$$

where $M(\vec{k}) \equiv M(\vec{k}, t{=}0)$. Note that because of translational invariance $\langle M(\vec{k},t)\rangle = 0$ for $\vec{k} \neq 0$.

$C(k,t)$ is just the kind of single variable correlation function we have discussed in section 5.1; $\vec{k}$ simply acts as a fixed label. We can therefore immediately apply the results obtained above; from eqs. (5.20), we obtain the Laplace transform in the form

$$C(kz) = \frac{i\beta^{-1}}{z + i\Sigma(kz)}\, \chi(k) , \tag{5.44}$$

where

$$\Sigma(kz)\, \chi(k) = \sigma(kz) = \beta\, \langle \dot{M}(k)| Q \frac{i}{z - QLQ} Q | \dot{M}(k) \rangle \tag{5.44a}$$

and

$$\chi(k) = \int \frac{d\omega}{\pi}\, \chi''(k\omega)/\omega = \beta \langle M(k)| M(k) \rangle \tag{5.44b}$$

as always. There is no term $\Omega(k)$ since $M(k,t)$ is purely odd under time reversal.

What is the projector? In direct application of (5.13), we can write it in the form

$$P(\vec{k}) = |M(\vec{k})\rangle \beta\chi^{-1}(k) \langle M(\vec{k})| = 1 - Q(\vec{k}) , \tag{5.45a}$$

so that $Q(\vec{k})$ removes magnetizationfluctuations of wave vector $\vec{k}$. But there is no need here to single out the wave vector $\vec{k}$. Without paying a price, we can sum over all wave vectors, and use

$$P = V \int \frac{d\vec{k}}{(2\pi)^3} \; P(\vec{k}) = 1 - Q \qquad (5.45b)$$

in eq. (5.44a). The reason is translational invariance in space. In an isotropic system fluctuations of different wave vectors do not couple, or $\langle M(\vec{k}) | M(\vec{k}') \rangle = 0$ if $\vec{k} \neq \vec{k}'$. Even if we use Q of (5.45b) in (5.44a), only the part $Q(\vec{k})$ does in fact contribute. Q has the intuitive advantage, however, that it removes magnetization fluctuations of arbitrary wavelength from the Liouvillian, and the mathematical advantage that Q is a translationally and rotationally invariant operator, in an isotropic infinitely extended system.

[Small digression for the weary: It is customary, in similar contexts, to use the discrete wave vectors of box quantization, and define collective coordinates then by

$$M_{\vec{k}} = \sum_a 2\mu s^a \; e^{i\vec{k} \cdot \vec{r}^a}$$

in place of (5.43b). For these, one obtains, for example,

$$\beta \langle M_{\vec{k}} | M_{\vec{k}'} \rangle = V \chi(k) \, \delta_{\vec{k}, \vec{k}'}$$

Since in this text we always work with continuous wave vectors, we have used them here as well, inserting factors V here and there to make the equations come out right. The advantage is that Fourier transforms of correlation functions, as usually defined here, contain no factors V, see (5.43c). The orthogonality appears now in the horrid form

$$\beta \langle M(\vec{k}) | M(\vec{k}') \rangle = \frac{(2\pi)^3}{V} \; \delta(\vec{k} - \vec{k}') \, \chi(k) = \text{"}\delta_{\vec{k}, \vec{k}'}\text{"} \; \chi(k),$$

but physical expressions are V-less. The projector, given by (5.45b), can be written in real space as

$$P = 1-Q = \int d\vec{r} \int d\vec{r}\,' \, | M(\vec{r}) \rangle \, \beta \, \chi^{-1}(\vec{r}-\vec{r}\,') \, \langle M(\vec{r}\,')| \, . \tag{5.46}$$

And $\sigma(\vec{k}z)$ of (5.44a) is directly the continuous Fourier-Laplace transform, as usually defined, of the local function

$$\gamma(\vec{r}-\vec{r}\,', z) = \beta \, \langle \dot{M}(\vec{r}) | Q \, e^{-iQLQt} Q | \dot{M}(\vec{r}\,') \rangle \, . \tag{5.47}$$

End of digression.]

We return to eqs. (5.44). First, remember that the magnetization operator obeys the continuity equation

$$\partial_t M(\vec{r}, t) + \vec{\nabla} \cdot \vec{j}\,(\vec{r}, t) = 0 \text{ or } \dot{M}(\vec{k}) = i\vec{k}\,\vec{j}\,(\vec{k}) \, , \tag{5.48}$$

where the magnetization current has the Fourier component

$$\vec{j}\,(\vec{k}) = V^{-\frac{1}{2}} \sum_{a} (2\mu s^{a}/m) \, \vec{p}^{\,a} \, e^{i\vec{k}\cdot\vec{r}^{\,a}} \tag{5.48a}$$

at $t = 0$. Because of this conservation law, $\sigma(kz)$ is of order $k^2$, and can be written in the form

$$\sigma(kz) = k_i \, d_{ij}(kz) k_j \, , \tag{5.49a}$$

with

$$d_{ij}(kz) = \beta \, \langle j_i(\vec{k}) | \, Q \, \frac{i}{z - QLQ} \, Q | j_j(\vec{k}) \rangle \, . \tag{5.49b}$$

The point is more than formal. The current density is a local operator, its correlation function falls off rapidly with distance, and thus $d_{ij}(kz)$ should be smooth and quite finite as $k \to 0$. Since $\chi(k)$ is finite in the same limit, approaching the spin magnetic susceptibility as $k \to 0$, the function

$$D(kz) = (\hat{k}_i\, d_{ij}(kz)\, \hat{k}_j)\, \chi^{-1}(k) \quad \text{where } \hat{k} = \vec{k}/k \tag{5.50}$$

is also finite as $k \to 0$. Inserting eqs. (5.49) and (5.50) in (5.44) we recover eq. (2.101), or

$$C(kz) = \frac{i\beta^{-1}}{z + ik^2 D(kz)}\, \chi(k)\ , \tag{5.51}$$

but with a somewhat better explanation for the factor $k^2$ than we could give in section 2.11. By the argument given there, after eq. (2.101), $C(kz)$ will thus have a hydrodynamic pole at

$$z = -ik^2 D(0,0) + O(k^4)\ , \tag{5.52}$$

provided that $D(kz)$ is well-behaved for small k <u>and</u> z, since (5.52) relies on making a Taylor expansion about the point $k = 0$ and $z = 0$. Now is it well behaved?

For the k-dependence, we have just settled the matter. Indeed, in the limit as $k \to 0$, the tensor $d_{ij}(kz)$ must be diagonal since the only k-independent second rank tensor is $\delta_{ij}$. Thus

$$\lim_{k \to 0} d_{ij}(\vec{k}z) = d(z)\delta_{ij} , \tag{5.53}$$

$$d(z) = \frac{1}{3}\beta \lim_{k \to 0} \langle \vec{j}(k) | Q \frac{i}{z - QLQ} Q | \vec{j}(k) \rangle . \tag{5.53a}$$

Now to the z-dependence. We know, of course, that D(kz) or d(z) is analytic in the upper half z-plane. What is in question, however, is analyticity at $z = 0$, i.e. on the real axis. That this is not a trivial matter one can see even from eq. (5.51) itself. Namely, in the limit of infinite wavelength, the correlation function C(kz) has a pole at $z = 0$ since

$$\lim_{k \to 0} C(kz) = \frac{i}{z} \beta^{-1} \chi . \tag{5.54}$$

And of course, d(z) is the $k \to 0$ limit of a kind of correlation function itself. However, as we have seen often, even just now when writing $\Sigma(kz) = k^2 D(kz)$, the pole (5.54) is there because the magnetization is a conserved density. (5.54) is an expression of the fact that

$$\lim_{k \to 0} C(k, t) = \beta^{-1} \chi \qquad \text{does not decay,} \tag{5.55}$$

as a consequence of the conservation law.

On the other hand, d(kz) is a correlation function of the current density, and the latter is <u>not</u> a conserved quantity. To understand this further, remember that if a function decays as $e^{-\gamma t}$, then its Laplace transform has a pole at $z = -i\gamma$. More generally, if

$$\lim_{t \to \infty} e^{\gamma t} f(t) < \infty ,$$

then $f(z) = \int_0^\infty dt\, e^{izt} f(t)$ is analytic not only in the upper half of the complex z-plane, but in addition in a strip of the lower half plane, of width $\gamma$; see fig. 5.1.

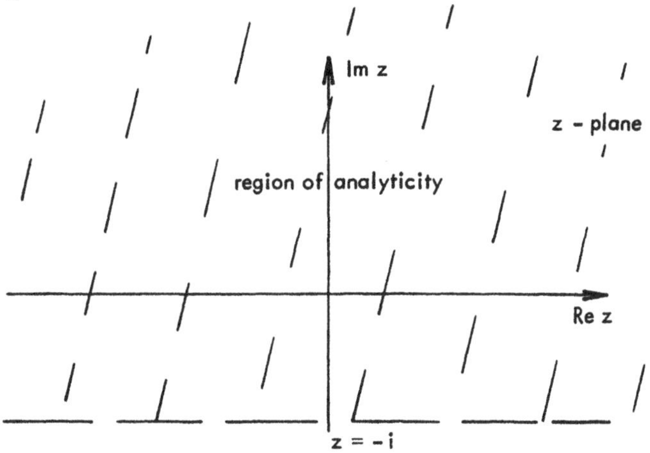


Fig. 5.1

In particular, f(z) is then analytic at z = 0, and can be expanded about this point.

There is good reason to expect that this is the behavior of the memory function d(z), that its time equivalent

$$d(t) = \lim_{k \to 0} \frac{1}{3} \langle \vec{j}(k) | Q\, e^{-iQLQt}\, Q | \vec{j}(t) \rangle \tag{5.56}$$

decays for large times as $e^{-t/\tau}$ where $\tau$ is a short microscopic life time. More generally, that D(kt) decays at least as $e^{-t/\tau(k)}$ where $\tau = \tau(k \to 0)$ is finite, and determined by the microscopic interactions. The reason for this behavior is apparent from eq. (5.56) itself. Infinite lifetimes result from local conservation laws as we have just seen, and described in physical terms in section 2.1. Complicated many-body systems are far too chaotic for infinitely slow relaxation to occur "accidentally". Now from the representation (5.56) we see that d(t) is a time correlation function which has, however, the conserved degree of freedom, M, projected out of its dynamics. In other words, the full Liouville operator L has a set of very small eigenvalues, of order $k^2$ for small k, whose eigenmodes are described by M(k), but the projector Q of eq. (5.45b) removes all of these, and QLQ no longer contains them. Consequently, $e^{-iQLQt}$ describes only microscopically rapid decay, and d(z) or D(kz) must be analytic in z, near z = 0.

We have been wordy with this statement, partly because it is at the very basis of much that is to follow, and partly because it would be difficult indeed, and probably impossible, to present a rigorous mathematical proof of the statement. Let us, instead, ask what could reasonably go wrong. There are two major possibilities. First, there could be conserved densities in addition to the one considered here, namely $M(\vec{r}t)$, densities whose slow fluctuations would therefore still be contained in (5.56). So long as these are few in number as is usually the case, they can be easily accounted for by applying the multivariate formalism of section 5.2, and thus projecting them out as well. We will encounter this possibility. While in our present example, spin diffusion in liquid $He^3$, there are indeed the additional conserved densities of mass, energy, and momentum, these do not in fact couple to the magnetization because of symmetry, and they do not have to be considered therefore. The other major possibility for failure is that there might be a second fundamental mechanism, apart from conservation laws, which enforces infinitely slow decay of certain collective fluctuations. Indeed, such a mechanism is in force in systems of broken symmetry, and we will analyze this case later.

Two further remarks are in place here. The first is to point out that we are concerned here with chaotic systems at finite temperature. At extremely low temperatures the thermal motion becomes very sluggish, transition processes become frozen out, and a host of long-lived "elementary excitations" results. With these, we will not be concerned in this text. The second remark is to point out that if the variable $\int d\vec{r}\, M(\vec{r}, t)$ is conserved, the variable $[\int d\vec{r} M(\vec{r}, t)]^2$ say, is conserved as well. The impact of such "mode-mode-coupling" terms, or in Fourier space of variables like $M(\vec{k}-\vec{k}', t)\, M(\vec{k}', t)$, on our considerations is subtle. Suffice it to say this: there are indications that in two dimensions, such "multi-phonon" modes, which are still contained in QLQ, are so strongly coupled to the single phonon modes which we consider, that they completely destroy the analyticity of the single phonon memory function D(kz), and hydrodynamics does not exist. In three dimensions, phase space considerations show that the "mode-mode-coupling" is weak, and does not alter our arguments in any essential way. (These multi-phonon modes are responsible for the infamous "long time tails" of memory functions, or weakly non-analytic contributions $\sim z^{1/2}$ to the function D(kz). See, for example, Ernst et al. (1970), Dorfman and Cohen (1970), and Zwanzig (1972)). Only very close to second order phase transitions, where the amplitude of spontaneous fluctuations increases tremendously, are these processes of major importance, and there they change the structure of hydrodynamics. See Kadanoff and Swift (1968) and Kawasaki (1970), whose treatment of dynamical phenomena near critical points is in a language similar to ours.

That is enough for Cassandra. It may be well to summarize the main argument: (1) It is always possible to write the correlation function of a variable $M(\vec{r}, t)$ in the form (5.44), where the memory function $\Sigma(kz)$ is a projected ("irreducible", as the diagram people would say) correlation function.

(2) If $M(\vec{r}, t)$ is the local density of a conserved quantity, $\Sigma(kz) = k^2 D(kz)$ is of order $k^2$ as $k \to 0$, resulting in a long-lived hydrodynamic mode with a lifetime $\tau(k) \to \infty$ as $k \to 0$. (3) If $M(\vec{r}, t)$ is the only conserved density in the system, or at least if no other conserved variable couples to M because of symmetry, then $D(kz)$ contains only microscopically rapid processes, and is smooth near $z = 0$. In this case, $C(kz)$ is described by a simple Lorentzian diffusion process for small k and z, with the diffusion coefficient given by $D = D(0, 0)$. Apart from the hydrodynamic pole at $z = -ik^2 D$, $C(kz)$ will, of course, have further poles $z = \omega(k) - i\gamma(k)$. However, their lifetimes $\gamma^{-1}(k)$ are finite as $k \to 0$, and the strength of these non-hydrodynamic modes is of relative order $k^2$ and vanishes as $k \to 0$.

There is one slightly puzzling matter to clear up. From the general symmetry and positivity properties (5.31) and (5.32) it is easy to see that $D(0,0) = D$ is real and positive, and that the diffusion coefficient, using (5.50) and (5.53), can be written in the form

$$D\chi = \frac{1}{3k_B T} \lim_{k \to 0} \int_0^\infty dt \, \langle \vec{j}(k) | Q e^{-iQLQt} Q | \vec{j}(k) \rangle . \tag{5.57a}$$

However, our previous result (2.29a) would appear in the form

$$D\chi = \frac{1}{3k_B T} \lim_{\epsilon \to 0} \lim_{k \to 0} \int_0^\infty dt \, e^{-\epsilon t} \langle \vec{j}(k) | e^{-iLt} | \vec{j}(k) \rangle \tag{5.57b}$$

Note the convergence factor, and the absence of projectors Q. In fact the two expressions are completely equivalent.

The proof is based upon the formula

$$\varphi(z) = \sigma(z) - \frac{i}{z} \sigma(z) \chi^{-1} \varphi(z) , \tag{5.58}$$

where $\sigma(z)$ is as given in (5.20c), and $\varphi(z)$ by the same expression without the projectors,

$$\varphi(z) = \beta \langle \dot{A}| \frac{i}{z - L} |\dot{A}\rangle. \tag{5.59}$$

Eq. (5.58) is a rigorous algebraic identity. It holds provided only that $\langle \dot{A}|A\rangle = 0$ so that $Q|\dot{A}\rangle = |\dot{A}\rangle$. It has first been given by Mori (1965) and, independently, by Berne et al. (1966), and we leave the simple proof to the reader.

In our present case $A \equiv M(k)$, and since $\dot{M}(k) = i\vec{k}\cdot\vec{j}(k)$, both $\sigma(kz)$ and $\varphi(kz)$ are of order $k^2$. If we write $\sigma(kz) = k^2 d(kz)$ and $\varphi(kz) = k^2 e(kz)$, we find from (5.58) that

$$e(kz) = \frac{1}{1 + \frac{i}{z} k^2 d(kz)\chi^{-1}(k)} \; d(kz) \tag{5.60}$$

and therefore, that

$$\lim_{k\to 0} d(kz) = \lim_{k\to 0} e(kz) . \tag{5.60a}$$

In other words,

$$\lim_{k\to 0} \langle \vec{j}(k)|Q \frac{i}{z-QLQ} Q|\vec{j}(k)\rangle = \lim_{k\to 0} \langle \vec{j}(k)| \frac{i}{z-L} |\vec{j}(k)\rangle. \tag{5.61}$$

This shows the equivalence of (5.57a) and (5.57b). Notice, however, that the limit as $k\to 0$ has to be performed while keeping $z$ finite; (5.60) makes that amply clear. Only after $k\to 0$, may we let $z = i\epsilon\to 0$ to obtain the diffusion coefficient according to (5.57b).

The integrand in (5.57a) falls off, at any k, within microscopic times since the slow fluctuations are projected out. There is therefore no convergence problem even as $k \to 0$, and we can simply set $\epsilon = 0$. However, the non-projected correlation function in (5.57b) develops according to the full dynamics; it does, therefore, contain contributions which do not decay as $k \to 0$. What eq. (5.61) says is that the strength of these terms is of order $k^2$, and so if the k-limit is performed first, they do not contribute to the integral. This is why in computing transport coefficients from Kubo formulae, using normal non-projected correlation functions, one has to eliminate the spatial variation before one eliminates the time dependence.

# CHAPTER 6

# BROWNIAN MOTION

Investigating a suspension of plant pollen dispersed in water, the botanist Robert Brown observed under the microscope that the tiny particles executed an irregular fluctuating motion. That was in 1829, and throughout the 19th century there were people for whom this trembling behavior was an aspect of the mysterious vis vitae. Why not; it had at first been observed only on particles of biological origin, after all. Less mystically inclined pragmatists noticed then, however, that the particles of an ink suspension were trembling as well. The physical theory of Brownian motion was first considered by Einstein and Smoluchowsky, and later further developed by Langevin and others. It continues to be a fascinating subject of statistical physics. For details see the review by S. Chandrasekhar (1943) and other articles reprinted in the collection of classical papers on stochastic processes by N. Wax (1954). In this chapter we shall deal with the theory of Brownian motion from the dynamical point of view taken earlier, dispensing with the usual a priori introduction of purely stochastic elements. Our treatment is presented as a simple and physical application of the formalism given in chapter 5, rather than a full and exhaustive theory of Brownian motion. A similar but more thorough analysis has been given by Lebowitz and Resibois (1965), see also the references cited there.

## 6.1 The Momentum Autocorrelation Function

If a heavy Brownian particle moves with velocity $\vec{v}$ through a stationary fluid of light particles, it experiences a frictional force, $-\zeta\vec{v}$, where $\zeta$ is the friction coefficient. The B-particle therefore slows down according to the equation

$$\partial_t \vec{v}(t) + \zeta\vec{v}(t) = 0\,, \tag{6.1}$$

which is easily solved: $\vec{v}(t) = e^{-\zeta t}\,\vec{v}(0)$. At equilibrium with the fluid, the B-particle will be at rest on the average, but spontaneous velocity fluctuations occur. The mean square velocity is given by the equipartition theorem as

$$\langle \frac{M}{2} v^2 \rangle = \frac{3}{2} k_B T\,, \tag{6.2}$$

where M is the mass of the B-particle. Combining, somewhat naively, the eqs. (6.2) and (6.1) we find that we can describe the decay of spontaneous velocity fluctuations by

$$\langle \vec{v}(t)\cdot\vec{v}(0)\rangle = e^{-\zeta t}\,\langle \vec{v}(0)\cdot\vec{v}(0)\rangle = e^{-\zeta t}\,3k_BT/M\,. \tag{6.3}$$

This result, eq. (6.3), is better than our "derivation" of it. It is this equation which we will first discuss.

Obviously, eq. (6.1) cannot be microscopically correct. The frictional force on the B-particle is ultimately due to frequent collisions with the molecules of the fluid, and can appear as a velocity-proportional continuous drag only on a time scale which is very long compared to the frequency and duration of individual collisions. If we want to derive eq. (6.3) from the microscopic dynamics, we have thus to keep in mind that this equation can hold, at best, only for times long compared to the rapid processes which characterize molecular motion of the fluid particles. Because it is so heavy, the B-particle itself changes its velocity on a much slower time scale. This is the circumstance--wide separation of time scales--for which the memory function method is most admirably suited.

In setting up a microscopic theory of Brownian motion, let us introduce one simplification from the start, by treating both the fluid particles and the B-particle as mass points. While this is certainly not a good description of those large colloidal particles which are visible under a light microscope, the diffusion of heavy mass points contains the essence of the Brownian problem, and generalizations to particles of more complicated structure present no difficulties of principle. One feature of the simplification is that the B-particle is described by only one pair of canonical position and momentum coordinates, $\vec{r}^{\,o}$ and $\vec{p}^{\,o}$. Denoting by $\vec{r}^{\,\alpha}$ and $\vec{p}^{\,\alpha}$ the position and momentum of the α-th fluid particle, we can write the Hamiltonian of the system in the form

$$H = \sum_{\alpha} \frac{p^{\alpha 2}}{2m} + \frac{1}{2} \sum_{\alpha\neq\beta} v(|\vec{r}^{\,\alpha\beta}|) + \frac{p^{o2}}{2M} + \sum_{\alpha} u(|\vec{r}^{\,o} - \vec{r}^{\,\alpha}|) , \tag{6.4}$$

where $\vec{r}^{\,\alpha\beta} \equiv \vec{r}^{\,\alpha} - \vec{r}^{\,\beta}$. The first two terms give the energy of the host fluid, the third term the kinetic energy of the B-particle, and the last term its interaction with each of the fluid particles. In keeping with the assumption of mass points, we will take the interaction potentials v(r) and u(r) as spherically symmetric. We can also write down the Liouville operator easily enough, in the form

$$L = L_f + \delta L , \tag{6.5}$$

where

$$iL_f = \sum_{\alpha} \frac{p^{\alpha}}{m} \frac{\partial}{\partial r^{\alpha}} - \frac{1}{2} \sum_{\alpha\neq\beta} \nabla v(r^{\alpha\beta}) \left(\frac{\partial}{\partial p^{\alpha}} - \frac{\partial}{\partial p^{\beta}}\right) \tag{6.5a}$$

is the Liouvillian of the fluid alone, and

$$i\delta L = \frac{p^{o}}{M} \frac{\partial}{\partial r^{o}} - \sum_{\alpha} \nabla u(r^{o} - r^{\alpha}) \left(\frac{\partial}{\partial p^{o}} - \frac{\partial}{\partial p^{\alpha}}\right) \tag{6.5b}$$

contains the interaction, apart from the first term which describes the freely streaming B-particle. Since quantum mechanics adds nothing of interest, we treat this as a classical system. Also, we have left out vector arrows, which would occur in obvious places, to unclutter the expressions a little.

Instead of the velocity, let us consider the momentum $\vec{p}^{\,o}(t)$ of the B-particle and its autocorrelation function

$$C_{ij}(t) = \langle p_i^o(t)\, p_j^o(0)\rangle = \langle p_i^o | e^{-iLt} | p_j^o \rangle \,. \tag{6.6}$$

It is averaged over an ensemble in which the B-particle is at equilibrium with the host fluid. In this state of affairs, no direction is distinguished, and thus evidently

$$C_{ij}(t) = \delta_{ij} C(t) \,. \tag{6.7}$$

From equilibrium statistical mechanics--namely, the equipartition theorem- we easily obtain the quantity called $\chi$ in chapter 5, namely

$$\beta^{-1} C_{ij}(0) \equiv \chi_{ij} = M\delta_{ij} \,. \tag{6.8}$$

The memory function, denoted as $\gamma(t)$ in chapter 5, is isotropic as well, and formally given by

$$\gamma_{ij}(t) = \frac{1}{k_B T} \langle \dot{p}_i^o | Q\, e^{-iQLQt}\, Q | \dot{p}_j^o \rangle = \gamma(t)\, \delta_{ij} \,, \tag{6.9}$$

in straightforward application of the formalism of sections 5.1 and 5.2. The projector removes fluctuations of the B-momentum. It can be formally written as

$$Q = 1 - P = 1 - |\vec{p}^{\,o}\rangle \frac{1}{Mk_B T} \langle \vec{p}^{\,o}| \,. \tag{6.10}$$

Note that $\langle \dot{p}_i^o\; p_j^o \rangle = 0$ from time reversal invariance; therefore, the frequency matrix $\omega_{ij}$, given in (5.28), vanishes. Defining a Laplace transform $C(z)$ as usual, and

$$\sigma(z) = \frac{1}{m} \int_0^\infty dt\, e^{izt} \gamma(t) \,, \tag{6.11}$$

we obtain, as a simple application of eqs. (5.20) or (5.28),

$$C(z) = \frac{i}{z - i\frac{m}{M}\sigma(z)} Mk_BT \tag{6.12}$$

We have inserted the factor $1/m$ in (6.11) simply so as to make the factor appearing in front of $\sigma(z)$ in (6.12) dimensionless.

By comparison of (6.12) to eq. (5.51) and in view of the discussion given there, the further procedure is now already obvious. The role of the factor $k^2$ in hydrodynamic collective modes is now taken over by the mass ratio, $m/M$. For an infinitely massive B-particle, $m/M = 0$, and $C(z)$ contains one pole at $z = 0$. This pole, of course, describes the fact that a particle of infinite mass does not move so that $\vec{p}^{\,o}(t)$ is time independent, and therefore

$$C(t) = \text{const.} \quad \text{or} \quad C(z) = \frac{i}{z}\ \text{const.} \quad \text{as } m/M \to 0 \ . \tag{6.13}$$

As the mass ratio becomes finite, this pole will shift into the lower half frequency plane, to a position given by the vanishing of the denominator of (6.12). To order $m/M$ this position will thus be given by

$$z = -i\frac{m}{M}\sigma(z) \simeq -i\frac{m}{M} \lim_{m/M\to 0} \sigma(0) \equiv -i\zeta \ , \tag{6.14}$$

provided that $\sigma(z)$ is well-behaved near $z = 0$, even as $m/M \to 0$ (which, to again point out this fact, the correlation function $C(z)$ is not). Taking the proviso for granted, we thus find that for small frequencies

$$C(z) = \frac{i\,Mk_BT}{z+i\zeta} \qquad \text{or} \qquad C(t) = Mk_BT\, e^{-\zeta t} \ , \tag{6.15}$$

where the second equation holds, correspondingly, for long times. This result is, of course, eq. (6.3) since $\vec{v} = \vec{p}^{\,o}/M$ and thus $\langle \vec{v}(t) \cdot \vec{v}(0) \rangle = 3\, C(t)\, M^{-2}$. From (6.14) it is easy to see that $\zeta$ is real and positive. The question of the validity of (6.3) has thus been reduced to whether or not $\sigma(z)$ is, even for small $m/M$, smooth near $z = 0$. We therefore have to examine the frequency spectrum of $\sigma(z)$ or, equivalently, the relaxation times contained in $\gamma(t)$ as given by eq. (6.9).

In considering a similar question in section 5.4, we pointed to the projectors Q which remove the slow processes from the dynamics of $\gamma(t)$. Here, we can carry the analysis a little further by considering the Liouville operator (6.5) in some detail. Let us split it in the form

$$L = L^{o} + L_B + L_{f\to B}\ , \tag{6.16}$$

where

$$L^{o} = L_f - i \sum_\alpha \nabla u(r^{o}-r^{\alpha}) \frac{\partial}{\partial p^{\alpha}} \tag{6.16a}$$

contains, in addition to the fluid Liouvillian, the action of the B-particle on the host fluid, while

$$L_{f\to B} = i \sum_\alpha \nabla u(r^{o}-r^{\alpha}) \frac{\partial}{\partial p^{o}} \tag{6.16b}$$

contains the reverse, the forces which the fluid exerts on the B-particle;

$$L_B = -i \frac{p^{o}}{M} \frac{\partial}{\partial r^{o}} \tag{6.16c}$$

describes the freely moving B-particle. What is the order of magnitude of those terms?

Note that the B-particle is intrinsically slower than the lighter fluid particle. The relative order of magnitude of their momenta can be inferred from the equipartition result, namely

$$\langle p_i^{o\,2} \rangle = Mk_BT\ , \quad \text{while} \quad \langle p_i^{\alpha\,2} \rangle = mk_BT \tag{6.17}$$

for each vector component. Therefore $p^{\alpha}/p^{o} \approx (m/M)^{1/2}$ as an order of magnitude estimate. Consequently, $L_{f\to B}$ and $L_B$ are smaller than $L^{o}$, by a factor $(m/M)^{1/2}$. Symbolically,

$$\frac{L_B + L_{B\to f}}{L^o} \sim \left(\frac{m}{M}\right)^{1/2} \tag{6.18}$$

The reader can sharpen this argument by formulating the whole problem in dimensionless momenta and positions, measuring momenta with respect to the respective mean thermal momenta $(mk_BT)^{1/2}$ and $(Mk_BT)^{1/2}$, and scaling positions to the range of the interparticle forces, say. In any case,

$$L \to L^o \quad \text{as} \quad m/M \to 0 . \tag{6.19}$$

Since it is also true that $L^o$ does not act on the Brownian coordinates, so that

$$L^o|\vec{p}^o\rangle = 0 \quad \text{or} \quad L^oP = 0 \quad \text{or} \quad L^o = L^oQ , \tag{6.20}$$

we can therefore write $\gamma(t)$ of eq. (6.9), in the limit $M\to\infty$, as

$$\lim_{M\to\infty} \gamma(t) \equiv \gamma_\infty(t) = \frac{1}{3k_BT} \langle\dot{\vec{p}}^o|e^{-iL^ot}|\dot{\vec{p}}^o\rangle , \tag{6.21}$$

where we have used the fact that $\langle\dot{p}^o|p^o\rangle = 0$ or $\langle\dot{p}^o|Q = \langle p^o|$. Note further that

$$\dot{\vec{p}}^o \equiv iL\vec{p}^o = -\sum_{\alpha=1}^{N} \vec{\nabla}\, u(r^o - r^\alpha) \equiv \vec{F} \tag{6.22}$$

is the force which the fluid particles exert on the B-particle. The zero mass-ratio memory function can therefore be written in the final form

$$\gamma_\infty(t) = \frac{1}{3mk_BT} \langle \vec{F}_\infty(t)\cdot \vec{F}_\infty(0)\rangle , \tag{6.23}$$

where the time-dependence of $\vec{F}_\infty(t)$ is given by $L^o$. $F_\infty(t)$ is, inother words, the force which the host fluid exerts on an <u>infinitely heavy</u> B-particle. In this limit then, $\gamma_\infty(t)$ contains only the microscopically rapid dynamical processes characteristic of the internal motion of a fluid of light particles, in the presence of a B-particle which is held fixed in space, and consequently $\gamma_\infty(t)$ will rapidly decay to zero. Its Laplace transform, $m\,\sigma(z)$, will therefore be analytic in z, near $z = 0$, which was the question we wanted to analyze.

In summary then, we have shown that velocity correlations of a very heavy B-particle decay exponentially, irreversibly, with a friction constant $\zeta$ which is of order of the mass ratio m/M, and given by (6.14) and (6.23) as

$$\zeta = \frac{m}{M}\zeta_\infty + O((m/M)^2) \; ,$$

$$\zeta_\infty = \frac{1}{3mk_BT} \int_0^\infty dt \, \langle \vec{F}_\infty(t) \cdot \vec{F}_\infty(0) \rangle \; . \qquad (6.24)$$

This formula was first derived by Kirkwood (1945) by quite different methods. Its value for practical computations should be apparent. To calculate $\zeta_\infty$, and thus gain numerical information about the motion of a B-particle, one "only" has to solve Newton's equations for the fluid particles in the presence of a fixed force center. The corresponding Hamiltonian is (6.4), with the kinetic energy $p_o^2/2M$ omitted. With the methods of computer molecular dynamics this calculation can be accomplished for a sufficiently large number of fluid particles, say one thousand. One monitors, as functions of time, the positions of all fluid particles, and obtains from these the force $\vec{F}_\infty(t)$ exerted at time t on the force center. To calculate the equilibrium-averaged time correlation function $\gamma_\infty(t)$ of (6.23), it would in principle be necessary to solve Newton's equations again and again, for arbitrary initial conditions over which the average is performed. Instead, taking advantage of the presumed ergodicity of the dynamics, one calculates $\gamma_\infty(t)$ by performing a time average over a particular phase space trajectory, i.e., as

$$\gamma_\infty(t) = \frac{1}{3k_BT} \cdot \frac{1}{T_o} \int_0^{T_o} d\tau \, \vec{F}_\infty(t+\tau)\, \vec{F}_\infty(\tau) \; , \qquad (6.25)$$

where the time $T_o$ must be chosen so large that the time integral (6.25) becomes independent of $T_o$. $\zeta_\infty$ is then obtained by calculating the time integral over the function $\gamma_\infty(t)$. In an unpublished report, A. Rahman has perfomed such a calculation of the Brownian friction coefficient. For details on similar calculations, see Berne and Forster (1971) and references quoted there.

As mentioned, the theory of Brownian motion is, formally, perfectly analogous to the hydrodynamic theory of, for example, spin diffusion. The formal role of the wavenumber k is now played by the mass ratio m/M. A comparison of (6.15), in the form

$$C(z) = \frac{i\,Mk_BT}{z + i\frac{m}{M}\zeta_\infty}\,, \tag{6.26}$$

with the hydrodynamic spin diffusion result (2.104) or with the transverse momentum density correlation function in a fluid, eq. (4.39), indicates that $\zeta_\infty$ plays the formal role of a dissipative transport coefficient, analogue of the spin diffusion coefficient D or the shear viscosity $\eta$. Eq. (6.24) can be recognized as a Kubo-type formula. Indeed, by an argument which is entirely parallel to that given in section 5.4 and based on the general relation (5.58), we can, in the limit of infinite B-mass M, omit the projectors from eq. (6.9), and write the friction constant in the form

$$m\zeta_\infty = \frac{1}{3k_BT} \lim_{\epsilon\to 0}\lim_{M\to\infty}\int_0^\infty dt\, e^{-\epsilon t}\,\langle \vec{F}(t)\cdot\vec{F}(0)\rangle\,, \tag{6.27}$$

where the time dependence is that generated by the full Liouvillian (6.5), and the order of the limits matters. This expression corresponds to eq. (5.57b), for example. Finally, the "absorptive response function"

$$\chi''(\omega;M) = \frac{1}{3}\int_{-\infty}^{\infty} dt\, e^{i\omega t}\,\langle \frac{i}{2}[\vec{p}^{\,o}(t),\vec{p}^{\,o}(0)]_{P.B.}\rangle \tag{6.28}$$

is given, at low frequencies and large M, by

$$\chi''(\omega;M) = \frac{m\,\omega\,\zeta_\infty}{\omega^2 + (\frac{m}{M}\,\zeta_\infty)^2} , \tag{6.29}$$

which follows by taking the real part of C(z), at $z = \omega$. Therefore, we can finally write $\zeta_\infty$ in the (almost) standard form

$$m\,\zeta_\infty = \lim_{\omega \to 0}\,[\,\lim_{M \to \infty} \omega\chi''(\omega;M)] , \tag{6.30}$$

which compares with eqs. (2.57) or (4.53).

There is one interesting, if as yet not dramatic, difference between the theory of Brownian motion as presented here, and hydrodynamic spin diffusion as indicated in chapter 2 and particularly in section 5.4. In either case, there is a hydrodynamic, dissipative pole at small z, given by

$$z = -i\,\Sigma(z) \approx -i\,\Sigma(0) . \tag{6.31}$$

The pole exists because

$$\Sigma(z) = \sigma(z)\,\chi^{-1} \tag{6.32}$$

is slowly varying near $z = 0$, and becomes arbitrarily small as either $k \to 0$ as for the spin diffusion memory function, or as $M \to \infty$ for the Brownian momentum memory function. In the case of spin diffusion, the reason for this behavior was the conservation law which forced $\sigma(z)$ to vanish as $k \to 0$ while $\chi$ remained finite in this limit. In our theory of Brownian motion, $\sigma(z)$ remains finite even in the limit as $M \to \infty$ but this time, $\chi^{-1} = 1/M$ vanishes in this limit. Both possibilities lead to irreversible hydrodynamic behavior. The difference in its origin is the reason why, in the expression (6.30) for the transport coefficient, there is no factor M in front of $\chi''$ which would correspond to the factor $1/k^2$ appearing in eq. (2.57), for example.

We could, of course, make this difference go away, formally, by scaling the B-momentum to the Brownian mass M, i.e., by studying the dimensionless momentum variable $\vec{\xi}^{o}(t) = \vec{p}^{o}(t)/\sqrt{Mk_BT}$ and its correlations. The small factor m/M in $\Sigma(z)$ would then be shifted from $\chi^{-1}$ to $\sigma(z)$, and it would thus appear in a fashion entirely reminiscent of spin diffusion theory. We find it preferable to work with the physical momentum whose time derivative, the force on the B-particle, is finite even as $M \to \infty$, but whose fluctuations in thermal equilibrium, given by the "susceptibility" $\chi(M) = MkT$, grow without bound as $M \to \infty$, thereby leading to hydrodynamically slow irreversible decay of time correlations. Those hydrodynamic modes which we discussed in chapters 2 and 4, owed their existence to microscopically exact conservation laws. In later chapters we will discuss examples of modes whose existence is, conversely, due to infinitely large static fluctuations, as they occur for certain degrees of freedom in ferromagnets, in superfluids, and in other systems in which spontaneous order is present.

## 6.2 The Generalized Langevin Equation

Having related the irreversible decay of the Brownian momentum autocorrelation function to underlying microscopic principles, let us now return to the simple phenomenological relaxation equation (6.1). As it stands, this equation can obviously not be microscopically correct. All of the complicated microscopical detail of fluid particles zooming hither and thither, bumping into the B-particle whenever the mood strikes them, cannot be completely lost on the B-particle's behavior. As our derivation of eq. (6.3) makes clear, what eq. (6.1) does contain is the gross effect of these collision, that which one sees when observing the B-particle with poor time resolution--eq. (6.3) was shown to hold at low frequencies when more rapidly fluctuating processes are ironed out. If one observes on a finer time scale, one has to insert into (6.1), in addition to the average friction force

$-\zeta M\vec{v}(t)$, a rapidly oscillating force $\vec{f}(t)$, due to the quickly changing fluid particle environment which the B-particle sees in fact. Instead of eq. (6.1), the motion of the B-particle is thus more correctly described by

$$\partial_t \vec{v}(t) + \zeta \vec{v}(t) = \frac{1}{M}\vec{f}(t) \ . \tag{6.33}$$

The average effect on the B-particle is contained in the friction term; the remainder, $\vec{f}(t)$, thus vanishes on the average,

$$\langle \vec{f}(t) \rangle = 0 \ . \tag{6.33a}$$

It is also uncorrelated with the B-particle velocity, i.e.,

$$\langle f_i(t) \cdot v_i \rangle = 0 \ . \tag{6.33b}$$

This must be true for eq. (6.33) to result in the exponential decay law for the velocity autocorrelation function, eq. (6.3), which we have just shown to be correct, at sufficiently low frequency. Finally, since the force $\vec{f}(t)$ is due to the extremely rapid and chaotic motion of the fluid particles, which collide with each other often between collisions with the B-particle, and thus quickly loose the memory of their earlier behavior, $\vec{f}(t)$ will be correlated only over extremely short times, and we should be able to represent the stochastic nature of $\vec{f}(t)$ by

$$\langle f_i(t) \cdot f_i(t') \rangle = 2\zeta M k_B T \ \delta(t-t')\delta_{ij} \ . \tag{6.33c}$$

The factor in this equation, $2\zeta M k_B T$, is a necessary consequence of the requirement that eqs. (6.33) describe a stationary process. The reader will have little difficulty in convincing himself of this fact. Kubo has called eq. (6.33c) the Second Fluctuation-Dissipation Theorem. Eqs. (6.33) constitute P. Langevin's stochastic (since the force, $\vec{f}(t)$, is assumed to be a random time-dependent process) theory of Brownian motion. In the next section, we will draw some simple and standard conclusions from this theory. First, however, let us derive it microscopically.

In such a derivation, we want again to begin with the Liouville equation which, for an arbitrary real dynamical variable A(t), is

$$\partial_t A(t) = iL\, A(t) \quad \text{or} \quad A(t) = e^{iLt} A \ , \tag{6.34}$$

and which describes a microscopically reversible system in purely mechanical motion. All forces, in this microscopic description, are mechanical rather than stochastic. That the forces on the B-particle behave at long times, in an effectively stochastic manner with short memory must be a limiting consequence of the Liouville equation. For convenience, let us consider (6.34a) as an equation for the bra $\langle A(t)|$ introduced in chapter 5, and take a Laplace transform so that

$$\langle A(z)| = \langle A| \frac{i}{z-L} \ , \tag{6.35}$$

where

$$\langle A(z)| = \int_0^\infty dt\, e^{izt} \langle A(t)| = \int_0^\infty dt\, e^{izt} \langle A| e^{-iLt} \ .$$

The Liouville operator, L, acts to the left in this equation. Introducing the projector onto A, P = 1-Q, and following in detail the derivations which have led us from eq. (5.9) to eq. (5.20), we can easily manipulate (6.35) into the "equation of motion" form

$$[z-\Omega+i\,\Sigma(z)]\langle A(z)| = i\langle A| + i\langle f_A(z)| \ , \tag{6.36}$$

where $\Omega$ and $\Sigma(z)$ are as given in (5.19) or (5.20), and

$$\langle f_A(z)| = \langle A| \frac{1}{z-LQ}\, LQ = \langle \dot{A}|Q \frac{i}{z-QLQ}\, Q \ . \tag{6.36a}$$

The second and more symmetric eq. (6.36a) follows from the first by little cosmetic operations whose proof is simple. Eq. (6.36) is the Laplace transform of an equation of motion for the "bra" $\langle A(t)|$ which is of the form, for $t > 0$,

$$(\partial_t + i\Omega)\langle A(t)| + \int_0^t d\tau\, \Gamma(t-\tau)\, \langle A(\tau)| = \langle f_A(t)| \ , \tag{6.37}$$

where $\Sigma(z)$ is the Laplace transform of $\Gamma(t)$ which is given in eq. (5.21a), and

$$\langle f_A(t) | = \langle \dot{A} | Q \, e^{-iQLQt} \, Q \,. \tag{6.37a}$$

This equation, which has first been derived by Mori (1965), is of a form reminiscent of Langevin's approximate equation (6.33), and it has therefore been called the "generalized Langevin equation". It is fully rigorous, and could also have been obtained directly from the Liouville equation (6.34), without the (convenient) detour via the Laplace transform. It separates the dynamics into three parts: (a) an instantaneous oscillation term given by $\Omega$ which describes the motion of $\langle A(t) |$ due to its intrinsic dynamics; $\Omega$ vanishes whenever the degree of freedom has a definite signature under time reversal; (b) a time-lagged linear memory term giving the effect of past behavior of A(t), on its present time rate of change; this term is due to the coupling of A to all other degrees of freedom (to the "bath"), accordingly giving the flow of information about the state of motion of A(t) into the bath and, with some time delay, back to A(t) itself; (c) finally, there is a driving force $f_A(t)$ external to the degree of freedom A, which summarizes the direct influence of internal bath dynamics on the behavior of A(t). These words are meant to illustrate the separation accomplished by eq. (6.37) in an intuitive fashion; to verify them in detail would require a great deal of more involved analysis. Note, however, that the left hand side of eq. (6.37) contains quantities, namely $\Omega$ and $\Gamma(t)$, which fully determine the dynamical behavior of the averaged auto-correlation function $C_A(t) = \langle A(t) | A(0) \rangle$ as shown in eq. (5.21). The remaining influence on the time rate of change of the variable A(t) itself, is a microscopic force $f_A(t)$ which evolves in time according to the modified Liouville operator QLQ from which thermal fluctuations of the degree of freedom A are removed. In a case in which the latter are the only slow fluctuations expected in the system, $f_A(t)$ will change in time on a much faster microscopic scale than A(t) itself does.

Equation (6.37) is, of course, a generalization of Langevin's equation (6.33), introducing memory effects into the latter, and giving a semi-explicit expression for the random force, f(t). This expression does verify two fundamental properties: First, because of the projector Q in (6.37a), the random force $f_A(t)$ is, at all t, uncorrelated with the initial value of the variable A, i.e.,

$$\langle f_A(t) | A \rangle = 0 , \tag{6.38a}$$

which is analogous to eq. (6.33b). Second, from the expression (6.37a) we can calculate the force autocorrelation function, and find that it is given by

$$\begin{aligned} \langle f_A(t) | f_A(t') \rangle &= \langle \dot{A} | Q \; e^{-iQLQ(t-t')} \, Q | \dot{A} \rangle \\ &= \Gamma(t-t') \, C(0) = kT \, \gamma(t) \, . \end{aligned} \tag{6.38b}$$

This fundamental and rigorous result is the generalization, of course, of the second fluctuation-dissipation theorem, eq. (6.33c).

Before returning to Brownian motion, let us quickly note that this formal framework easily applies to a simultaneous consideration of several variables $A_i(t)$ , i = 1 ... N, as well. As in section 5.2, for the multivariate case eq. (6.37) is simply read as a matrix equation,

$$[\partial_t \delta_{ik} + i\Omega_{ik}] \langle A_k(t) | + \int_0^t d\tau \, \Gamma_{ik}(t-\tau) \langle A_k(\tau) | = \langle f_i(t) | , \tag{6.39}$$

where the frequency matrix $\Omega_{ij}$ and the memory matrix $\Gamma_{ij}(t)$ are those given in eqs. (5.28), and the "random forces" $f_i(t)$ are given by

$$| f_i(t) \rangle = Q \, e^{iQLQt} \, Q | \dot{A}_i \rangle , \tag{6.39a}$$

in which Q is the projector of eq. (5.27). The reader will find the derivation of (6.39) straightforward. We have given our considerations as equations of motion for "bras" $\langle A(t) |$, in order to tie them more directly into what we did in sections 5.1 and 5.2. Since one is accustomed to identifying dynamical variables with kets--not that it makes any difference--and since that is the way these equations often appear in the literature, they are equivalent to

$$(\partial_t \delta_{ik} - i\Omega^*_{ik})\, |A_k(t)\rangle + \int_0^t d\tau\, \Gamma^*_{ik}(t-\tau)\, |A_k(\tau)\rangle = |f_i(t)\rangle, \tag{6.40}$$

obtained from (6.39) by complex conjugation. The fundamental properties, eqs. (5.38), have their obvious generalizations,

$$\langle f_i(t) | A_j \rangle = 0, \tag{6.41a}$$

$$\langle f_i(t) | f_j(t') \rangle = \Gamma_{ik}(t-t')\, C_{kj}(0) \equiv k_B T\, \gamma_{ij}(t-t'), \tag{6.41b}$$

where

$$\gamma_{ij}(t) = \frac{1}{k_B T} \langle \dot{A}_i | Q\, e^{-iQLQt}\, Q | \dot{A}_j \rangle \tag{6.42a}$$

is the unnormalized memory matrix, and

$$C_{ij}(0) = k_B T\, \chi_{ij} = \langle A_i | A_j \rangle \tag{6.42b}$$

the static "susceptibility" matrix.

The equations of this section want to be used with some caution. The dynamical variable, A(t), is a purely mechanical, as yet unaveraged, quantity, and in a complex system it is bound to fluctuate wildly. It is, in many circumstances, a bit artificial to separate from this motion the more regular part which is observed on the average close to thermal equilibrium. While our derivation of equations like (6.37) makes no explicit reference to the microscopic state of the system from which this motion begins, it is clear that with the projector defined via a thermal ensemble, the separation accomplished by (6.37) will be sensible only for initial states which are in fact close to equilibrium. Under different circumstances, e.g., in a turbulent system, alternative definitions of the scalar product involved in the projector P would be more adequate. Note that the abstract formalism leaves us, for example, free choice as to the temperature at which we want to evaluate the projectors, and thus the memory functions, but that physical considerations will of course require choosing that temperature at which the system as a whole is actually held; only then can we expect the force f(t) to be reasonably random. By multiplying eq. (6.37) with A and averaging, one sees then immediately that, in view of (6.38a) the memory function equation for the equilibrium-averaged time correlation function is regained.

## 6.3 Diffusion of a Heavy Particle

When the mass M of the B-particle is large, we saw that the memory function for the momentum auto-correlation function, $\Sigma(z) = \frac{m}{M}\sigma(z)$ in (6.12), is dominated by rapid processes, and can be replaced by its limit as $z \to 0$,

$$\frac{m}{M}\sigma(z) \simeq \frac{m}{M}\sigma(0) = \zeta , \tag{6.43}$$

so long as we are interested in much smaller frequencies, or longer times, than those contained in $\sigma(z)$ or $\gamma(t)$. (6.43) corresponds to approximating $\gamma(t)$ by

$$\gamma(t) = 2m\,\sigma(0)\,\delta(t) , \tag{6.44}$$

where the factor 2 is there because the time integral in (6.11) extends only over positive times. If we therefore apply the general framework of section (6.2), i.e., eq. (6.37), to the momentum $\vec{p}(t)$ of the Brownian particle, we find that the general and rigorous equation

$$\partial_t \vec{p}(t) + \frac{1}{M}\int_0^t d\tau\, \gamma(t-\tau)\, \vec{p}(\tau) = \vec{f}(t), \tag{6.45}$$

with $\gamma(t)$ as given in eq. (6.9), reduces for long times to the stochastic equation

$$\partial_t \vec{p}(t) + \zeta\vec{p}(t) = \vec{f}(t) , \tag{6.46}$$

where the force is not correlated with the initial momentum,

$$\langle f_i(t)\, p_j \rangle = 0 , \tag{6.46a}$$

and where the force autocorrelation function, given in general by

$$\langle f_i(t)\, f_j(t')\rangle = k_B T\ \gamma(t-t')\ \delta_{ij}$$

as in (6.41b), can for long time differences be replaced by

$$\langle f_i(t)\, f_j(t')\rangle = 2\zeta\, M k_B T\ \delta(t-t')\ \delta_{ij} \tag{6.46b}$$

in accordance with (6.44), if high frequency fluctuations are not of interest. We thus obtain the Langevin theory of eqs. (6.33). It will be applicable whenever the mass ratio m/M is sufficiently small since the B-particle relaxes on a scale which is slower by a factor m/M than the microscopic lifetimes which characterize the time dependence of $\gamma(t)$.

A quantity of great experimental interest which we can now calculate, is the mean distance $\Delta r(t)$ which the B-particle traverses in time t. The calculation is standard, see Chandrasekhar (1943). From (6.46) we obtain

$$\vec{p}(t) - \vec{p}(0)\, e^{-\zeta t} = \int_0^t d\tau\ e^{-\zeta(t-\tau)}\, \vec{f}(\tau). \tag{6.47}$$

Further, since $\partial_t \vec{x}(t) = \vec{p}(t)/M$, a second integration yields the position $\vec{x}(t)$ of the B-particle which was initially at $\vec{x}(0)$ with momentum $\vec{p}(0)$:

$$\vec{x}(t) - \vec{x}(0) = \vec{p}(0)\, \frac{1-e^{-\zeta t}}{M\zeta} + \frac{1}{M\zeta} \int_0^t d\tau\ \vec{f}(\tau)\ [1-e^{-\zeta(t-\tau)}]. \tag{6.48}$$

The mean distance can then easily be obtained,

$$[\Delta r(t)]^2 = \langle (\vec{x}(t) - \vec{x}(0))^2 \rangle, \tag{6.49}$$

by squaring (6.48) and using (6.46 a,b) and the equipartition theorem, eq. (6.8). In the limit of very long times, $t \gg \zeta^{-1}$, one finds

$$[\Delta r(t)]^2 = 6\, Dt\,, \tag{6.50}$$

where the diffusion coefficient D is given in terms of the friction constant by

$$D = \frac{k_B T}{M\zeta} = \left(\frac{k_B T}{m}\right) \frac{1}{\zeta_\infty} \quad . \tag{6.50a}$$

Equation (6.50a) is often called an Einstein relation. Because of (6.3), it is equivalent to the equation

$$D = \frac{1}{3} \int_0^\infty dt \, \langle \vec{v}(t) \cdot \vec{v}(0) \rangle . \tag{6.51}$$

As it stands, eq. (6.33) does not, of course, specify all statistical properties which concern the B-particle. To fully characterize the stochastic force, we have to know, in addition to the variance (6.33), its higher moments $\langle f(t_1)\, f(t_2)\, f(t_3) \rangle$, $\langle f(t_1)\, f(t_2)\, f(t_3)\, f(t_4) \rangle$ etc. It is usually assumed--though our derivation has not as yet shown that to be true--that the random force f(t) has, at any time, a Gaussian distribution. This is most easily stated in terms of a generating functional, in the form

$$\langle e^{i \int_0^\infty dt \vec{U}(t) \vec{f}(t)} \rangle = e^{-\lambda \int_0^\infty dt \, \vec{U}^2(t)} . \tag{6.52}$$

where $\lambda = Mk_BT\zeta$. By taking derivatives with respect to the arbitrary function $\vec{U}(t)$, we obtain (6.33c) from (6.52), and in addition

$$\begin{aligned} \langle f(1)\, f(2)\, f(3)\, f(4) \rangle &= \langle f(1)\, f(2) \rangle \langle f(3)\, f(4) \rangle \\ &+ \langle f(1)\, f(3) \rangle \langle f(2)\, f(4) \rangle + \langle f(1)\, f(4) \rangle \langle f(2)\, f(3) \rangle , \end{aligned} \tag{6.53}$$

for example, where $f(1) \equiv f_{i_1}(t_1)$. With this assumption of a Gaussian random force, the Langevin theory is equivalent to a Fokker-Planck equation for the distribution function of Brownian particles which we will consider in the next section.

## 6.4 The Fokker-Planck Equation

We have so far concentrated on small Brownian momentum fluctuations, described by the momentum auto-correlation function. A more exhaustive treatment of Brownian motion has to consider the distribution function $\hat{f}(rpt)$, normally introduced so that $\hat{f}(rpt)\, d\vec{r}\; d\vec{p}$ is the number of B-particles in the small volume element $d\vec{r}\; d\vec{p}$. Since we are interested only in extremely dilute suspensions of B-particles in which these move independently of one another, we can again consider a system which contains a single B-particle, and interpret $\hat{f}(rpt)\; d\vec{r}\; d\vec{p}$ as the probability of finding the B-particle in $d\vec{r}\; d\vec{p}$, at time t. From $\hat{f}(rpt)$ all quantities of statistical interest concerning the behavior of the B-particle can be obtained, among them results we have already given. It is of importance therefore to have an equation of motion for $\hat{f}$. The requisite equation is due to A. D. Fokker and M. Planck who in 1917 derived, on the basis of stochastic considerations, their famous equation,

$$(\partial_t + \frac{\vec{p}\cdot\vec{\nabla}}{m})\, \hat{f}(rpt) \;=\; \zeta\; \vec{\partial}\cdot (\vec{p} + Mk_BT\; \vec{\partial})\; \hat{f}(rpt)\,, \tag{6.54}$$

where $\vec{\partial} \equiv \partial/\partial\vec{p}$. In this chapter, we will derive this equation from statistical mechanics and the purely mechanical Liouville equation. For specific applications of eq. (6.54), see e.g. Chandrasekhar (1943).

To start, we need an operator or dynamical variable, $f(rpt)$, whose average over some Gibbs ensemble represents the probability density $\hat{f}(rpt)$. The requisite quantity can evidently be written in the form

$$f(\vec{r}\vec{p},t) \;=\delta(\vec{r}-\vec{r}^{\,o}(t))\;\; \delta(\vec{p} - \vec{p}^{\,o}(t))\;, \tag{6.55}$$

where as before, $\vec{r}^{\,o}(t)$ and $\vec{p}^{\,o}(t)$ are the microscopic position and momentum of the B-particle, at time t. In full thermal equilibrium, in a uniform system,

$$\langle f(rpt)\rangle_{eq.} = \frac{1}{V}\,(\frac{2\pi M}{\beta})^{-3/2}\;\; e^{-\beta\frac{p^2}{2M}}\;\;, \tag{6.56}$$

indicating that the particle is anywhere in the volume V with constant probability, and that its velocity-distribution is Maxwellian. The quantity of greatest interest is the equilibrium-averaged correlation function

$$C(rpt,\, r'p't') \;=\; \langle\, f(rpt)\; f(r'p't')\rangle_{eq.} - \;\langle f(rpt)\rangle_{eq.}\langle f(r'p't')\rangle_{eq.}\;, \tag{6.57}$$

which gives the conditional probability of finding, at time t, the B-particle at point $\vec{r}$ with momentum $\vec{p}$ if we know that at time t', it was at point $\vec{r}'$ and moved with momentum $\vec{p}'$. We shall therefore study this function. For convenience, we will use the notation "1" $\equiv \{\vec{r}, \vec{p}\}$ and the bracket notation introduced earlier, so that (6.57) appears in the form

$$C(11', t-t') = \langle f(1,t) | f(1',t') \rangle = \langle f(1) | e^{-iL(t-t')} | f(1') \rangle \qquad (6.58)$$

where L is the Liouville operator of eq. (6.5), and $f(1) \equiv f(1, t=0)$. Because of the connection between fluctuations and linear response, knowledge of C is, of course, equivalent to knowledge of the probability density $\hat{f} = \langle f \rangle_{\text{non-eq.}}$ for any initial state which is only slightly displaced from equilibrium--more precisely, for any state which can be produced by acting with arbitrary external forces on the B-particle. Let us, then, apply the formalism of chapter 5 to the calculation of C.

We can easily calculate the equal time value of C. It is given by

$$C(11', t=0) = \beta^{-1} \chi(11') = \delta(11') \langle f(1) \rangle, \qquad (6.59)$$

where $\delta(11') = \delta(\vec{r} - \vec{r}')\delta(\vec{p} - \vec{p}')$. The frequency matrix $\omega$ of eq. (5.28) is also easily written down since

$$\omega(11') = i\beta \langle \dot{f}(1) | f(1') \rangle = i \langle [f(1), f(1')]_{P.B.} \rangle_{eq.}$$

so that, evaluating the Poisson bracket, we obtain

$$\omega(11') = i[\nabla \partial' - \partial \nabla'] \langle f(1) \rangle \delta(11') = -i\beta \frac{p\nabla}{M} \langle f(1) \rangle \delta(11')$$

and therefore

$$\Omega(11') = \omega(1\bar{1}) \chi^{-1} (\bar{1}1') = -i \frac{p\nabla}{M} \delta(11'). \qquad (6.60)$$

Here, $\chi^{-1}$ is the matrix inverse of $\chi$, and an integral $\int d\bar{1} = \int d^3\bar{r} \int d^3\bar{p}$ over the barred variables has been implied. The reader will have no difficulty going through the steps which lead to (6.60). This simple result gives, of course, the free streaming of the B-particle whose equation of motion can now, in analogy to the general equation (5.21) or (5.28), be written in the form

$$(\partial_t + \frac{p\nabla}{M}) C(11',t) + \int_0^t d\tau\, \Gamma(1\bar{1};t-\tau) C(\bar{1}1',\tau) = 0. \tag{6.61}$$

The difficult part in this equation is, of course, the memory function, given in general by

$$\Gamma(11',t) = \gamma(1\bar{1},t)\, \chi^{-1}(\bar{1}1') = \gamma(11',t)/\beta\langle f(1')\rangle, \tag{6.62}$$

where

$$\gamma(11',t) = \beta\langle f(1)| LQ\, e^{-iQLQt}\, QL| f(1')\rangle . \tag{6.62a}$$

This is the first equation in which occurs the projection operator Q. Since we are now concerned with correlations of all B-particle properties, summarized in f(1), the projector is now, in accordance with eq. (5.27), formally given by

$$P = 1-Q = \int d1 \int d1'\, |f(1)\rangle \beta\chi^{-1}(11')\, \langle f(1')| \tag{6.63a}$$

or, inserting $\chi$ from (6.59), it is defined by

$$PA = \int d1\, f(1)\, \frac{\langle f(1)A\rangle_{eq.}}{\langle f(1)\rangle_{eq}} , \tag{6.63b}$$

for any microscopic property A. For any property $B(r^o p^o)$ which depends only on the coordinates of the B-particle, but not on the host fluid variables, it is obvious that $P|B\rangle = |B\rangle$ or that $Q|B\rangle = 0$.

Let us now insert the Liouville operator L of (6.16) into (6.62a), and consider the end parts $QL|f\rangle$ of this expression. It is easy to see that the part $L^o$ of L does not contribute,

$$L^o | f(1')\rangle = 0 \qquad \text{and} \qquad \langle f(1)| L^o = 0, \tag{6.64}$$

since $L^o$ acts only on the host particles. The Brownian streaming term does not contribute either since

$$L_B|f(1)\rangle = |-i\frac{\overset{\circ}{p}\overset{\circ}{\nabla}}{M}f(1)\rangle = i\frac{p\nabla}{M}|f(1)\rangle,$$

so that

$$Q\,L_B|f(1)\rangle = 0\,. \tag{6.65}$$

Therefore, we are left with

$$\begin{aligned} QL|f(1)\rangle &= QL_{f\to B}|f(1)\rangle \\ &= i\,Q|\sum_{\alpha}\vec{\nabla}U(r^{\circ}-r^{\alpha})\frac{\partial}{\partial\vec{p}^{\,\circ}}f(1)\rangle = i\frac{\partial}{\partial\vec{p}}Q|\vec{F}f(1)\rangle, \end{aligned} \tag{6.66}$$

where we have inserted the force $\vec{F}$ on the B-particle according to eq. (6.22), and replaced the internal momentum gradient $\partial/\partial p^{\circ}$ by minus the external gradient $\partial/\partial p$ which holds when it acts on f(1). With these manipulations, eq. (6.62a) for the memory function appears in the form

$$\beta^{-1}\gamma(11',t) = \partial_i\partial_i'\langle F_i f(1)|Q\,e^{-iQLQt}Q|F_i f(1')\rangle\,. \tag{6.67}$$

We expect, from our experience in section 6.1, that eq. (6.61) simplifies when the B-particle becomes very heavy, $M\to\infty$. Let us therefore consider this limit. In eq. (6.18) we indicated that in this limit, the full Liouville operator L reduces to $L^{\circ}$, the Liouvillian for the host fluid in the presence of a fixed B-particle. $L^{\circ}$, given explicitly in (6.16a), acts only on the fluid particles so that the projectors in the exponential in (6.67) can be omitted. It is also not difficult to show that the remaining two projectors in (6.67) can be omitted as well since, for example,

$$\langle f(1)|\vec{F}f(1')\rangle = \delta(11')\langle f(1)\rangle\langle\vec{F}\rangle = 0\,,$$

so that $P|\vec{F}f(1')\rangle = 0$. Consequently, we can write (6.67), in the limit as $M\to\infty$, as

$$\beta^{-1}\gamma_{\infty}(11',t) = \partial_i\partial_i'\langle F_i f(1)|e^{-iL^{\circ}t}|F_i f(1')\rangle. \tag{6.68}$$

And since $L^o$ acts only on the force $\vec{F}$ but not on the density $f(1')$, we can independently perform the static average over the f-product in (6.68), and finally obtain

$$\gamma_\infty(11',t) = \vec{\partial}\,\vec{\partial}'\,\langle f(1)\,f(1')\rangle m\,\gamma_\infty(t)\,, \tag{6.69}$$

where $\gamma_\infty(t)$ is the force autocorrelation function already encountered in eq. (6.23). With (6.59), we finally obtain

$$\Gamma_\infty(11',t) = -\frac{m}{M}\,\gamma_\infty(t)\,\vec{\partial}(\vec{p} + Mk_BT\,\vec{\partial})\,\delta(11') \tag{6.70}$$

and therefore, for large M,

$$(\partial_t + \frac{\vec{p}\cdot\vec{\nabla}}{M})\,C(11',t) = \frac{m}{M}\,\vec{\partial}(\vec{p} + Mk_BT\vec{\partial})\int d\tau\,\gamma_\infty(\tau)C(11',t-\tau) \tag{6.71}$$

The collision operator on the right hand side is explicitly small of order m/M, and vanishes as $M \to \infty$. For large B-mass M therefore, $C(\tau)$ will contain relaxation times of order $(m/M)^{-1}$ which are much larger than those rapid microscopic times contained in $\gamma_\infty(t)$ which characterize the chaotization of the host fluid. Under these circumstances, we can replace $\gamma_\infty(t)$ in (6.71) by an instantaneous $\delta$-function in time, i.e. replace (6.71) by the Markoffian equation

$$(\partial_t + \frac{\vec{p}\cdot\vec{\nabla}}{M})\,C(11',t) = \zeta\vec{\partial}\cdot(\vec{p} + Mk_BT\,\vec{\partial})\,C(11',t)\,, \tag{6.72}$$

if we are interested in times longer than those contained in $\gamma_\infty(t)$. This, of course, is the Fokker-Planck equation (6.54), written in terms of the conditional probability density $C(11',t)$, where

$$\zeta = \frac{m}{M}\int_0^\infty dt\,\gamma_\infty(t) = \frac{m}{M}\,\zeta_\infty \tag{6.72a}$$

is the same friction constant we encountered in section 6.1, eq. (6.24). Note that by contrast to the eq. (6.54), (6.72) describes a mathematically well-defined object whose initial value is a fully specified equilibrium average, namely eq. (6.59). Suspicious readers might sharpen our somewhat hurried derivation by introducing dimensionless canonical momentum variables as suggested after eq. (6.18), and discussing the order in (m/M) of the various terms somewhat more carefully than we have done here.

As the quickest of all uses of eq. (6.72), we might rederive the momentum autocorrelation function since

$$\langle \vec{p}(t)\cdot\vec{p}(t')\rangle = \int d\vec{p} \int d\vec{p}' \int d\vec{r} \int d\vec{r}'\ \vec{p}\cdot\vec{p}'\ C(11',t), \tag{6.73}$$

as is obvious from the definition of C in (6.57). We can perform all but one of those integrals directly on (6.72). Namely, the function

$$g_i(p,t) = \int d\vec{p}' \int d\vec{r} \int d\vec{r}'\ \ p_i'\ C(11',t) \tag{6.74}$$

fulfils the equation

$$\partial_t g_i(p,t) = \zeta\ \vec{\partial}\,(\vec{p} + Mk_BT\ \vec{\partial})\ g_i(p,t), \tag{6.75a}$$

and (6.59) prescribes the initial condition to be

$$g_i(p,0) = (2\pi M/\beta)^{-3/2}\ p_i\ e^{-\beta \frac{p^2}{2M}} . \tag{6.75b}$$

This is an eigenfunction of the Fokker-Planck operator, and thus (6.73) is solved by

$$g_i(p,t) = e^{-\zeta t}\ g_i(p,0)\ , \tag{6.76}$$

from which our previous result (6.15) follows immediately.

# CHAPTER 7

# BROKEN SYMMETRY

We have analyzed, in considerable detail, collective fluctuations in a variety of many-body systems, concentrating on the region of small wavenumber k or large wavelength $\lambda = 2\pi/k$ where these fluctuations involve cooperative behavior of extremely large numbers of particles. We have paid particular attention to those modes whose lifetime $\tau(k)$ becomes infinite as $k \to 0$. These hydrodynamic modes contain the most informative characterization of the macroscopic dynamic properties of the many-body system. Spin diffusion in an itinerant ($He^3$) or isotropic Heisenberg paramagnet, heat diffusion, normal sound and the transverse shear diffusion in a normal one-component fluid were our examples. In all these instances, the existence of hydrodynamically long-lived modes was a direct consequence of microscopic conservation laws. The reader may want to recall the discussion given in section 2.1 in connection with fig. 2.1, and the more formal analysis presented in section 5.4. A variety of additional fluctuation phenomena can be treated in the same manner. In a binary fluid, for example, there is a long-lived mode associated with concentration fluctuations, see Martin (1965). In a weakly anharmonic crystal in which umklapp processes can be neglected, the phonon gas conserves both the energy and the heat current densities, and a second sound mode results, see Kwok and Martin (1966). In a molecular fluid in which the intermolecular forces are only weakly non-central, one obtains long-lived collective angular momentum fluctuations, see Ailawadi, Berne and Forster (1971).

There is, however, another category of cooperative hydrodynamic processes which are not the direct result of a microscopic continuity equation. Spin conservation alone told us that the spin density in a paramagnet undergoes a diffusion process. However, in the ferromagnetic regime, described by the same Hamiltonian with obviously the same conserved quantities, there are long-lived propagating spin waves. There are also hydrodynamic spin waves in an antiferromagnet which are collective excitations of the staggered magnetization $\vec{N}(r,t)$, a quantity which is not microscopically conserved. In a nematic liquid crystal, a molecular liquid whose molecules are aligned on the average, there are cooperative fluctuations of the molecular orientation which decay infinitely slowly as $k \to 0$. And finally, second sound in superfluid Helium is due to fluctuations of the phase of the macroscopic wave function, manifestly not a conserved quantity.

The fundamental reason why all these modes are nevertheless infinitely long-lived as $k \to 0$, is the spontaneous breaking of a continuous symmetry and the presence of long-ranged order in all of these systems.

The mechanism is physically very simple, really. Consider a ferromagnet, for example. In this system, which we take to be composed of spins attached to the sites of some regular lattice, the forces between two spins favor parallel alignment. Even if these forces act only over short microscopic distances, they succeed for sufficiently low temperature, in the ferromagnetic regime, in aligning all spins on the average along some common direction. Which direction? Well, assuming here that the lattice itself is quite passive, that it exerts no orienting influence on the spins, the direction of preferential orientation is completely arbitrary. At least in the absence of external fields, the free energy is independent of that direction. But if it therefore costs no energy at all to rotate the system as a whole, it will cost very little energy to produce a fluctuation of very long wavelength in which the direction of alignment varies in a sinusoidal fashion in space, as shown in fig. 7.1 . Consequently,

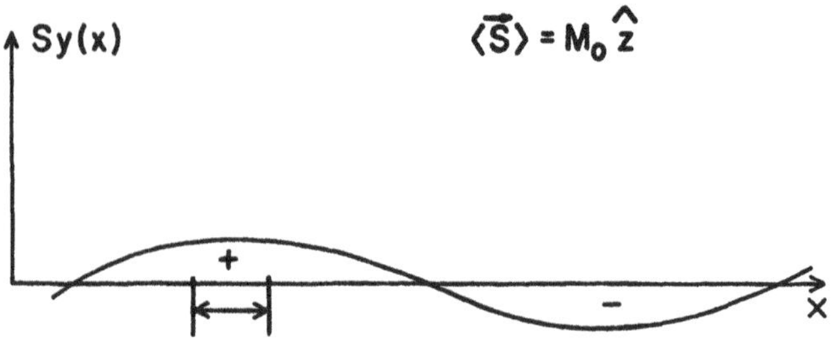


Fig. 7.1

the likelihood for such a fluctuation to arise spontaneously, measured by $\langle |S_y(k)|^2 \rangle$, must go to infinity as its wavelength becomes infinitely long, and normally one finds

$$\chi_\perp(k) = \beta\langle |S_y(k)|^2 \rangle = \frac{M_o^2}{Rk^2} \qquad \text{as } k = \frac{2\pi}{\lambda} \to 0 \,, \tag{7.1}$$

where R is some positive constant, and $M_o^2$ is inserted for normalization. Now $\beta^{-1}\chi_\perp(k)$ is, of course, the spatial Fourier transform of the static correlation function

$$S_\perp(\vec{r}-\vec{r}') = \langle S_y(\vec{r})S_y(\vec{r}')\rangle - \langle S_y(\vec{r})\rangle\langle S_y(\vec{r}')\rangle \,, \tag{7.2}$$

where, since we have taken the average overall alignment in the z-direction, $\langle S_y \rangle = 0$ actually. (7.1) thus corresponds to the statement that for large $|\vec{r}-\vec{r}'|$,

$$S_\perp(\vec{r}-\vec{r}') = \frac{k_BT}{R} \cdot \frac{M_o^2}{4\pi|\vec{r}-\vec{r}'|} \,. \tag{7.3}$$

The small-k-divergence (7.1) of the state susceptibility $\chi_\perp(k)$ is an expression of the fact that correlations of the transverse-spin density, $S_y(\vec{r})$, are of infinitely long range.

Now suppose a fluctuation as shown in fig. 7.1 has arisen at $t = 0$, spontaneously or by external means. How will it decay? The answer is: very slowly.

The point is that in any region which is small compared to $\lambda$, such as indicated by arrows in fig. 7.1, the spins are essentially uniformly aligned — not in the z-direction, but any direction is fine as far as the internal forces among these spins are concerned. There is no locally active restoring force as it were. The <u>only</u> reason why $S_y(\vec{r},t)$ relaxes at all is that half a wavelength away, the alignment direction is different, and so the spins must rotate to finally reach the equilibrium state of uniform alignment. In a sense, equilibration requires transport, over a distance of the order $\lambda/2$, of the "information about the local alignment direction." As $\lambda \to \infty$, the time required for this process must go to infinity as well, as some power of $\lambda$.

The reader will notice the similarity of the argument given here to that presented for conserved densities on page 11. It is true, of course, that in an isotropic ferromagnet of the type considered here, the spin density $S_y(\vec{r}\ t)$ is also conserved. However, the present argument has nothing whatsoever to do with that property which is not shared by the "symmetry-restoring" variables" in antiferromagnets, superfluids, or liquid crystals. It would have been more pedagogic to pick one of these for an example were it not for the suggestion that most everybody knows ferromagnets so much better.

## 7.1 Long-Ranged Correlations and Slow Modes

For an introductory formal understanding of the connection between long-ranged correlations and hydrodynamic modes, consider a dynamical variable A(rt), which is <u>not</u> conserved, such as the local molecular orientation in a liquid crystal. This means that the microscopic equation of motion is of the general form

$$\partial_t A(r,t) + \vec{\nabla}\vec{j}^{\,A}(r,t) = S(r,t)$$

or

$$\partial_t A(k,t) - i\vec{k}\,\vec{j}^{\,A}(k,t) = S(k,t)\,. \tag{7.4}$$

While this separation of $\dot{A}$ into a streaming-term, $\vec{\nabla}\vec{j}$, and a source term, S, is not unique, the point is, of course, that the latter is not the divergence of a local density, i.e. that S(k,t) is finite as $k \to 0$:

$$S = \lim_{k\to 0} S(k,0) = \frac{1}{\sqrt{V}} \int d\vec{r}\; S(r,t) \neq 0$$

or

$$\frac{d}{dt} \frac{1}{\sqrt{V}} \int d\vec{r}\, A(r,t) \Big|_{t=0} \equiv S \neq 0\,. \tag{7.5}$$

Now the correlation function, defined as usual by

$$C_{AA}(kz) = \beta^{-1} \int \frac{d\omega}{\pi i}\, \chi''_{AA}(k\omega)/\omega(\omega - z)\,, \tag{7.6}$$

can always be written in the form (5.20),

$$C_{AA}(kz) = \frac{i\beta^{-1}\chi_A(k)}{z + i\;\sigma_A(kz)\,\chi_A^{-1}(k)}\,, \tag{7.7}$$

where the static susceptibility is given by

$$\chi_A(k) = \int \frac{d\omega}{\pi}\, \chi''_{AA}(k\omega)/\omega = \beta\,\langle A(k)|A(k)\rangle \tag{7.7a}$$

and the memory function by

$$\sigma_A(kz) = \beta\,\langle \dot{A}(k)|\, Q \frac{i}{z - QLQ} Q\,|\dot{A}(k)\,\rangle\,, \tag{7.7b}$$

with the projector Q defined as usually, and where we have assumed for simplicity that $\langle \dot{A}(k) | A(k) \rangle = 0$ as is usually the case.

If A(rt) where the density of a constant of the motion, we would have $\sigma(kz) \sim k^2$, and (7.7) would result in a hydrodynamic pole. For a non-conserved variable A, however, the limit

$$\lim_{k \to 0} \sigma_A(kz) = \beta \langle S | Q \frac{i}{z - QLQ} Q | S \rangle \neq 0 \tag{7.8}$$

is finite. If we were to make the (questionable) assumption that $\sigma(kz)$ in (7.7a) can be replaced by a constant, namely by

$$\lim_{z \to 0} \lim_{k \to 0} \sigma_A(kz) \equiv \sigma_A^o \tag{7.9}$$

which is always positive, then (7.7) would become

$$C_{AA}(kz) = \frac{i\beta^{-1} \chi_A(k)}{z + i \sigma_A^o \chi_A^{-1}(k)} \tag{7.10}$$

which one might hope to give an approximate description at small k and z. Assuming that the susceptibility $\chi_A(k)$ is finite as $k \to 0$ as so far it always has been, $\chi_A = \lim_{k \to 0} \chi_A(k)$, (7.10) would have a pole at

$$z = -i \sigma_A^o \chi_A^{-1} \equiv -i \tau_A^{-1} , \tag{7.11}$$

corresponding to a finite lifetime $\tau_A$. This result, which corresponds to an equation of motion

$$\partial_t \delta \langle A(rt) \rangle = - \frac{1}{\tau_A} \delta \langle A(rt) \rangle , \tag{7.12}$$

gives an adequate phenomenological description of a great many processes. For example, the Bloch equation for the magnetization in a non-conserving paramagnet (e.g. one with spin-lattice interaction) is an approximate equation of the type (7.10) or (7.12).

Suppose, however, that the variable $A(\vec{r}t)$, which is not a conserved density, has static correlations of infinite range (as a result of some broken continuous symmetry, or for whatever reason), i.e. that

$$\chi_A(k) = \frac{M_o^2}{R_A k^2} \qquad \text{as } k \to 0, \tag{7.13}$$

where the coefficient, $R_A M_o^{-2}$, must be positive since $\chi_A(k)$ is a positive quantity. ($M_o$ is inserted for later consistency.) The impact of (7.13) on dynamical properties is now a simple matter. Given (7.13), the normalized memory function $\Sigma_A(kz) = \sigma_A(kz)\,\chi_A^{-1}(k)$ has the same behavior at small k as it did in a conserving system even if $\sigma_A(kz)$ is now finite at small k and z. For example, in the spin-paramagnetic case (5.51), we had $\Sigma_M(kz) \sim k^2$ at small k because the continuity equation for the spin density led to $\sigma_M(kz) \sim k^2$ while $\chi_M(k)$ was finite as $k \to 0$. Now again we find $\Sigma_A(kz) \sim k^2$ in an ordered system not because the variable A is conserved which by assumption it is not, but because it is statically correlated over large distances, $\chi_A(k) \sim 1/k^2$. Otherwise, all is the same. By the argument which we have given repeatedly, see e.g. sections 2.11 or 5.4, we find a hydrodynamic pole in $C_{AA}(kz)$, at

$$z = -i\,\sigma_A^o\,(R_A/M_o^2)k^2, \tag{7.14}$$

rigorously to order $k^2$. And the absorptive part of the response function,

$$\chi''_{AA}(k\omega) = \omega \beta\, \mathrm{Re}\, C_{AA}(k, \omega + i\epsilon), \tag{7.15a}$$

is given by

$$\frac{1}{\omega}\chi''_{AA}(k\omega) = \frac{\sigma^{\circ}_A}{\omega^2 + [k^2\sigma^{\circ}_A R_A M_o^{-2}]^2} \qquad (7.15b)$$

for small k and $\omega$, corresponding to a Lorentzian spectral line centered at $\omega = 0$, and with a full width at half maximum which is of order $k^2$, namely $(2\sigma^{\circ}_A R_A M_o^{-2})k^2$. There is an important difference, however, between the fluctuations of a conserved density and the "symmetry-restoring mode" treated here. The total weight of $\chi''_{AA}$, i.e the area under $\chi''_{AA}(k\omega)/\omega$, diverges as $k \to 0$,

$$\int \frac{d\omega}{\pi} \frac{\chi''_{AA}(k\omega)}{\omega} = \chi_A(k) = \frac{M_o^2}{R_A k^2} \qquad \text{as } k \to 0, \qquad (7.16)$$

which is, of course, a restatement of eq. (7.13), and compatible as it must be with (7.15b). This property is characteristic of an ordered system. It is the reason, for example and as we shall see in Chapter 11, why light scatters so strongly from nematic liquid crystals. Note also that the Kubo-type expression for the transport coefficient $\sigma^{\circ}_A$, obtained from (7.15b) as

$$\sigma^{\circ}_A = \lim_{\omega\to 0}\,[\lim_{k\to 0}\ \omega\,\chi''_{AA}(k\omega)]\,, \qquad (7.17)$$

differs from eq. (4.54), say, by a factor $k^2$. The reader will notice the formal similarity of statements made in this section to results obtained in section 6.1 on Brownian momentum correlation where the role of $k^2$ was played by the mass ratio m/M. In particular, eq. (6.8) is analogous to (7.16), and eqs. (6.29) and (6.30) correspond to (7.15b) and (7.17), respectively.

The basic assertion of this section, eq. (7.13), clearly requires proof for every particular system to which we want to apply this scheme. First, however, it may be well to summarize the essential results we obtained in this preliminary discussion:

(a) The lifetime $\tau_A(k)$ of a dynamical mode $A(k,t)$ can generally be written as

$$\tau_A^{-1}(k) = \sigma'_A(k)/\chi_A(k), \tag{7.18}$$

where $\sigma'_A(k) = \text{Re}\,\sigma_A(k,i\varepsilon)$ is a generalized transport coefficient, and $\chi_A(k)$ a generalized susceptibility.

(b) If A is a conserved density, then $\sigma'_A(k) \sim k^2$ so that the mode is infinitely long-lived as $k \to 0$.

(c) If A is a "symmetry-restoring variable" with static correlations extending over an infinite range, then $\chi_A(k) \sim 1/k^2$, and again, the mode will have an infinite lifetime as $k \to 0$. Moreover, for small k this mode will be very strong.

This should be enough for an introduction. Now let us analyze these statements in more detail, and in particular, prove the assertion (7.13) on which they rest. What we will establish — the connection between broken symmetry, long-ranged correlations and long-lived modes — was first enunciated by J. Goldstone (1961) who showed that in relativistic quantum field theory, there must exist a Bose particle with vanishing rest mass whenever a continuous symmetry of the Lagrangian is spontaneously broken. See also Goldstone et al. (1962) and Katz and Frishman (1966). Applications of this theorem to the non-relativistic many-body problem have been discussed by Anderson (1958), Lange (1966), and others. The application of these ideas to hydrodynamic modes has received much less attention than to the elementary excitations observed at very low temperature, and less than it deserves.

## 7.2 Broken Symmetry in a Ferromagnet

The isotropic Heisenberg ferromagnet is the prototype of a system whose "symmetry is broken". What we mean by that is that a symmetry of the Hamiltonian— namely, rotational invariance — is not a symmetry of the state; the ferromagnetic state of the system specifies an axis of preferential spin orientation, and thus has a lower symmetry than its Hamiltonian.

The Heisenberg Hamiltonian, describing a system of interacting spins which are localized on the sites of a rigid lattice, is

$$H = -\frac{1}{2}\sum_{\alpha\neq\beta} J(|\vec{r}^{\,\alpha}-\vec{r}^{\,\beta}|)\ \vec{S}^{\alpha}\cdot\vec{S}^{\beta}. \tag{7.19}$$

$\vec{S}^{\alpha}$ is the spin operator at site $\vec{r}^{\,\alpha}$, and while spins at different sites commute, the three spin components at any given site have the usual angular momentum commutation relations

$$[S_i^{\alpha}, S_j^{\beta}] = i\hbar\epsilon_{ijk} S_k^{\alpha}\ \delta_{\alpha\beta}\ . \tag{7.20}$$

$J(|\vec{r}^{\,\alpha}-\vec{r}^{\,\beta}|)$ is the exchange interaction, dependent on the distance between sites but by assumption not on the orientation of the spins relative to the lattice. Therefore, the Hamiltonian (7.19) is invariant under common rotations of all spins relative to the fixed lattice. Mathematically, the total spin operator

$$\vec{S}^{\,tot} = \sum_{\alpha} \vec{S}^{\alpha} \tag{7.21}$$

commutes with the Hamiltonian,

$$[\vec{S}^{\,tot}, H] = i\hbar\frac{d}{dt}\vec{S}^{\,tot} = 0\ . \tag{7.22}$$

Read one way, this equation says, of course, that the total spin is conserved. Read the other way, it says that the Hamiltonian is rotationally invariant. Namely, the equation

$$[S_i^{tot}, S_i^{\alpha}] = i\hbar\epsilon_{ijk} S_k^{\alpha} \qquad (\text{all } \alpha) \tag{7.23}$$

identifies $\vec{S}^{tot}$ as the generator of spin rotations. The unitary transformation

$$U(\vec{\theta}) = \exp \frac{i}{\hbar} \vec{\theta} \cdot \vec{S}^{tot} \simeq 1 + \frac{i}{\hbar} \delta\vec{\theta} \cdot \vec{S}^{tot} \tag{7.24}$$

rotates all spins through the angle $|\vec{\theta}|$ about the axis along which $\vec{\theta}$ is oriented. The second eq. (7.24) holds for an infinitesimal rotation where one finds

$$U(\delta\vec{\theta})\ \vec{S}^{\alpha}\ U^{-1}(\delta\vec{\theta}) = \vec{S}^{\alpha} + \delta\vec{S}^{\alpha},$$

where

$$\delta\vec{S}^{\alpha} = \frac{i}{\hbar}[(\delta\vec{\theta} \cdot \vec{S}^{tot}), \vec{S}^{\alpha}] = (\delta\vec{\theta} \times \vec{S}^{\alpha}), \tag{7.25}$$

using (7.23). Because of (7.22) therefore, the Heisenberg Hamiltonian is invariant under the continuous group of transformations U,

$$U(\vec{\theta})\ H\ U^{-1}(\vec{\theta}) = H\,. \tag{7.26}$$

Despite this rotational invariance of H, a ferromagnet has a spontaneous, permanent magnetic moment. That is,

$$V\vec{M}_o = \langle \vec{S}^{tot} \rangle \neq 0\,, \tag{7.27}$$

where V is the volume. (We disregard here any domain structure, assuming we have succeeded in preparing a single-crystal, uniformly magnetized sample.) Now this means that the density matrix cannot commute with $\vec{S}^{tot}$. Namely, assume that the spontaneous moment is in the z-direction so that

$$\langle S_z^{tot} \rangle = VM_o \neq 0 \quad \text{but} \quad \langle S_x^{tot} \rangle = \langle S_y^{tot} \rangle = 0. \tag{7.28}$$

Then, since $[S_x^{tot}, S_y^{tot}] = i\hbar S_z^{tot}$, one has

$$\langle S_z^{tot} \rangle \equiv \mathrm{tr}\, \rho S_z^{tot} = (i\hbar)^{-1} \mathrm{tr}\rho [S_x^{tot}, S_y^{tot}]$$

$$= (i\hbar)^{-1} \mathrm{tr}[\rho, S_x^{tot}] S_y^{tot} = -(i\hbar)^{-1} \mathrm{tr}[\rho, S_y^{tot}] S_x^{tot} \neq 0. \quad (7.29)$$

Thus, the density matrix commutes neither with $S_x^{tot}$ nor with $S_y^{tot}$, even though the Hamiltonian does. This is no cause for panic. To describe the state of an oriented ferromagnet, which is what $\rho$ does describe, one needs such parameters as the temperature, and one has to specify the direction of spontaneous magnetization. As all spins are rotated about the x-axis, say, the state of the system changes, albeit in a rather trivial manner.

Of course, this means that the canonical ensemble $\rho \sim e^{-\beta H}$ is no longer a good ensemble for the spontaneously ordered system. Averaging over this ensemble would be to average, among other properties, over all directions of the total spin. That is fine in a paramagnet, and passes for a number of purposes in the ferromagnetic regime as well, but for other purposes, such as the calculation of $\langle \vec{S}^{tot} \rangle$, it would be a foolish thing to do. One could use $e^{-\beta H}$ to weigh states of different energy, but in addition one should specify that the trace is to be taken only over those states for which $\vec{S}^{tot}$ points in the z-direction. Formally, one would then have something like

$$\rho = \text{const.}\, \mathcal{P}(\vec{S}^{tot})\, e^{-\beta H}, \quad (7.30)$$

where the projection operator $\mathcal{P}(\vec{S}^{tot})$ eliminates all but those states for which $\vec{S}^{tot}$ points along $\hat{z}$. This is called a "restricted ensemble," and it is not very easy to work with.

A better procedure is to do what one does in the laboratory. To align a piece of ferromagnetic iron that has been collecting dust, one puts it in a magnetic field $\delta\vec{h} = \delta h \cdot \hat{z}$. When the field is turned off again the magnetization stays oriented along $\hat{z}$ for lack of a reason to turn to any other direction. Now in a constant external field $\delta\vec{h} = \delta h \cdot \hat{z}$ the canonical density matrix is

$$\rho(\delta\vec{h}) = e^{-\beta[H - Sz^{tot} \cdot \delta h]}/\text{tr of same}, \tag{7.31}$$

which commutes with $S_z^{tot}$, but not with $S_x^{tot}$ and $S_y^{tot}$. Then

$$\langle \vec{S}^{tot} \rangle (\delta\vec{h}) = \text{tr}\rho\,(\delta\vec{h})\,\vec{S}^{tot} = M_o(\delta h) \cdot \hat{z}\,. \tag{7.32}$$

Now of course, as $\delta h \rightarrow 0$, $M_o$ must remain finite in a ferromagnet,

$$\lim_{\delta h \to 0} M_o(\delta h) = M_o(T) \neq 0\,, \tag{7.32}$$

even though at $\delta h = 0$, the average (7.32) would vanish,

$$M_o(\delta h{=}0) = 0\,. \tag{7.33}$$

The external field, however small, gives the states with $\vec{S}^{tot} \sim \hat{z}$ a somewhat larger weight in the ensemble which, if the system is ferromagnetic, suffices to pull the spins into the z-direction. The content of eq. (7.32) is repeated in fig. 7.2.

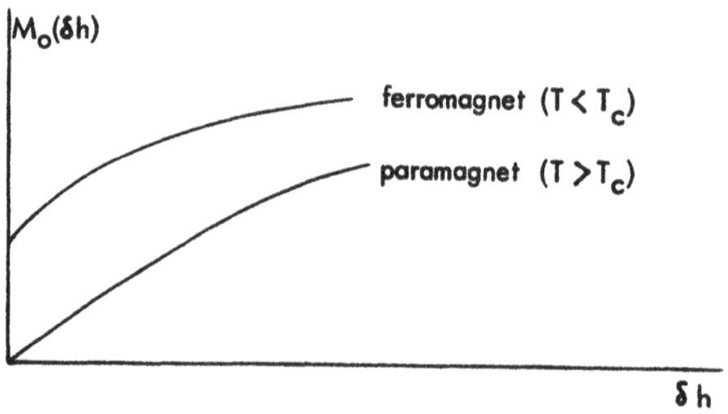


Fig. 7.2

It is, of course, one thing to state the fact of spontaneously broken symmetry as in eq. (7.32), and another to prove, given a Hamiltonian (7.19), that there is a range of temperatures $T < T_c$ for which (7.32) does in fact obtain. The proof is difficult, and will not be attempted here, all the more as the object of proof is so entirely plausible. Instead, let us draw what conclusions we can from the assertion that, even in the absence of an external field (rather, as $\delta h \to 0$) $\langle \vec{S}^{tot} \rangle = M_o \hat{z} \neq 0$.

First, let us use a soft-sell technique to show that $\chi_\perp(k)$ must diverge as a consequence. It is convenient to define a local magnetization operator by

$$\vec{M}(\vec{r}t) = \sum_\alpha \vec{S}^\alpha(t)\, \delta(\vec{r} - \vec{r}^\alpha) \qquad (7.34)$$

and its correlation function by

$$\chi''_{ij}(\vec{r}\,\vec{r}';t) = \langle (2\hbar)^{-1} [M_i(\vec{r}t), M_j(\vec{r}'\,0)] \rangle . \qquad (7.35)$$

The fact that the spins are on a lattice destroys translational invariance a little, but if we are interested in slow spatial variations the lattice structure doesn't matter. A Fourier transform can be defined by

$$\chi''_{ij}(\vec{k}\omega) = \frac{1}{V} \int_{-\infty}^{\infty} dt\, e^{i\omega t} \int dr \int dr'\, e^{-i\vec{k}\cdot(\vec{r}-\vec{r}')}\, \chi''_{ij}(\vec{r},\vec{r}';t). \quad (7.36)$$

Our soft-sell approach now is this: we take the sample off the shelf at $t = -\infty$, assuming that at that time it is magnetized in the z - direction. We then turn on, very slowly, a spatially varying external magnetic field which points in the x-direction. The time-dependent Hamiltonian in this field is

$$H(t) = H - \int dr\, M_x(r)\, \delta h_x(r)\, e^{\epsilon t} . \qquad (t<0) \quad (7.37)$$

What happens is that a non-zero magnetization in the x-direction will be produced. At $t = 0$, it is given by

$$\delta \langle M_x(k) \rangle = \chi_{xx}(k)\, \delta h_x(k), \qquad (7.38)$$

where

$$\chi_{xx}(k) = \int \frac{d\omega}{\pi} \chi''_{xx}(k\omega)/\omega \quad . \tag{7.38a}$$

Now what happens if the external field is uniform? Well, since we have no field present to make the system prefer the z-direction, the magnetization will simply flip over, completely. However, for a uniform field (7.38) becomes (cancelling a $\delta(\vec{k})$ on both sides)

$$\delta\langle M_x\rangle = \chi_{xx}(k=0)\,\delta h_x = M_o \quad . \tag{7.39}$$

Thus, $\delta\langle M_x\rangle$ must be finite even though $\delta h_x$ is infinitesimal, so that $\chi_{xx}(k)$ must be infinitely large at $k = 0$. Since from symmetry $\chi_{xx}(k)$ is an even function of $k$, the weakest singularity that will do the trick is normally

$$\chi_{xx}(k) = \frac{M_o^2}{Rk^2} \qquad \text{as } k \to 0 \, . \tag{7.40}$$

The same reasoning obviously gives a singularity in $\chi_{yy}(k)$. The divergence is thus obviously connected with the broken symmetry, the fact that at the slightest tickling the system can rotate from one direction ("state") to the other.

Note, on the other hand, that there is nothing very particular about $\chi_{zz}(k)$. In the limit as $k \to 0$, $\chi_{zz}(k)$ is finite since it does require effort to <u>increase</u> the magnetization in z-direction if the system is already spontaneously magnetized in the z-direction. And of course, $\chi_{zz}(k \to 0) = \left(\frac{\partial M}{\partial h}\right)_T$ is the normal susceptibility. In generalizeable language: the symmetry-breaking variable ($M_z$), whose non-vanishing expectation value expresses the fact of spontaneous order, has correlations of finite range, i.e. a finite susceptibility even at $k = 0$. The singularity occurs in the variables ($M_x$ and $M_y$) which change the spontaneous order, i.e. which tend to restore the full symmetry of the Hamiltonian.

## 7.3 The Bogoliubov Inequality

We have made the $k^{-2}$ singularity in $\chi_{xx}(k)$ plausible, but the arguments given would not rule out more bizarre possibilities like $k^{-1.3}$, for example. Now we can prove that the singularity is at least as $k^{-2}$.

We start by noticing that

$$\frac{1}{i\hbar}[S_y^{tot}, M_x(\vec{r})] = -M_z(\vec{r}') , \qquad (7.41)$$

which is eq. (7.23), expressing the fact that $S_y^{tot}$ generates spin rotations about the y-axis. Averaging (7.41), we immediately obtain

$$\frac{2}{V}\int dr \int dr'\, \chi''_{yx}(rr', t=0) = \int \frac{d\omega}{\pi}\, \chi''_{yx}(k=0,\omega) = -iM_o , \quad (7.42)$$

which is a sum rule that distinguishes the ferromagnetic ($M_o \neq 0$) from the paramagnetic ($M_o = 0$) range.

Next, we use the fact that the quantity

$$\langle A|B\rangle \equiv \beta^{-1}\int \frac{d\omega}{\pi}\, \frac{1}{\omega}\, \chi''_{AB}(k\omega) ,$$

defined for any two Hermitian variables A(rt) and B(rt), has all the properties of a scalar product in a Hilbert space as we pointed out in section 5.3. In particular, therefore, the Schwartz inequality holds,

$$\langle A|A\rangle\langle B|B\rangle \geqq |\langle A|B\rangle|^2 ,$$

which more explicitly reads

$$\int \frac{d\omega}{\pi}\, \frac{1}{\omega}\, \chi''_{AA}(k\omega) \int \frac{d\omega}{\pi}\, \frac{1}{\omega}\, \chi''_{BB}(k\omega) \geqq \left| \int \frac{d\omega}{\pi}\, \frac{1}{\omega}\, \chi''_{AB}(k\omega)\right|^2 . \quad (7.43)$$

This is one form of the celebrated Bogoliubov inequality (Hohenberg 1967). We use (7.43) with $A(rt) \equiv \dot{M}_y(rt)$ and $B(rt) \equiv M_x(rt)$. Since

$$\chi''_{\dot{M}_y M_x}(k\omega) = -i\omega\, \chi''_{M_y M_x}(k\omega) \equiv -i\omega\, \chi''_{yx}(k\omega)\,,$$

$$\chi''_{\dot{M}_y \dot{M}_y}(k\omega) = \omega^2 \chi''_{yy}(k\omega),$$

we obtain

$$\int \frac{d\omega}{\pi} \frac{\chi''_{xx}(k\omega)}{\pi} \equiv \chi_{xx}(k) > \frac{\left| \int \frac{d\omega}{\pi}\, \chi''_{yx}(k\omega) \right|^2}{\int \frac{d\omega}{\pi}\, \omega \chi''_{yy}(k\omega)} \quad . \tag{7.44}$$

For small k, the numerator becomes $M_o^2$ by the sum rule (7.42). In fact, for our lattice spin model, (7.42) holds for arbitrary k.

Now the total spin is a constant of the motion, see eq. (7.22), and thus the spin density obeys the continuity equation

$$\partial_t M_i(rt) + \nabla_j j^M_{ij}(rt) = 0, \tag{7.45}$$

so that

$$\omega^2 \chi''_{yy}(k\omega) = k_i k_j \chi''_{j_{yi} j_{yj}}(k\omega) \tag{7.45a}$$

is of order $k^2$. As a consequence, the denominator of eq. (7.44) is of order $k^2$ as well, and hence we find

$$\chi_{yy}(k) \gtrless \frac{M_o^2}{k^2 \cdot [\text{coefficient}]} \tag{7.46}$$

for small but finite k, where [coefficient] is a magnetization current correlation function. We conclude not only that $\chi_{xx}(k)$ is infinite at k = 0, but that it is at least as singular as $1/k^2$ which is quite a bit more. If we assume (as is in fact always found to be true, except at the critical point) that the singularity is no worse than $1/k^2$ either, i.e. that

$$\chi_{xx}(k) = \chi_{yy}(k) = \frac{M_o^2}{Rk^2} \qquad \text{as } k \to 0, \tag{7.47}$$

we also obtain a bound for the positive constant R, namely

$$R \leq \lim_{k\to 0} \frac{1}{k^2} \int \frac{d\omega}{\pi} \, \omega \, \chi''_{xx}(k\omega) \, . \qquad (7.48)$$

Our proof of (7.46) would be complete if we had explicitly demonstrated that the sum on r.h.s. of (7.48) is in fact finite. We have considered such sum rules before, and the reader is invited to evaluate (7.48). The answer (Lange 1966) is that the bound (7.48) is finite if the exchange forces are of finite range, i.e. if

$$\int d\vec{r} \; J(r) < \infty \, . \qquad (7.49)$$

## 7.4 The Goldstone Theorem

The connection between a symmetry of the Hamiltonian which is broken by the physical state, and the existence of a collective mode of low frequency for large wavelength, is of sufficient importance that the theme can stand an additional variation. In this section, therefore, let us formalize things a little, restricting the argument to translationally invariant systems however.

First, we need a symmetry that can be broken. We will call the corresponding Hermitian operator Q, and we assume that it is the integral of a local density $q(\vec{r}t)$,

$$Q = \int d\vec{r} \; q(\vec{r}t) \, . \qquad (7.50)$$

The assumption that Q commutes with the Hamiltonian,

$$[H, Q] = 0 \, , \qquad (7.51)$$

identifies Q as the generator of a continuous symmetry transformation $U(\phi) = \exp(i\phi Q/\hbar)$ under which the Hamiltonian is invariant, $U(\phi) H \, U^{+}(\phi) = H$. It also means that the observable $Q(t)$ is conserved,

$$\frac{d}{dt} Q(t) = 0 \quad \text{or} \quad \partial_t q(rt) + \vec{\nabla} \cdot \vec{j}^{q}(rt) = 0, \tag{7.52}$$

where $\vec{j}^{q}$ is the current corresponding to the quantity Q.

Now let us further assume that the symmetry operation Q transforms the local, Hermitian observable $A(\vec{r})$ into the observable $B(\vec{r})$, according to

$$\frac{1}{i\hbar} [Q, A(\vec{r})] = B(\vec{r}). \tag{7.53}$$

Then, if the state or ensemble $\rho$ is symmetric under Q, $[\rho, Q] = 0$, the average of B must vanish. If, on the other hand, we find a pair of operators A, B such that

$$\langle B(r) \rangle = \langle B \rangle \neq 0, \tag{7.54}$$

the thermodynamic state must break the Q-symmetry, i.e. $[\rho, Q] \neq 0$.

The consequences of these assumptions, meant to describe physically realized conditions of course, are interesting. First, equations (7.53) and (7.54) imply, of course, that

$$\int d\vec{r}' \left\langle \frac{1}{2\hbar} [q(\vec{r}t), A(\vec{r}'')] \right\rangle = \frac{i}{2} \langle B \rangle, \tag{7.55}$$

where t is arbitrary since Q(t) = Q(0) is time independent. (7.55) is the equation

$$\chi''_{q,A}(k=0, \omega) = i\pi \langle B \rangle \delta(\omega), \tag{7.56}$$

which is interesting if it also justifies the conclusion that

$$\lim_{k\to 0} \chi''_{qA}(k,\omega) = i\pi \langle B \rangle \delta(\omega). \tag{7.57}$$

For it is does, (7.57) signals the existence of a collective mode whose energy, $\omega(k)$, goes to zero as $k\to 0$, and/or whose lifetime, $\tau(k)$, goes to infinity as $k\to 0$. Something like

$$\chi''_{qA}(k\omega) = i\pi \langle B\rangle \delta(\omega - \omega(k))$$

or

$$\chi''_{qA}(k\omega) = i\langle B\rangle \frac{\tau(k)}{1 + (\omega\tau(k))^2} \tag{7.58}$$

or a combination of both, where however

$$\lim_{k\to 0} \omega(k) = 0 \quad \text{and} \quad \lim_{k\to 0} \tau(k) = \infty . \tag{7.59}$$

Further, because of the Bogoliubov inequality (7.43), namely

$$\int \frac{d\omega}{\pi\omega} \chi''_{AA}(k\omega) \equiv \chi_{AA}(k) \geq \frac{\left| \int \frac{d\omega}{\pi} \chi''_{qA}(k\omega) \right|^2}{\int \frac{d\omega}{\pi} \omega \chi''_{qq}(k\omega)} , \tag{7.60}$$

we would expect that $\chi_{AA}(k)$ diverges at least as $1/k^2$ since

$$\int \frac{d\omega}{\pi} \omega \chi''_{qq}(k\omega) = k^2 \int \frac{d\omega}{\pi} \frac{1}{\omega} \chi''_{iqiq}(k\omega) \tag{7.61}$$

must be of order $k^2$ on account of the conservation law (7.52). If the divergence is the minimal one, i.e.

$$\chi_{AA}(k) = \frac{\langle B\rangle^2}{R_A k^2} \qquad \text{as } k \to 0 , \tag{7.62}$$

then the "stiffness constant" $R_A$ is bounded by

$$R_A \leq \lim_{k\to 0} \frac{1}{k^2} \int \frac{d\omega}{\pi} \omega \chi''_{qq}(k\omega) . \tag{7.63}$$

Thus we find low-frequency "Goldstone modes" and long-ranged correlations, both as a result of the breaking of a continuous symmetry.

This is the formal structure. To be perfectly rigorous it would have to be fancied up some, see Katz and Frishman (1966) and Lange (1966). What the foregoing considerations show is that the "symmetry-breaking" variable B(r) is not necessarily the same as the "symmetry-restoring" variable A(r) (accepted language is not very meticulous about the distinction), and neither need to be the same as the symmetry operator Q or its density $q(\vec{r})$ as the ferromagnetic example might lead one to expect. Indeed, for the isotropic antiferromagnet, Q is again one or the other component of the total spin $\vec{S}^{tot}$ but A and B are components of the staggered magnetization density $\vec{N}(r)$. For a nematic liquid crystal, Q is the total angular momentum (whose density is not a local operator, unfortunately) while A and B are components of the molecular mass quadrupole moment, commonly called the "director". And for the superfluid, the broken symmetry is gauge invariance whose generator is the number operator N while the operator A is the phase of the quantum field $\psi(r)$. $\langle B \rangle^2$, by the way, is here the condensate density $n^o$ while the "stiffness constant" $R_A$ turns out to be the superfluid density which is usually named $\rho_s$.

## 7.5 Some Additional Considerations

This section would be in small print if one could do that on a typewriter. It contains some only loosely connected comments that some readers might want to work out in more detail while others will prefer skipping them.

The first concerns the justification of the "if" before eq. (7.57). The statement that we have, (7.56), is equivalently

$$\int \frac{d\omega}{\pi} \chi'' (k = 0, \omega) = i \langle B \rangle , \tag{7.64a}$$

$$\int \frac{d\omega}{\pi} \, \omega^n \, \chi''(k=0,\omega) = 0 \qquad (n \geq 1)\,, \tag{7.64b}$$

namely the broken symmetry, where $\chi'' \equiv \chi''_{qA}$, while the statement we need is

$$\lim_{k\to 0} \int \frac{d\omega}{\pi} \chi''(k,\omega) = i \, \langle B \rangle \,, \tag{7.65a}$$

$$\lim_{k\to 0} \int \frac{d\omega}{\pi} \, \omega^n \, \chi''(k,\omega) = 0 \qquad (n \geq 1)\,. \tag{7.65b}$$

Now the Fourier transform

$$f(k) = \int d\vec{r} \; e^{-i\vec{k}\cdot\vec{r}} \, f(r) \tag{7.66a}$$

of any function f(r) is uniform at k = 0, i.e.

$$\lim_{k\to 0} f(k) = f(k=0) = \int d\vec{r} \, f(r)\,, \tag{7.66b}$$

if f(r) falls of sufficiently rapidly at large r so that the integral exists, $\int d\vec{r} \, f(r) < \infty$. Consequently, the eqs. (7.64a) and (7.65a) are equivalent if only the function

$$f^{(0)}(r) = \int \frac{d\omega}{\pi} \chi''(r,\omega) = \hbar^{-1} \langle [q(r), A(0)] \rangle \tag{7.67}$$

vanishes rapidly as $r \to \infty$. This is true if q(r) and A(r) are both local operators which depend only on properties (position, momentum, spin, etc.) of those particles in a small neighborhood of $\vec{r}$, for then the commutator (7.67) vanishes if r is large enough. See the related comment on page 57.

Let us also look at the sum rule

$$f^{(1)}(r) = \int \frac{d\omega}{\pi} \, \omega \, \chi''(r,\omega) = \left(\frac{i}{\hbar}\right)^2 \left\langle \Big[ [H, q(r)], \, A(0) \Big] \right\rangle . \tag{7.68}$$

If all the forces in the system are of short range, then the energy density operator $\varepsilon(r)$ (whose integral is H) is also a local operator, namely one depending only on the properties of those particles which are with the potential range of the point $\vec{r}$. Under this assumption therefore, $[H, q(r)]$ commutes with $A(0)$ unless $\vec{r}$ is within a force range or so of $\vec{0}$, and $f^{(1)}(\vec{r})$ is short range as well so that its Fourier transform is well behaved near $k = 0$. Higher order sum rules submit to the same argument which makes it plausible that the basic Goldstone statement, namely eq. (7.65) or (7.57), is secured if the system in question is not plagued by Coulomb forces, say. More careful arguments are given by Lange (1966).

The second comment concerns the interpretation of eq. (7.57) in terms of the excitation spectrum of H. Throughout this text we have considered thermal averages at finite temperatures since the main interest in this text is dynamical, even hydrodynamical, properties at finite T. But everything said in section 7.4 applies equally if we are $T = 0$ where $\langle \ldots \rangle$ simply means the quantum-mechanical expectation value in the ground state $|0\rangle$ of the Hamiltonian H. In this case,

$$\chi''_{qA}(k\omega) = \frac{\pi}{\hbar} \sum_n \delta(\omega - \omega_n) \langle 0|q(-k)|n\rangle\langle n|A(k)|0\rangle - \begin{pmatrix} A \leftrightarrow q \\ \omega \rightarrow -\omega \end{pmatrix}, \tag{7.69}$$

where $E_n = \hbar\omega_n$ is the energy of the n-th excited state, and $q(k)$, $A(k)$ are Fourier amplitudes of the respective densities. Thus if the ground state breaks the Q-symmetry, $\langle 0|B|0\rangle \equiv \langle B\rangle \neq 0$, then eq. (7.57) signals the existence of a branch of elementary excitations whose energies, $\hbar\omega(k)$, vanish as $k \to 0$: there is no energy gap above the ground state. Again, only long-range forces would allow us to escape this conclusion.

In relativistic particle physics, the branch of excitations are seen, of course, as a particle, the Goldstone boson, and the absence of an energy gap carries over into the conclusion that this particle must have zero rest mass. Now symmetries and even more broken symmetries are of course very fashionable in particle theory, for many good and sufficient reasons, but the accompanying zero mass particles are a thorn on the rose: there simply aren't so many of those around. It is highly welcome therefore that long-range forces (read:gauge fields) can obliterate the zero-mass particle. This mechanism, called the Kibble-Higgs mechanism, is described in a superbly readable account by Kibble (1965).

The third comment concerns eq. (7.60) on which all hydrodynamic conclusions rest, and which we can write as

$$\chi_{AA}(k) \geq \frac{|\langle B \rangle|^2}{k^2 R} , \tag{7.70}$$

where

$$R = \lim_{k \to 0} \frac{1}{k^2} \int \frac{d\omega}{\pi}\, \omega\, \chi''_{qq}(k\omega) . \tag{7.71}$$

For many systems, one can prove, by methods similar to those hinted at after (7.68), that R is in fact finite, a consequence of the conservation of the local density q and valid for short-range forces. By (7.70) then, fluctuations of A are very strong, particularly at long wavelength. It is plausible, and easy to see for any particular system, that such fluctuations carry energy, for example, and if the total energy of the system is to remain finite the sum $\sum_k \chi_{AA}(k)$ must not diverge. However, while the sum is finite for a three-dimensional system if $\chi_{AA}(k) \sim k^{-2-\alpha}$, $\alpha < 1$, for a two-dimensional system

$$\sum_k \chi_{AA}(k) \geq \frac{|\langle B \rangle|^2}{R} \cdot 2\pi \int_0 \frac{dk}{k} \longrightarrow \infty \tag{7.72}$$

diverges at long wavelengths. The only escape is $\langle B \rangle = 0$, i.e. the absence of order in two dimensions which is destroyed by the strong fluctuations. Obviously, this brief remark is only meant to wet an occasional reader's appetite. He will find detail in Mermin (1966) or Hohenberg (1967); see also section 10.9 below.

The fourth and final comment concerns a phenomenological interpretation of eq. (7.62). We begin by noting that if A(r) couples to an external field $\alpha(r)$ in the form

$$\mathcal{H} = H - \int d\vec{r}\; A(r)\,\alpha(r) \;, \tag{7.73}$$

then the susceptibility $\chi_{AA}(r - r')$ is of course given by

$$\chi_{AA}(r - r') = \left.\frac{\delta \langle A(r) \rangle}{\delta\alpha(r')}\right|_{\alpha\to 0} \tag{7.74}$$

by the rules of statistical mechanics, see eq. (3.9). Now in the presence of a spatially varying field there will be gradients of the locally averaged property $\langle A(r)\rangle$ which contribute to the free energy. We would therefore write the free energy as

$$F + F_o + \frac{1}{2}\int d\vec{r}\; R'_A\, [\vec{\nabla}\langle A(r)\rangle]^2 - \int d\vec{r}\,\langle A(r)\rangle\,\alpha(r) \tag{7.75}$$

up to higher order derivatives of $\langle A \rangle$, where $F_o$ is the free energy of the undisturbed system, and depends only on temperature and pressure. Given $\alpha(r)$, the physical configuration $\langle A(r)\rangle$ is obtained by minimizing F which results in

$$\frac{\delta F}{\delta\alpha(r)} = -R'_A\,\nabla^2\langle A(r)\rangle - \alpha(r) = 0 \;. \tag{7.76}$$

If $\alpha(r)$ is now varied, we obtain by (7.74)

$$-R'_A\,\nabla^2\chi_{AA}(r-r') = \delta(r-r') \;, \tag{7.77}$$

which is eq. (7.62), with $R_A = R'_A\,\langle B\rangle^2$. The interpretation of (7.76) is, of course, that in response to the external force $\alpha(r)$ a compensating internal force field is set up which is given by

$$\alpha_{int}(r) = R'_A \nabla^2 \langle A(r) \rangle \tag{7.78}$$

This equation justifies calling $R_A$ a "stiffness constant". The second term in (7.75), is, of course, interpretable as an elastic contribution to the free energy, due to local variations of the order parameter. Note finally that if we omit the $\langle \ldots \rangle$ in (7.78) and compute the force correlation function, we find

$$\langle \alpha_{int}(r)\, \alpha_{int}(r') \rangle = -k_B T R'_A \nabla^2 \delta(r - r'). \tag{7.79}$$

The assertion (7.62) that $\chi(k) \sim k^{-2}$ rather than $k^{-2.3}$, which would also be compatible with (7.60), is thus seen as an expression of the prejudice that, while the order parameter correlations are of infinite range, the internal forces should be only locally correlated.

## 7.6 Hydrodynamic Goldstone Modes

Most of the attention which the Goldstone theorem has received in the literature has been in connection with elementary excitations. That is, with those eigenstates of the exact Hamiltonian which lie immediately above the (degenerate) ground state, and which one therefore observes at very low temperatures (Anderson 1964). In this book, by contrast, we are concerned with modes observed at finite temperatures when the system is in microscopically chaotic motion, modes which are coherent and long-lived even though microscopic collision times $\tau_c$ are so short that $\omega\tau_c \ll 1$. Can one make precise statements about these "hydrodynamic Goldstone modes"? If one were of an extremely critical disposition the answer would have to be no. More rigor can be applied to the discussion of elementary excitations than that of hydrodynamic modes. But given that second sound in superfluid Helium is among the modes we aim at, one should not give up so easily. In this section we will analyze, somewhat tentatively, just how convincingly the existence of hydrodynamic Goldstone modes can be established. The discussion is a little abstract, and the reader might want to return to it after he has read one or the other of the following chapters.

It seems best to begin by applying, without further ado, the memory function formalism which we will use again and again from now on, to the general example considered in section 7.4. We have one conserved density $q(rt)$ which should lead to a slow mode because it is conserved, and we have a symmetry-restoring variable, $A(rt)$, which should lead to a second slow mode because its static susceptibility $\chi_{AA}(k)$ diverges as $k \to 0$. Let us call these two densities

$$\{q(rt),\ A(rt)\} \equiv \{A_1(rt),\ A_2(rt)\} \tag{7.80}$$

to facilitate matrix notation. For definiteness, let us assume that both q and A are scalar densities, that q is even under parity and time reversal and that $A(rt)$ is even under parity, odd under time reversal. In this case, the susceptibility matrix $\chi_{\mu\nu}(k)$ cannot have off-diagonal elements, and to lowest order in k it can be written as

$$\chi_{\mu\nu}(k) = \begin{pmatrix} \chi_{qq}^{(0)} & 0 \\ 0 & \dfrac{\langle B\rangle^2}{R_A k^2} \end{pmatrix}, \tag{7.81}$$

which incorporates the most essential property of $A(rt)$, its infinite-range fluctuations. Within the framework of chapter 5, in particular the basic equation (5.28), the next object to consider is the unnormalized frequency matrix, $\omega_{\mu\nu}(k)$. Because of time reversal invariance, it has no diagonal elements, and its off-diagonal element is given by eq. (7.55) or (7.65), so that

$$\omega_{\mu\nu}(k) = \begin{pmatrix} 0 & i\langle B \rangle \\ -i\langle B \rangle & 0 \end{pmatrix} \qquad (7.82)$$

as $k \to 0$. If we could neglect dissipation, i.e. if we could neglect the memory function $\Sigma_{\mu\nu}$ in eq. (5.28), we could solve the matrix equation

$$\left[ z\,\delta_{\mu\lambda} - \omega_{\mu\kappa}\,\chi^{-1}_{\kappa\lambda} \right] C_{\lambda\nu} = i\beta^{-1}\,\chi_{\mu\nu} \qquad (7.83)$$

easily to obtain, for example,

$$C_{22}(kz) \equiv C_{AA}(kz) = i\beta^{-1}\,\frac{\langle B \rangle^2}{R_A k^2} \cdot \frac{z}{z^2 - \omega^2(k)}\,, \qquad (7.84)$$

and therefore, by taking the real part we would find the absorptive response function as

$$2\chi''_{AA}(k\omega)/\pi\omega = \frac{\langle B \rangle^2}{R_A k^2} \left[ \delta(\omega - \omega(k)) + \delta(\omega + \omega(k)) \right] \qquad (7.85)$$

where

$$\omega(k) = kc = k\,(R_A/\chi_{qq}(0))^{1/2}\,. \qquad (7.86)$$

Thus at long wave length there would be propagating waves.

Now of course, we cannot simply neglect $\Sigma_{\mu\nu}(kz)$. Its general expression is given in eq. (5.28), involving the projector $\hat{Q}$ which removes fluctuations of q and of A. The qq-element of the unnormalized memory function is thus given by

$$\sigma_{qq}(kz) = i\beta\,\langle \dot{q}(k) \,|\, \hat{Q}\,[z - \hat{Q}L\hat{Q}]^{-1}\,\hat{Q} \,|\, \dot{q}(k) \rangle\,, \qquad (7.87)$$

and because $q(rt)$ obeys the continuity equation (7.52), or

$$\dot{q}(k) = i\,\vec{k} \cdot \vec{j}^{\,q}(k)\,, \qquad (7.88)$$

$\sigma_{qq}(kz)$ is explicitly of order $k^2$. If we blithely assume that for low frequencies $\sigma_{qq}(kz)$ can be replaced by $\sigma_{qq}(k, z=0)$, which is real, we obtain to leading order in k

$$\sigma_{qq}(kz) \approx \sigma_{qq}(k,0) = k^2\gamma + k^4 \ldots \tag{7.89}$$

where $\gamma$ is a positive transport coefficient. We will rethink this replacement later. By the same token, however, we would also obtain

$$\sigma_{AA}(kz) \simeq \sigma_{AA}(0,0) \equiv \xi , \tag{7.90}$$

where $\xi$ is positive, and where there is no factor of k simply because A(rt) is, by assumption, not a conserved variable so that $\dot{A}(k=0)$ is finite. We will quickly dispose of the off-diagonal elements. $\sigma_{qA}(kz)$ would only contribute to leading order in k if it were of order k or less. Now it must be at least of order k because there is one factor k from the conservation law. However, since both q and A are assumed even under parity, $\sigma_{qA}(kz)$ must be an even function of k, and thus it must be of order $k^2$ which does not contribute to leading order in k.

With eqs. (7.81), (7.89) and (7.90) we thus obtain

$$\Sigma_{\mu\nu}(k,0) = [\sigma(k,0)\chi^{-1}(k)]_{\mu\nu} = \begin{pmatrix} k^2\gamma\chi_{qq}^{-1} & 0 \\ 0 & k^2\xi R_A/\langle B\rangle^2 \end{pmatrix}, \tag{7.91}$$

where the factor $k^2$ in $\Sigma_{11}$ is due to q-conservation while the factor $k^2$ in $\Sigma_{22}$ comes from the long-ranged correlations. If we insert (7.91) into the general memory function equation (5.28), we can solve for the correlation functions $C_{\mu\nu}(kz)$, and we obtain, for example,

$$C_{AA}(kz) = i\beta^{-1}\,\frac{\langle B\rangle^2}{R_A k^2}\cdot\frac{z+ik^2\gamma\chi_{qq}^{-1}}{z^2-c^2k^2+izk^2\Gamma}, \tag{7.92}$$

where

$$\Gamma = \gamma\chi_{qq}^{-1}(0) + \xi R_A/\langle B\rangle^2. \tag{7.93}$$

Thus we find a well-developed mode, propagating with the velocity c given in (7.86), and decaying with a hydrodynamically slow lifetime

$$\tau(k) = [\frac{1}{2}k^2 \Gamma]^{-1} \qquad \text{as } k \to 0 \ . \tag{7.94}$$

The correlation function $C_{AA}(kz)$ is in all aspects analogous to the sound wave contribution to $C_{nn}(kz)$ in normal fluids, see eq. (4.42a), even though the physical origin of the propagating waves is very different in the two cases. For a physical example of the formal theory just given, see the treatment of antiferromagnetic spin waves in chapter 9.

If we have not made a mistake in the derivation of eq. (7.92), this result should be asymptotically rigorous for small wave vectors and frequencies. Let us, therefore, inquire into the validity of the recipes which we have used to obtain (7.92).

There is, of course, nothing to criticize in the basic memory function equation (5.28) which we have used. It does not say very much, but what it says is always true. The physical statements on which the recipe is based, are (1) the fact that the A-susceptibility diverges as $k \to 0$, (2) the assumption that for small frequency z, the memory functions $\sigma_{\mu\nu}(kz)$ are smooth functions of z, and (3) the assumption that as functions of k, the memory functions $\sigma_{\mu\nu}(k, i0)$ behave normally, so that they are not, for example, singular as $k \to 0$. If these assertions are granted, then the conclusions, including the existence of a hydrodynamic mode as exhibited in (7.92), are inescapable. The reader might again profit, at this point, from a glance over the arguments given in section 5.4.

We have given a proof of the $k^{-2}$-singularity in $\chi_{AA}(k)$ in section 7.4, and some further discussion in section 7.5. As the present discussion should have made clear, it is this singularity which gives rise to the hydrodynamic "Goldstone mode" (whether the $k^{-2}$-singularity arises from broken symmetry as it usually does, or from some other mysterious cause). We will thus accept it here as given.

Is $\sigma_{\mu\nu}(kz)$ a smooth function of z at small z? We have discussed a very similar question in section 5.4. There, we pointed out that a many-body system is so chaotic that almost all processes decay within short microscopic times $\tau$ which are determined by the strength of the microscopic interactions. Slow macroscopic decay is the unusual behavior which will normally happen only for special reasons: conservation laws, broken symmetry. Now the dynamical operator which determines the relaxation of $\sigma_{\mu\nu}(kz)$ is not the Liouvillian L but the projected Liouvillian $\hat{Q}L\hat{Q}$, see eq. (5.28) or eq. (7.87), from which the presumed hydrodynamic fluctuations have been removed. Those processes that are still contained in $\hat{Q}L\hat{Q}$ should thus relax locally and rapidly. We can codify this physical expectation in the assertion that

$$\hat{Q}iL\hat{Q} > \tau_c^{-1} . \tag{7.95}$$

If this is true, then

$$q_{\mu\nu}(kz) = -\beta \langle \dot{A}_\mu(k) | \hat{Q} [iz - \hat{Q}iL\hat{Q}]^{-1} \hat{Q} | \dot{A}_\nu(k) \rangle \tag{7.96}$$

will be analytic and smooth in the range $|z|\tau_c << 1$. These words do not constitute a formal proof, and indeed a formal proof is likely to be forbiddingly difficult. However, the words will convince a reasonable physicist. Moreover, they are words that enter the derivation of hydrodynamics in normal systems (see e.g. section 5.4) exactly as they do in systems of broken symmetry. What eq. (7.95) expresses is simply that in either system local equilibrium is established on a rapid microscopic time scale because the processes which establish local equilibrium do not require transport over macroscopic distances, of either the conserved quantities or of the "information about the local order".

Turning to the third assumption mentioned above, note that in our derivation it is important that

$$\xi = \sigma_{AA}(0, i0) = \beta \langle \dot{A}(k) | \hat{Q} [iL\hat{Q}]^{-1} \hat{Q} | \dot{A}(k) \rangle \tag{7.97}$$

is finite as $k \to 0$ even though we know that

$$\chi_{AA}(k) = \beta \langle A(k) | A(k) \rangle = \frac{\langle B \rangle^2}{R_A k^2} \tag{7.98}$$

diverges in this limit. How convincing can this assertion be made? Again, we will not attempt to prove from first principles that $\xi$ is finite. Instead, we will attempt to demonstrate its plausibility. ("A demonstration convinces a reasonable man; a proof, a stubborn one." Kac 1967. Even reasonable men might argue, of course, whether the following tentative considerations qualify for a demonstration rather than an expression of faith.)

We have already noticed that (7.98) indicates the presence of infinite-range correlations,

$$\chi_{AA}(\vec{r}-\vec{r}\,') = \frac{\langle B \rangle^2}{R_A} \cdot \frac{1}{4\pi |\vec{r}-\vec{r}\,'|} \tag{7.99}$$

for large $|\vec{r}-\vec{r}\,'|$. In a chaotic thermal system with short-range forces, this behavior is startling and only plausible if there is a reason for it. Broken symmetry has been seen to be that reason. It is therefore reasonable to assume that all long-ranged correlations can be attributed solely to the broken symmetry, in our case to the necessarily strong fluctuations of the one symmetry-restoring variable $A(r)$. Mathematically, this assumption states that for any two local variables $A_\mu(r)$, $A_\nu(r)$, the modified susceptibilities

$$\tilde{\chi}_{\mu\nu}(\vec{r}-\vec{r}\,') = \beta \langle A_\mu(\vec{r}) | \hat{Q}_A | A_\nu(\vec{r}\,') \rangle \text{ are of finite range.} \tag{7.100}$$

Here, $\hat{Q}_A$ is the projector which removes the symmetry-restoring fluctuations, and can be written in the form

$$\hat{Q}_A \equiv 1 - \hat{P}_A = 1 - \int d\vec{r} \int d\vec{r}\,' \, |A(\vec{r})\rangle \, \beta \chi^{-1}_{AA}(\vec{r}-\vec{r}\,') \langle A(\vec{r}\,')| , \tag{7.101}$$

if there is, as in the presently considered abstract case, only one broken symmetry (Q) and one symmetry-restoring variable (A).

The assertion (7.100) is partly a prescription to define and find the proper order parameter variable $A(\vec{r})$, and partly, of course, a physical assumption subject, in principle, to microscopic verification. It implies that the Fourier transforms

$$\tilde{\chi}_{\mu\nu}(k) = \chi_{\mu\nu}(k) - \chi_{A_\mu A}(k)\,\chi^{-1}_{AA}(k)\,\chi_{AA_\nu}(k) \tag{7.102}$$

have finite limits as $k \to 0$ even though this may not be true for the individual terms in (7.102). It also implies, incidentally, that $\lim \tilde{\chi}_{\mu\nu}(\vec{k} \to 0) = \chi_{\mu\nu}(\vec{k}=0)$ so that the value of the limit is independent of the path in $\vec{k}$-space along which the limit is taken. The assertion (7.100) thus states that while there are long-ranged correlations in the system they are of the simplest possible kind. They can be attributed to a single variable, $A(\vec{r})$. Applied to a superfluid, for example, (7.100) states that there is one "mode", the "superfluid component", which introduces long-ranged correlations into the system; that correlations in all other degrees of freedom however, summarized in the "normal component", are of finite range as they are in a normal chaotic system. But that is forging ahead a little.

Returning to eq. (7.97) now, we notice that

$$\sigma'_{AA}(k, i0) \leq \beta\tau_c \,\langle \dot{A}(k) |\, \hat{Q} \,| \dot{A}(k) \rangle \tag{7.103}$$

because of (7.95), where the projector, $\hat{Q}$, removes not only the long-ranged fluctuations of the symmetry-restoring variable $A(r)$, but those of the conserved variable $q(r)$ as well. This latter fact does not change the conclusion. We can now answer the question posed above affirmatively. $\sigma_{AA}(k, i0)$ is finite as $k \to 0$

because it does not involve the fluctuations of the symmetry-breaking variable A(r) which are explicitly projected out. The same holds true, of course, for other components of the memory matrix $\sigma_{\mu\nu}(k, i0)$. In particular, by the same token we would assert that the damping constant $\gamma$ of (7.89) should be finite because it is given by fluctuations of the current $\vec{j}^{q}(r)$ from which, however, the long-ranged correlations of the order parameter have been removed.

Arguments which are as wordy as these are likely to admit of exceptions. From the naive, but valid, derivation which lead us to the hydrodynamic expression (7.92), it is clear that the latter holds only for sufficiently small frequencies $\omega$ and wave numbers k. In line with our further considerations, it should hold so long as

$$\omega\tau_c << 1 \qquad \text{and } k\ell << 1 \,, \tag{7.104}$$

where $\tau_c$ is the collision time which characterizes microscopic relaxation, while $\ell$ is a length which characterizes the spatial decay of the functions $\tilde{\chi}_{\mu\nu}(\vec{r}-\vec{r}\,')$ of eq. (100). One may identify $\ell$ as a mean free path. It is the length over which a particle, or a phonon or other excitation, may propagate before collisions that it suffers along the way destroy all correlations with its initial state. At very low temperatures, collisions become infrequent and ineffective, and both $\tau_c$ and $\ell$ become so large that the range of validity of hydrodynamics shrinks. For superfluid Helium, hydrodynamic second sound is only observed at very low frequencies. As $T \to 0$, and if there is not strong scattering by impurities and such, $\tau_c \to \infty$ so to speak, and the modes that one observes in a variety of systems are not collision-dominated hydrodynamic modes ($\omega\tau_c \ll 1$) but collisionless elementary excitations ($\omega\tau_c \gg 1$).

# CHAPTER 8

## HYDRODYNAMIC SPIN WAVES IN FERROMAGNETS

We have now all the machinery ready to discuss collective and, in particular, hydrodynamic modes in all reasonable systems, whether these modes owe their existence to microscopic conservation laws, like normal sound, or whether they are "Goldstone modes", like second sound in superfluids. The rest is therefore just a series of exercises. In this chapter, we will continue to consider magnets and their characteristic collective modes: spin waves. The treatment given here is not the only one possible but it is the one that I like best since it makes most apparent how few the tricks really are which nature uses to keep theoreticians busy, and experimentalists happy. A similar treatment, in spirit if not in detail, has been given by Halperin and Hohenberg (1969a) and by Lubensky (1971). It is important to point out that we are not dealing with the spin wave excitations (Feynman 1972, Anderson 1964) which form the low-lying excited states above the perfectly ordered ground state of the Hamiltonian, and which determine, for example, the thermodynamic properties of ferromagnets at low temperatures. While there is an obvious relation, our treatment aims, instead, at modes at finite temperatures when local equilibrium is rapidly established within a microscopically short time $\tau$, and when the order is incomplete. The modes discussed here, and restricted to the hydrodynamic frequency domain $\omega << \tau^{-1}$, may be given the omnibus label "renormalized".

These modes can be investigated without explicit reference to the microscopic Hamiltonian, and our results hold consequently for systems more general than those for which the Heisenberg Hamiltonian (7.19) is appropriate. The important properties which we will use, however, are that the total spin $\vec{S}^{tot}$ is conserved, which excludes "spin-lattice" interactions or spin-orbit forces in itinerant magnets, and that the forces are of sufficiently short range. For definiteness, we will refer to the Heisenberg Hamiltonian (7.19), but as before we shall disregard the essentially trivial complications which the lattice structure introduces.

## 8.1 Symmetry Properties

We did our general symmetry analysis in section 3.2. For the isotropic ferromagnet, there are two small new points: the independence of rotations of either the spins or the lattice, and the presence of spontaneous magnetization. These are easily dealt with. Rotations of the lattice, but not the spins, leave the system physically invariant. Such rotations affect the wave vector $\vec{k}$ but not the direction of $\vec{M}$, and thus the spin correlation function of eq. (7.36) is a scalar function of $\vec{k}$, in other words

$$\chi''_{ij}(\vec{k}\omega) \text{ is a function of } k^2 \text{ only.} \tag{8.1}$$

Still, of course, $\chi''_{ij}(k\omega)$ is a tensor under spin rotations. To construct such a tensor, we have available only $\vec{M}_o$, the spontaneous magnetization, from which three tensors can be built: $\delta_{ij}$, $M_{oi}M_{oj}$, and $\epsilon_{ijk}M_{ok}$. Therefore, $\chi''_{ij}$ can be expressed in invariant form as

$$\chi''_{ij}(k\omega) = \frac{\langle M_i\rangle\langle M_j\rangle}{|\langle\vec{M}\rangle|^2}\,\chi''_\ell(k\omega) + \left(\delta_{ij} - \frac{\langle M_i\rangle\langle M_j\rangle}{|\langle\vec{M}\rangle|^2}\right)\chi''_t(k\omega) + i\epsilon_{ijk}\frac{\langle M_k\rangle}{|\langle\vec{M}\rangle|}\,\chi''_a(k\omega), \tag{8.2}$$

where the three functions, $\chi''_\ell$, $\chi''_t$, $\chi''_a$ can only depend on $|\langle\vec{M}\rangle|$ which is itself a function of the temperature.

Time reversal symmetry is changed a little since $\langle\vec{M}\rangle$ changes sign under time reversal as was indicated in eq. (3.22). One finds altogether that

$$\chi''_\ell(k\omega) \text{ and } \chi''_t(k\omega) \text{ are real, odd functions of } \omega,$$

$$\chi''_a(k\omega) \text{ is a real, even function of } \omega. \tag{8.3}$$

It is also not difficult to see that

$$\omega\chi''_\ell(k\omega) \geqq 0 \quad \text{and} \quad \omega\chi''_t(k\omega) \geqq 0,$$

$$\chi''^2_t(k\omega) \geqq \chi''^2_a(k\omega), \tag{8.4}$$

as a consequence of the stability restrictions of eq. (3.28).

It is customary to introduce circularly polarized components of $\chi''_{ij}$ by

$$\chi''_\perp(k\omega) = \chi''_t(k\omega) + \chi''_a(k\omega). \tag{8.5a}$$

This is evidently the transform of

$$\chi''_\perp(r,r';t) = \langle\frac{1}{2\hbar}[M^+(r,t), M^-(r',0)]\rangle, \tag{8.5b}$$

where

$$M^\pm = \frac{1}{\sqrt{2}}(M_x \pm i\, M_y), \tag{8.5c}$$

if $\langle\vec{M}\rangle$ points in the positive z-direction.

Apart from the three components $M_i(rt)$ of the magnetization whose fluctuations are clearly hydrodynamic, we also have to consider the magnetic energy density, $\epsilon(rt)$, which obeys a conservation law and will therefore lead to a hydrodynamic mode. For the isotropic Heisenberg magnet $\epsilon(rt)$ can be defined by

$$\epsilon(\vec{r}t) = -\frac{1}{2}\sum_{\alpha}\delta(\vec{r}-\vec{r}^{\alpha})\sum_{\beta(\neq\alpha)} J(|r^{\alpha}-r^{\beta}|)\vec{S}^{\alpha}(t)\cdot\vec{S}^{\beta}(t). \qquad (8.6)$$

It is obviously a scalar under spatial as well as spin rotations. Consequently, it can only couple to the z-component of $\vec{M}$, or in invariant form

$$\chi''_{\epsilon M_i}(\vec{k}\omega) = \frac{\langle M_i\rangle}{|\langle\vec{M}\rangle|}\ \chi''_{\epsilon M}(k\omega), \qquad (8.7)$$

where $\chi''_{\epsilon M}(k\omega)$ is easily shown to be a real and odd function of $\omega$, and depends only on $k^2$. In the paramagnetic regime no pseudovector is available to construct $\chi''_{\epsilon M_i}$, and therefore this function vanishes. This is why in our discussion in chapter 2 we could disregard energy fluctuations completely when discussing the spin dynamics. Even in the ferromagnet the cross correlation function is bounded by

$$[\chi''_{\epsilon M}(k\omega)]^2 \leq \chi''_{M_3M_3}(k\omega)\ \chi''_{\epsilon\epsilon}(k\omega). \qquad (8.8)$$

Here and henceforth we take the spontaneous magnetization in the z-direction: $\langle\vec{M}\rangle = M_0\,\hat{z}$.

## 8.2 Undamped Spin Waves

As a consequence of (8.2) and (8.7) there is no coupling between the two sets $\{M_x, M_y\}$ and $\{M_z, \epsilon\}$ of hydrodynamic variables. We will first consider the former whose fluctuations are by far the more interesting ones. What follows is a straightforward application of the formalism developed in chapter 5. To calculate the matrix of correlation functions

$$C_{\mu\nu}(k,t) = \langle A_\mu(k,t)\,|\,A_\nu(k,0)\rangle, \qquad (8.9a)$$

whose Laplace transform is given by

$$C_{\mu\nu}(kz) = \beta^{-1} \int \frac{d\omega}{\pi i} \chi''_{\mu\nu}(k\omega)/\omega(\omega - z) \tag{8.9b}$$

and where

$$A_1 \equiv M_x \quad \text{and} \quad A_2 \equiv M_y , \tag{8.9c}$$

we need the three matrices $\chi_{\mu\nu}(k)$, $\omega_{\mu\nu}(k)$, and $\sigma_{\mu\nu}(kz)$ which enter the general equations (5.28).

The transverse spin susceptibility has already been given in eq. (7.47), namely

$$\chi_{11}(k) = \chi_{22}(k) = \frac{M_o^{\ 2}}{R k^2} (1 + O(k^2)), \tag{8.10}$$

while $\chi_{12}(k) = \chi_{21}(k) = 0$ from time reversal symmetry.

The frequency matrix $\omega_{\mu\nu}$ of eq. (5.28) is also easy to obtain. Its diagonal elements vanish because of time reversal symmetry, and using eqs. (5.28) and (7.42), the off-diagonal element is given by

$$\omega_{12}(k) = -\omega_{21}(k) = \int \frac{d\omega}{\pi} \chi''_{M_x M_y}(k\omega) = i M_o. \tag{8.11}$$

For the Heisenberg system, in which there is no orbital motion, (8.11) holds for any k which is smaller than a vector of the inverse lattice. For more general conserving magnets (8.11) holds at least for sufficiently small k, i.e., to order $k^2$.

From (8.10) and (8.11) the normalized frequency matrix $\Omega_{\mu\nu}(k)$ is therefore given by, to order $k^4$,

$$\Omega_{\mu\nu}(k) = [\omega(k)\chi^{-1}(k)]_{\mu\nu} = \frac{Rk^2}{M_o} \begin{pmatrix} 0 & i \\ -i & 0 \end{pmatrix} . \tag{8.12}$$

Sine to order $k^2$ the damping matrix $\Sigma_{\mu\nu}(kz)$ of eq. (5.28) can be omitted, as will be seen below, we obtain the eigenfrequencies of collective modes from

$$\det[z - \underset{\approx}{\Omega}(k)] = 0, \tag{8.13}$$

which leads to the dispersion relation

$$z = \pm (R/M_o)k^2 + O(k^4). \tag{8.14}$$

Thus there is a propagating hydrodynamic mode in the system, called a spin wave, with the a bit unusual characteristic that its group velocity increases with decreasing wavelength. Unusual? As collective excitations go yes, but one might say that spin waves have the same dispersion relation as the de Broglie waves which correspond to free massive particles.

The eigenmodes corresponding to spin waves are found by diagonalizing the matrix $\Omega$. The reader who goes through this exercise will convince himself that the eigenmodes are given by the operators $M^{\pm}$ defined in (8.5c). For example, $\chi''_{\perp}$ as defined in (8.5a) is given by

$$\chi''_{\perp}(k\omega) = -M_o\, \delta(\omega - k^2 R/M_o) \tag{8.15}$$

in the approximation which omits the damping terms.

## 8.3 Damped Spin Waves

In order to include damping, we have to consider the memory matrix $\sigma_{\mu\nu}(kz)$, generally defined in eq. (5.28), e.g., by

$$\sigma_{11}(kz) = \beta\langle \dot{M}_x(k)|\, Q \frac{i}{z-QLZ}\, Q\, |\, \dot{M}_x(k)\rangle, \tag{8.16}$$

where L is the quantum mechanical Liouvillian, and the projector Q (which we do not need explicitly) removes fluctuations of the variables $\vec{M}$ and $\epsilon$. Noting that the magnetization is conserved,

$$\dot{M}_i(rt) + \nabla_j j_{ij}(rt) = 0 \quad \text{or} \quad \dot{M}_i(kt) = i\, k_j\, j_{ij}(kt), \tag{8.17}$$

$\sigma_{11}$ is seen to be at least of order $k^2$. On the basis of all the words which fill sections 5.4 and 7.6 we can replace $\sigma(kz)$ by $\sigma(k,0)$, rigorously to order $k^4$, and thus parametize the memory function by

$$\sigma_{11}(k0) = \sigma_{22}(k0) = k^2\sigma_\perp, \tag{8.18a}$$

$$\sigma_{12}(k0) = -\sigma_{21}(k0) = -k^2\mu M_o, \tag{8.18b}$$

where both $\sigma_\perp$ and $\mu$ are real by the general symmetry and reality requirements on the memory matrix, but where only $\sigma_\perp$ is constrained to be positive. We have put into (8.18b) an explicit factor of $M_o$ because $\sigma_{12}(k,0)$ must be an odd function of $M_o$ from time reversal, and vanishes in the paramagnetic regime. The reader will remember that the symmetry and reality properties of the memory functions $\sigma_{\mu\nu}(\vec{k}z)$ are precisely the same as those of the correlation functions $C_{\mu\nu}(\vec{k}z)$ which can be conveniently inferred from eqs. (8.9) and (8.2).

Given $\underset{\approx}{\omega}$, $\underset{\approx}{\sigma}$, and $\underset{\approx}{\chi}$, each to lowest order in k, we thus obtain the matrix

$$z\delta_{\mu\nu} - \Omega_{\mu\nu}(k) + i\Sigma_{\mu\nu}(k,z=0) = i\beta^{-1}\chi_{\mu\lambda}C^{-1}_{\lambda\nu}(kz)$$

$$= \begin{pmatrix} z + ik^4\sigma_\perp RM_o^{-2} & -ik^2(RM_o^{-1})(1+\mu k^2) \\ ik^2(RM_o^{-1})(1+\mu k^2) & z + ik^4\sigma_\perp RM_o^{-2} \end{pmatrix} . \tag{8.19}$$

We will henceforth omit the $\mu$-term in the off-diagonal matrix element which is here seen to be a higher order (in k) correction, and which will be discussed below. The poles of $C_{\mu\nu}(kz)$ are then easily seen to be given by

$$z = z(k) = \pm (R/M_o)k^2 - i\sigma_\perp (R/M_o^2)k^4, \tag{8.20}$$

indicating two propagating spin waves whose damping is of order $k^4$. The result that the lifetime of spin waves if proportional to the fourth power of their wavelength $\lambda$ is seen to be due to the double cause: the fact that the spin density $\vec{M}(rt)$ is conserved contributes a power $\lambda^2$ as it does in a paramagnet, see section 5.4; and the additional fact of the broken symmetry and long-ranged spin order contributes another power of $\lambda^2$ to the hydrodynamic lifetime $\tau(\lambda)$ of spin waves.

From (8.19) we obtain the complex correlation functions, namely

$$C_{11}(kz) = C_{22}(kz) = i\beta^{-1} \frac{M_o^2}{Rk^2} \cdot \frac{z + \frac{i}{2}\gamma_\perp k^4}{z^2 - r^2k^4 + izk^4\gamma_\perp}, \tag{8.21a}$$

$$C_{12}(kz) = -\frac{\beta^{-1}M_o}{z^2 - r^2k^4 + izk^4\gamma_\perp}, \tag{8.21b}$$

where we have introduced the natural abbreviations

$$r = R/M_o \quad \text{and} \quad \gamma_\perp = 2\sigma_\perp R/M_o^2. \tag{8.22}$$

And even if these formulae are a little unwieldy we will also write down the absorptive correlation functions. From the real part of $C_{11}$ one obtains

$$\chi''_t(k\omega)/\omega = \frac{1}{2}\frac{M_o^2}{Rk^2} \cdot \frac{\gamma_\perp k^4(\omega^2 + r^2k^4)}{[\omega^2 - r^2k^4]^2 + [\omega k^4\gamma_\perp]^2}, \tag{8.23a}$$

while from the imaginary part of $C_{12}$ one finds

$$\chi''_a(k\omega)/\omega = \frac{\omega k^4\gamma_\perp M_o}{[\omega^2 - r^2k^4]^2 + [\omega k^4\gamma_\perp]^2}. \tag{8.23b}$$

The reader will want to compare these results with those that we obtained in chapter 2 for the same system but in the paramagnetic regime, namely $\chi''_a = 0$ and

$$\chi''_t(k\omega)/\omega = \chi \frac{k^2D}{\omega^2 + (k^2D)^2} \quad \text{(paramagnet)}. \tag{2.55b}$$

The comparison is most instructive if we separate (8.21a) or (8.23a) into the two pole contributions, namely

$$\chi''_t(k\omega)/\omega = \frac{1}{2}\frac{M_o^2}{Rk^2}\left[\frac{k^4\gamma_\perp/2}{(\omega-rk^2)^2+(k^4\gamma_\perp/2)^2}+\frac{k^4\gamma_\perp/2}{(\omega+rk^2)^2+(k^4\gamma_\perp/2)^2}\right], \tag{8.24}$$

which is mathematically equivalent to (8.23a).

Instead of the single Lorentzian line in the paramagnet, which is centered at $\omega = 0$, for the ferromagnet we obtain a split line, consisting of two Lorentzians which are symmetrically displaced from $\omega = 0$ and centered at $\omega = \pm rk^2$. For very small wave vector k these

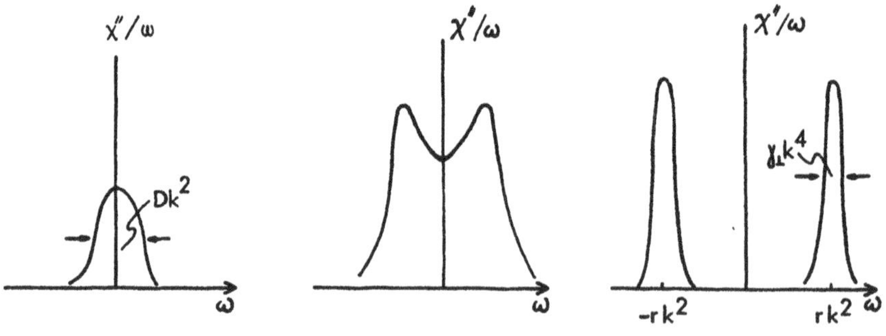


a) paramagnet small k

b) ferromagnet moderate k

c) ferromagnet small k

Fig. 8.1

two peaks are well separated since their full width at half maximum is given by $\gamma_\perp k^4$. By contrast, the width of the paramagnetic peak is much wider, and given by $2Dk^2$. In an ideal world in which the magnetic response function $\chi''(k\omega)$ could be measured at arbitrarily small wave vector k one would therefore expect to find a result of the form given in fig. 8.1c, in the ferromagnetic region. If one did not find this kind of spectrum one would have cause for great surprise and concern since our result, eq. (8.24), is based only on very general principles and should be rigorous for asymptotically small k.

Alas, the world is not ideal. The best experimental method to measure $\chi''_t(k\omega)$ is by inelastic magnetic neutron scattering, and since the wave vector K of thermal neutrons is large ($K \approx 1\ \mathring{A}^{-1}$ for T = 300°K), it is difficult to measure very small momentum transfer $|\vec{k}| = |\vec{K}_{in} - \vec{K}_{out}|$. Nevertheless, for ferromagnetic iron and nickel precise neutron scattering experiments have been performed (Collins et al. 1969, Minkiewcz et al. 1969) from which $\chi''_t(k\omega)$ can be obtained, as indicated in the scattering formula (2.30) or (A.20), for wave vectors k as small as $k \approx 0.04\ \mathring{A}^{-1}$. The picture that emerges bears out eq. (8.24) in the ferromagnetic, and (2.55b) in the paramagnetic regions. (In real iron and nickel one has to include the effects of long-ranged dipole-dipole interactions which we have omitted.) Note that "small k" means, of course, that $k\xi << 1$ where $\xi$ is the range of the static correlation functions $\tilde{\chi}(r)$ as discussed briefly in section 7.6. As the transition temperature $T_c$ is approached, either from below or above, the correlation length $\xi$ becomes very large, going to infinity at $T_c$. Near $T_c$, therefore, we would expect fig. 8.1c to be replaced by fig. 8.1b as is indeed observed. In the critical region, very close to $T_c$, the correlation length is so large that our hydrodynamic analysis breaks down. It has to be replaced by the more subtle considerations of "dynamical scaling" (Halperin and Hohenberg 1969b, Stanley 1971) which will not be treated here.

## 8.4 Sum Rules and Kubo Formulae

The hydrodynamic result (8.23) has several of those properties with which we have become familiar in simpler systems. For example, (8.23a) exhausts the thermodynamic sum rule

$$\int \frac{d\omega}{\pi} \frac{1}{\omega} \chi''_t(k\omega) = \frac{M_o^2}{Rk^2} \qquad \text{(ferromagnet)} \tag{8.25a}$$

to relative order $k^2$, whereas for the paramagnet we had

$$\lim_{k\to 0} \int \frac{d\omega}{\pi} \frac{1}{\omega} \chi''_t(k\omega) = \chi \qquad \text{(paramagnet)}, \tag{8.25b}$$

where $\chi$ is the magnetic spin susceptibility. Since this sum rule, in either case, just states the value of $C_t(k,t)$ at $t = 0$, it is evident that the memory function procedure of deriving $\chi''$ will always give results which exhaust the thermodynamic sum rule.

Equations (7.42) or (8.11) are reflected in the sum rule

$$\lim_{k\to 0} \int \frac{d\omega}{\pi} \chi''_a(k\omega) = M_o \qquad \text{(ferromagnet)} \tag{8.26}$$

which is also fulfilled, again by construction. Eq. (8.26) expresses the fact of broken symmetry while (8.25a) describes the related occurrence of long-ranged order. In an isotropic paramagnet $\chi''_a$ vanishes.

Given the Hamiltonian it would be an easy matter to derive further sum rules by the procedure generally described in section 3.4. For example, it would be useful to know the numerical value of

$$c_\infty^2 = \lim_{k\to 0} \frac{1}{k^2} \int \frac{d\omega}{\pi} \omega\chi''_t(k\omega)$$

$$= \lim_{k\to 0} \frac{1}{k^2 V} \int d\vec{r} \int d\vec{r}\,' e^{-i\vec{k}(\vec{r}-\vec{r}\,')} \langle \frac{i}{\hbar} [\dot{M}_y(\vec{r}), M_y(\vec{r}\,')] \rangle, \tag{8.27}$$

since the general equation (7.63), applied to the present case, would then give an upper bound for the stiffness constant R, namely

$$R \leq c_\infty^2. \tag{8.28}$$

Of course, the hydrodynamic result (8.23a) is of no use here since for it the integral $\int \omega \chi_t''(k\omega)$ diverges. However, for the simple itinerant system which we treated in chapter 2 the sum rule (8.27) is simply the f-sum rule, and it results in $c_\infty^2 = n/m$. (Remember that in the present chapter we have set the magnetic moment per particle equal to 1.) For the isotropic Heisenberg ferromagnet, described by the Hamiltonian (7.19), the sum rule (8.27) can also be calculated, and expressed in the form

$$c_\infty^2 = \frac{1}{6} \sum_{\beta(\neq\alpha)} (r^{\alpha\beta})^2 J^{\alpha\beta} \langle \vec{S}^\alpha \cdot \vec{S}^\beta - S_y^\alpha S_y^\beta \rangle . \tag{8.29}$$

This is, of course, reminiscent of the corresponding sum rule for the momentum density in simple fluids, eq. (4.76). (8.29) shows that $c_\infty^2$ is determined by short-range correlation in the system.

From the hydrodynamic result (8.23a) one extracts a Kubo-type relation for the spin wave damping constant $\gamma_\perp$, or rather $\sigma_\perp$. This relation is

$$\sigma_\perp = \lim_{\omega\to 0} \lim_{k\to 0} \frac{\omega}{k^2} \chi_t''(k\omega), \tag{8.30}$$

which looks, of course, just like the corresponding expressions (4.53) which give the transport coefficients in a simple fluid. The reader might amuse himself by showing that (8.30) is in fact the same expression as (8.18a); in doing so, he will need the master formula (5.58) in order to eliminate the projection operator Q which is contained in the microscopic expression (8.16) for $\sigma_{11}(kz)$.

That projector is useful, however, in that it suggests that the transport coefficient $\sigma_\perp$ is determined by local and rapid fluctuations. Now locally a ferromagnet and a paramagnet are not very different. If we therefore compare eq. (8.30) for the ferromagnet with the corresponding equation which follows from (2.55b) for the paramagnetic regime, namely

$$D\chi = \lim_{\omega\to 0} \lim_{k\to 0} \frac{\omega}{k^2} \chi''_t(k\omega) \qquad \text{(paramagnet)}, \tag{8.31}$$

we would expect that $\sigma_\perp$ and $D\chi$ should be of the same order of magnitude. This kind of estimate would have to be corrected, of course, for the normal Arrhenius-law temperature dependence of the transport coefficients, and even then it would completely break down in the critical region near $T_c$, the transition temperature. Much should be said about this question which is, however, well outside the scope of this book.

## 8.5 Corrections

This section deals with a minor subtlety. We have omitted, from all equations following (8.19), the off-diagonal element $k^2 \mu M_0$ of the memory matrix. If we had carried it along, we would have found a spin wave dispersion relation of the form

$$z = \pm \epsilon(k) - i\,\Gamma(k) = \pm r k^2 (1 + \mu k^2) - \frac{i}{2} \gamma_\perp k^4 \tag{8.32}$$

which corrects the spin wave "energy" by a small term of relative order $k^2$. That is the only effect of this term. Including it, we can write the eqs. (8.21) in the form

$$C_{11}(kz) = i\beta^{-1} \chi_{11}(k) \left[\frac{1/2}{z - \epsilon(k) + i\Gamma(k)} + \frac{1/2}{z + \epsilon(k) + i\Gamma(k)}\right], \tag{8.33a}$$

$$C_{12}(k) = i\beta^{-1} \chi_{11}(k) \left[\frac{i/2}{z - \epsilon(k) + i\Gamma(k)} - \frac{i/2}{z + \epsilon(k) + i\Gamma(k)}\right]. \tag{8.33b}$$

Note, however, that the term $\mu k^2$ which stems from $\sigma_{12}(k, i0)$ is not the only one which would contribute to the $k^4$-correction of the spin wave energy. A similar contribution would come from the static susceptibility

$$\chi_{11}(k) = \frac{M_o^2}{Rk^2}(1 + xk^2 + k^4 \ldots), \tag{8.34}$$

as the reader will easily show. Terms of this sort might be interesting for some purposes but they are outside of what is usually termed hydrodynamics. As is easily seen from our eqs. (8.33), our previous results are systematic in that they determine both real and imaginary parts of the eigenfrequencies to lowest non-vanishing order in k, and the corresponding pole strengths also to lowest non-vanishing order. The reader might want to convince himself explicitly that the z-dependence of the memory functions $\sigma_{\mu\nu}(k,z)$ would only give rise to higher order corrections as well, in this sense.

## 8.6 "Longitudinal" Fluctuations

In the isotropic paramagnet, the magnetization $\vec{M}$ does not couple to the energy density $\epsilon$. If there are no other conserved quantities in the system, both $\vec{M}$ and $\epsilon$ undergo independent Brownian diffusion processes, and their respective correlation functions $\omega^{-1}\chi''_{MM}$ and $\omega^{-1}\chi''_{\epsilon\epsilon}$ are Lorentzians in the hydrodynamic regime of small frequency and wave vector.

Apart from the emergence of spin waves in the transverse variables $M_x$ and $M_y$, the most noteworthy thing that happens in the ordered ferromagnetic state is that the energy density can now couple to the longitudinal component $M_z$ of the magnetization. For example, the static susceptibility matrix has, at k = 0, the components

$$\chi_{MM} = \frac{\partial\langle M\rangle}{\partial h}\Big|_{T,h=0}, \quad \chi_{\epsilon\epsilon} = T\frac{\partial\epsilon}{\partial T}\Big|_{h=0},$$

$$\chi_{M\epsilon} = \chi_{\epsilon M} = T\frac{\partial M_o}{\partial T}\Big|_{h=0}, \tag{8.35}$$

in the absence of an external magnetic field h. There are, however, no very dramatic manifestations of this coupling. The frequency matrix vanishes from time reversal, and to lowest order the memory matrix is of order $k^2$ due to spin and energy conservation, and it is real and symmetric. Instead of containing one diffusive Lorentzian, the longitudinal correlation function $\omega^{-1} \chi''_{M_z M_z}(k\omega)$ now emerges as a sum of two Lorentzians, both centered at $\omega = 0$, both with a width of order $k^2$ as $k \rightarrow 0$, both with positive strength. The combined strength of the two Lorentzians is, of course, given by the longitudinal magnetic susceptibility and is thus finite even as $k \rightarrow 0$, in contrast to what we found for the transverse spin wave amplitudes. A more detailed analysis of these longitudinal fluctuations is left as an exercise.

# CHAPTER 9

## HYDRODYNAMIC SPIN WAVES IN ANTIFERROMAGNETS

It is always tempting to speculate (Weidlich 1971) on the applicability of models of ferromagnets to the decision behavior of social groups. Each spin lining up with its neighbors, is that not reminiscent of the adaptive behavior of (some) individuals in (some) societies? If so, then the society which reflects antiferromagnetic behavior would be of a peculiarly antagonistic sort indeed: In an ideal antiferromagnet, each spin watches carefully where its nearest neighbors point--and then points in the exactly opposite direction. In one dimension, this results in an array like this:

Fig. 9.1

In a three-dimensional simple cubic lattice, each spin that points up is surrounded by six nearest neighbor spins which point down. In an antiferromagnet, therefore, things look like two interpenetrating sublattices, the spins in each of which are aligned as in a ferromagnet, but in opposite directions for the two sublattices. Of course, at finite temperature $T < T_c$ the antialignment is not perfect, just as the alignment in a ferromagnet is not. However, on the average each spin in an antiferromagnet is surrounded by more antialigned than aligned spins.

The Hamiltonian for the simplest isotropic Heisenberg antiferromagnet is identical in form to eq. (7.19), namely

$$H = -\frac{1}{2} \sum_{\alpha \neq \beta} J^{\alpha\beta}\, \vec{S}^{\alpha} \cdot \vec{S}^{\beta} . \tag{9.1}$$

However, while the nearest neighbor coupling energy J in a ferromagnet is positive, favoring parallel alignment, J is negative in antiferromagnetic systems. Consequently, states of antiparallel orientation of nearest neighbors are energetically favored, and at sufficiently low temperature the system tends to be antiferromagnetic. This is a crude picture, but for our purposes it will suffice.

Now we have characterized the ferromagnetic state by the fact that, even in equilibrium, it has a spontaneous magnetization, $\langle \vec{M} \rangle \neq 0$. In the normal antiferromagnet, $\langle \vec{M} \rangle = 0$ obviously; there is no net macroscopic magnetic moment. Rather, what is non-zero on the average is the "staggered magnetization" $\vec{N}$ . Instead of adding all spins with the same sign as we did to obtain the total magnetization,

$$\vec{M}^{tot}(t) = \sum_{\alpha} \vec{S}^{\alpha}(t), \tag{9.2}$$

we now add them with opposite signs for the two sublattices,

$$\vec{N}^{tot}(t) = \sum_{\alpha} \eta^{\alpha}\, \vec{S}^{\alpha}(t), \tag{9.3}$$

where $\eta^{\alpha}$ has opposite sign for nearest neighbors: $\eta^{\alpha} = 1$ for all spins on one sublattice, and $\eta^{\alpha} = -1$ for all spins on the other. Then, while the ferromagnetic order can be characterized by

$$\langle \vec{M}^{tot} \rangle \neq 0 \quad \text{and} \quad \langle \vec{N}^{tot} \rangle = 0 \quad \text{(ferromagnet)}, \tag{9.4}$$

the antiferromagnetic order is described by

$$\langle \vec{M}^{tot} \rangle = 0 \quad \text{and} \quad \langle \vec{N}^{tot} \rangle \neq 0 \qquad \text{(antiferromagnet)}. \tag{9.5}$$

Just as we defined the local magnetization operator by

$$\vec{M}(r,t) = \sum_{\alpha} \vec{S}^{\alpha}(t)\, \delta(r-r^{\alpha}), \tag{9.6}$$

we can now define the local density of the staggered magnetization by

$$\vec{N}(r,t) = \sum_{\alpha} \eta^{\alpha}\, S^{\alpha}(t)\, \delta(r-r^{\alpha}). \tag{9.7}$$

It will, on the average, have a certain non-zero value $N_o$ in some direction which we choose to be the z-direction,

$$\langle \vec{N}(rt) \rangle = N_o\, \hat{z}, \tag{9.8a}$$

$$\langle \vec{M}(rt) \rangle = 0. \tag{9.8b}$$

## 9.1 Long-Range Order

Now matters are not so simple that we could simply replace $\vec{M}$ by $\vec{N}$ everywhere, and be done with. There are two important differences for the antiferromagnet:

(a) As in the ferromagnet, two continuous symmetries are broken, namely rotations of all spins about the x and y axes. The generators of these symmetries are $M_x^{tot}$ and $M_y^{tot}$ as before, while the symmetry-restoring variables are now $N_x$ and $N_y$.

(b) As before, the total magnetization is microscopically conserved,

$$\frac{d}{dt} M^{tot} = \frac{1}{i\hbar} [M^{tot}, H] = 0 \tag{9.9a}$$

and

$$\partial_t M_i(r,t) + \nabla_i j_{ij}(rt) = 0, \tag{9.9b}$$

since this is simply a consequence of the isotropy of the Hamiltonian. However, the total staggered magnetization, $\vec{N}^{tot}$, is not conserved,

$$[\vec{N}^{tot}, H] = (i\hbar/2) \sum_{\alpha\neq\beta} J^{\alpha\beta}(\eta^\alpha - \eta^\beta)\, \vec{S}^\alpha \times \vec{S}^\beta \neq 0. \tag{9.10}$$

Let us now define correlation functions as always by

$$\chi''_{AB}(\vec{k}\,\omega) = \frac{1}{V} \int_{-\infty}^{\infty} dt \int d\vec{r} \int d\vec{r}\,' \; e^{i\omega t - i\vec{k}\cdot(\vec{r}-\vec{r}\,')} \langle \frac{1}{2\hbar}[A(\vec{r},t), B(\vec{r}\,',0)] \rangle \tag{9.11}$$

and pursue the consequences of the broken symmetry. The generator property of $\vec{M}^{tot}$ is expressed by

$$\frac{1}{i\hbar}[M_i^{tot}, N_j(r')] = \epsilon_{ijk} N_k(r'), \tag{9.12}$$

which because of (9.8) leads to

$$\langle \frac{1}{i\hbar}[M_x^{tot}, N_y(r')] \rangle = N_o ,$$

$$\langle \frac{1}{i\hbar}[M_y^{tot}, N_x(r')] \rangle = -N_o ; \tag{9.13}$$

or equivalently

$$\lim_{k\to 0} \int \frac{d\omega}{\pi} \chi''_{M_x N_y}(\vec{k}\,\omega) = i N_o ,$$

$$\lim_{k\to 0} \int \frac{d\omega}{\pi} \chi''_{M_y N_x}(\vec{k}\,\omega) = -i N_o , \tag{9.14}$$

where, for the simple Heisenberg antiferromagnet considered here, (9.14) holds in fact at any k. Comparing (9.13) with (7.53) and proceeding step by step to (7.62) we obtain the now familiar divergence at small k in the static susceptibilities,

$$\chi_{N_x N_x}(k) = \chi_{N_y N_y}(k) = \frac{N_o^2}{R' k^2}(1 + O(k^2)), \qquad (9.15)$$

where R' is a positive constant. (9.15) is the antiferromagnetic equivalent of the equation (8.10) for the corresponding ferromagnet. (9.15) indicates that fluctuations of the transverse staggered magnetization are correlated over large spatial distances, and it is the main reason why antiferromagnetic spin waves exist, as we will see. As in (7.63) we also obtain a bound for the antiferromagnetic stiffness constant R', namely

$$R' \leq \lim_{k \to 0} \frac{1}{k^2} \int \frac{d\omega}{\pi} \; \omega \; \chi''_{M_x M_x}(\vec{k}, \omega), \qquad (9.16)$$

which is to say

$$R' \leq c_\infty^2, \qquad (9.17)$$

where $c_\infty^2$ is given by the expression (8.29) which makes only implicit reference to the specific (ferro- or antiferromagnetic) order present.

The two static functions $\chi_{N_x N_x}(k)$ and $\chi_{N_y N_y}(k)$, which are equal by symmetry, are the only ones which must be singular as $k \to 0$, according to the Bogoliubov inequality argument discussed in section 7.3. For all other susceptibilities, we shall assume that they are nice and well-behaved as $k \to 0$. Indeed, if for example $\chi_{M_x M_x}(k)$ were infinite as $k \to 0$, it would mean that an infinitesimal, uniform magnetic field in the x-direction would produce a finite response $\delta\langle M_x \rangle$. That, however, plainly cannot be so since in an antiferromagnet there is no spontaneous net magnetization, and since it does cost a finite amount of energy, and thus a finite external field, to create one.

## 9.2 Symmetry Considerations

If we want to construct a hydrodynamic fluctuation theory, we have to explicitly consider both variables $\vec{M}(r,t)$ and $\vec{N}(r,t)$: $\vec{M}(r,t)$ because it obeys a microscopic conservation law, and $\vec{N}(rt)$ because its static susceptibility diverges as a result of broken symmetry. We also ought to include the conserved energy density as a further hydrodynamic variable but we will disregard it for the moment. We then have three sets of correlation functions, namely

$$\chi''_{N_iN_j}(\vec{k}\omega) \ ; \ \chi''_{M_iM_j}(\vec{k}\omega) \ ; \ \chi''_{N_iM_j}(\vec{k}\omega) \ . \tag{9.18}$$

First, on the basis of the same argument of the independence of spatial and spin rotations that led to (8.1), it is clear that

$$\text{all} \quad \chi''(\vec{k}\,\omega) \ \text{ are functions of } k^2 \text{ only.} \tag{9.19}$$

Moreover, all three sets of $\chi''$'s are tensors, and since there are only three tensors which can be built from the only vector in the system, namely $\langle\vec{N}\rangle$, all $\chi''$'s must be of the invariant form

$$\chi''_{ij}(k\omega) = n_in_j\,\chi''_\ell + (\delta_{ij} - n_in_j)\,\chi''_t + i\,\epsilon_{ijk}n_k\,\chi''_a, \tag{9.20}$$

where $n_i = \langle N_i\rangle/|\langle\vec{N}\rangle|$, and where the $\chi''$'s on the right hand side are functions of k, ω, and of course the temperature. Since $\chi''_{N_iM_j}(\vec{k},\omega) = -\chi''_{M_jN_i}(-\vec{k},-\omega)$, this leaves us with 9 functions $\chi''$ to be determined. Noting that $\langle\vec{N}\rangle$ changes sign under time reversal we find, in analogy to (8.3), that

$$\begin{aligned}&\text{the three } \chi''_\ell\,(k\omega)\text{'s are real, odd in } \omega,\\ &\text{the three } \chi''_t(k\omega)\text{'s are real, odd in } \omega,\\ &\text{the three } \chi''_a(k\omega)\text{'s are real, even in } \omega.\end{aligned} \tag{9.21}$$

With $\langle\vec{N}\rangle$ pointing in the z-direction, we have found so far that there are no cross correlations between the two sets of variables $\{N_x, N_y; M_x, M_y\}$ and $\{N_z, M_z\}$.

However, there is one more symmetry. Clearly, it does not matter which sublattice we count as ↑, and which as ↓. In other words, replacing all $\eta^a$'s in (9.7) by $-\eta^a$ cannot change anything. This operation, which is a kind of parity, replaces $\{\vec{N}, \vec{n}, \vec{M}\}$ by $\{-\vec{N}, -\vec{n}, \vec{M}\}$. Therefore $\chi''_{N_iN_j}$ and $\chi''_{M_iM_j}$ must be even in $\vec{n}$ so that

$$\chi''_{a,NN} = \chi''_{a,MM} = 0, \tag{9.22}$$

while $\chi''_{N_iM_j}$ must be odd in $\vec{n}$ so that

$$\chi''_{\ell,NM} = \chi''_{t,NM} = 0. \tag{9.23}$$

Consequently, the invariant representation (9.20) becomes

$$\chi''_{N_iN_j}(k\omega) = n_in_j\ \chi''_{\ell,NN}(k\omega) + (\delta_{ij} - n_in_j)\ \chi''_{t,NN}(k\omega), \tag{9.24a}$$

$$\chi''_{M_iM_j}(k\omega) = n_in_j\ \chi''_{\ell,MM}(k\omega) + (\delta_{ij} - n_in_j)\chi''_{t,MM}(k\omega), \tag{9.24b}$$

$$\chi''_{N_iM_j}(k\omega) = \chi''_{M_iN_j}(k\omega) = i\,\epsilon_{ijk}\, n_k\ \chi''_a(k\omega). \tag{9.24c}$$

This leaves five functions to be determined. The six variables $\vec{N}$ and $\vec{M}$ are now broken down into four sets,

$$\{N_x, M_y\}\ ;\ \{N_y, M_x\}, \tag{9.25}$$

and

$$\{M_z\}\ ;\ \{N_z\}, \tag{9.26}$$

between which there is no coupling.

Here we are only interested in the dynamics of the first two variables, $N_x$ and $M_y$. The second set, $\{N_y, M_x\}$, is of course completely equivalent. The longitudinal spin density $M_z$ is conserved but its static susceptibility, $\chi_{M_zM_z}(k) = \chi_\ell$ as $k \to 0$, is finite. Since $M_z$ is not coupled to any of the other hydrodynamic variables, its autocorrelations will, for small k and $\omega$, display the usual diffusion structure,

$$\chi''_{M_zM_z}(k\omega) = \frac{\omega k^2 D_\ell}{\omega^2 + (k^2 D_\ell)^2}\, \chi_\ell\ , \tag{9.27}$$

where $D_\ell$ is a longitudinal spin diffusion coefficient. The derivation of (9.27) follows in all detail that given for the paramagnet in section 5.4, and needs no further discussion.

The fourth variable, $N_z$, is neither conserved, nor is there any reason for its static correlations to be unusually large. $N_z$ is therefore not a hydrodynamic variable, and its fluctuations decay within microscopic times. The spectrum of these fluctuations, as of those of all other non-hydrodynamic variables, is not of a universally predictable form, and it is not particularly interesting. Only near the critical point would we have to reconsider this assessment. Just above $T_c$ the system is still macroscopically isotropic, and $\chi_{N_zN_z}(k)$ must equal $\chi_{N_xN_x}(k)$ which goes to infinity as $k \to 0$ and $T \to T_c$. However, we will continue to follow our overall policy of avoiding the peculiar difficulties that arise in nearly critical systems; see Stanley (1971).

Now we can also quickly deal with the energy density $\epsilon(r,t)$ which, as in the ferromagnet, can be defined by

$$\epsilon(rt) = -\frac{1}{2} \sum_\alpha \delta(r - r^\alpha) \sum_{\beta(\neq\alpha)} J^{\alpha\beta}\, \vec{S}^\alpha(t) \cdot \vec{S}^\beta(t). \tag{9.28}$$

$\epsilon(rt)$ is conserved and should therefore be included in the roster of hydrodynamic variables. Now the two functions $\chi''_{\epsilon N_i}(k\omega)$ and $\chi''_{\epsilon M_i}(k\omega)$ are vectors in spin space, and since $\vec{n}$ is the only vector available, they must be of the form

$$\chi''_{\epsilon N_i}(k\omega) = n_i \quad \chi''_{\epsilon N}(k\omega), \tag{9.29a}$$

$$\chi''_{\epsilon M_i}(k\omega) = n_i \quad \chi''_{\epsilon M}(k\omega). \tag{9.29b}$$

However, since both $\epsilon$ and $\vec{M}$ are even under the "parity" operation, $\eta^\alpha \to -\eta^\alpha$ and $\vec{n} \to -\vec{n}$, $\chi''_{\epsilon M}$ vanishes in fact.

Thus the energy density $\epsilon$ is only coupled to $N_z$. But $N_z$ is not a hydrodynamic variable and therefore, for small k and $\omega$, $N_z$ is not qualitatively distinguished from the trillions of other rapidly fluctuating degrees of freedom which contribute to the overall energy fluctuations. In other words, $\epsilon$ is not coupled to any other hydrodynamic variable. And whenever this is so, one gets a simple diffusion law at small k and $\omega$,

$$\chi''_{\epsilon\epsilon}(k\omega) = \frac{\omega k^2 D_T}{\omega^2 + (k^2 D_T)^2} \, T c_h, \tag{9.30}$$

where $D_T$ is the coefficient of thermal diffusion, and $V c_h = (\partial E/\partial T)_h$ the specific heat at constant and vanishing external magnetic field.

After these many preliminaries, which will doubtlessly have bored some but hopefully helped others, we can proceed in our quest for hydrodynamic spin waves in the Heisenberg antiferromagnet.

## 9.3 Undamped Spin Waves

We are now ready to apply the memory function machinery of chapter 5 to the two variables $A_\mu(r,t)$ where

$$A_1(rt) \equiv N_x(rt) \text{ and } A_2(rt) \equiv M_y(rt). \tag{9.31}$$

As always, we will analyze the matrix of Kubo correlation functions,

$$C_{\mu\nu}(r,r';t) = \langle A_\mu(rt) \mid A_\nu(r',0)\rangle, \tag{9.32}$$

defined as in (5.37) so that the Fourier-Laplace transforms are related to $\chi''$ by

$$C_{\mu\nu}(k,z) = \beta^{-1} \int \frac{d\omega}{\pi i} \frac{\chi''_{\mu\nu}(k\omega)}{\omega(\omega-z)} . \tag{9.33}$$

This matrix obeys the general "equation of motion"

$$[z\underline{1} - \underline{\Omega}(k) + i\underline{\Sigma}(kz)]_{\mu\lambda} C_{\lambda\nu}(kz) = i\beta^{-1} \chi_{\mu\nu}(k), \tag{9.34}$$

whose individual terms are generally identified in (5.28). Let us now consider these terms one by one.

The susceptibility matrix $\chi_{\mu\nu}(k)$ is diagonal since both $N_x$ and $M_y$ have the same negative signature under time reversal. For sufficiently small k the diagonal elements are given, according to our considerations in section 9.1, by

$$\chi_{11}(k) = \frac{N_o^2}{R'k^2} \quad \text{and} \quad \chi_{22}(k) = \chi_\perp , \tag{9.35}$$

where $\chi_\perp = (\partial M_x / \partial h_x)_{T,h=0}$ is the thermodynamic spin susceptibility which is finite and, of course, positive.

The unnormalized frequency matrix $\omega_{\mu\nu}(k)$ is generally given by

$$\omega_{\mu\nu}(k) = \int \frac{d\omega}{\pi} \chi''_{\mu\nu}(k\omega) \tag{9.36}$$

and can thus be read off eq. (9.14) above:

$$\omega_{12}(\vec{k}) = -\omega_{21}(-\vec{k}) = i N_o, \tag{9.37}$$

while the diagonal elements vanish from time reversal symmetry. Consequently the normalized frequency matrix is of the form

$$\Omega_{\mu\nu}(k) = [\omega(k)\chi^{-1}(k)]_{\mu\nu} = \begin{pmatrix} 0 & iN_o/\chi_\perp \\ -i R'k^2/N_o & 0 \end{pmatrix} . \tag{9.38}$$

Anticipating that the memory matrix $\Sigma_{\mu\nu}(kz)$ is negligible to lowest order in k we find the mode dispersion relations by solving

$$\det[z\underset{\approx}{1} - \underset{\approx}{\Omega}(k)] = 0, \tag{9.39}$$

which results in

$$z = \pm ck \quad \text{where} \quad c^2 = R'/\chi_\perp \ . \tag{9.40}$$

Thus there are oscillatory modes, spin waves, propagating with a speed c which is determined by the stiffness constant $R'$ and the magnetic susceptibility $\chi_\perp$ . Note the contrast to the result (8.14) which holds for the corresponding ferromagnet.

It may be helpful to rephrase these equations and results in terms of macroscopic non-equilibrium fluctuations $\delta\langle A_\mu(rt)\rangle$ which are set up by small external fields $\delta a_\mu(r)$ that are turned on for negative times. The general results of chapter 3, especially eq. (3.13), tell us that the linear dynamic response to such adiabatic external fields is given by

$$\delta\langle A_\mu(kz)\rangle = C_{\mu\nu}(kz)\,\beta\,\delta a_\nu(k). \tag{9.41}$$

Thus the equation of motion (9.34) is also an equation of motion for the macroscopic fluctuations $\delta\langle A_\mu\rangle$. If we omit the dissipative matrix $\Sigma_{\mu\nu}$, insert our expression (9.38) for $\underset{\approx}{\Omega}$, and reconvert things into real space and time, these equations become

$$\partial_t\,\delta\langle M_y(rt)\rangle - R'\nabla^2\,\delta\langle N_x(rt)\rangle/N_o = 0, \tag{9.42a}$$

$$\partial_t\,\delta\langle N_x(rt)\rangle/N_o - \chi_\perp^{-1}\,\delta\langle M_y(rt)\rangle = 0. \tag{9.42b}$$

The first equation is in agreement, of course, with the conservation of magnetization. According to (9.42b) an inhomogeneity of the magnetization sets in motion a fluctuation of the staggered magnetization whose non-uniformity, in turn, drives the magnetization. In effect the spin system acquires "elastic" properties because of the long-range order. While in the paramagnet a spin fluctuation ambles back to equilibrium by diffusive Brownian motion, in the antiferromagnet it can feed into a "conserved" current, $\delta\vec{j} = -R'\,\vec{\nabla}\,\delta\langle N\rangle N_o^{-1}$. Consequently, the fluctuation will overshoot as it returns to equilibrium, and a propagating "elastic" wave results.

## 9.4 Spin Wave Damping

We now include the memory function, $\Sigma_{\mu\nu}(kz)$. It is generally given, according to eq. (5.28), by

$$\Sigma_{\mu\nu}(kz) = \begin{pmatrix} (R'k^2/N_o^2)\,\sigma_{11}(kz) & \chi_\perp^{-1}\,\sigma_{12}(kz) \\ (R'k^2/N_o^2)\,\sigma_{21}(kz) & \chi_\perp^{-1}\,\sigma_{22}(kz) \end{pmatrix} \tag{9.43}$$

in terms of the unnormalized matrix

$$\sigma_{\mu\nu}(kz) = \beta\langle \dot{A}_\mu(k)\,|\,Q\,\frac{i}{z-QLQ}\,Q\,|\dot{A}_\nu(k)\rangle\ . \tag{9.44}$$

In (9.43) we have inserted the inverse matrix $\chi_{\mu\nu}^{-1}(k)$, to lowest order in k.

To extract useful information from the general formula (9.44) the reader will want to recall all the words used, for example, in section 5.4. First to the projector Q in (9.44). It removes the fluctuations of the variables $A_1 = N_x$ and $A_2 = M_y$. However, there is no reason to be so restrictive. Because symmetry forbids all the other hydrodynamic variables from coupling to $N_x$ and $M_y$, we can define Q by

$$Q = 1 - V\int\frac{d^3k}{(2\pi)^3}\sum_{\mu,\nu}|A_\mu(k)\rangle\beta\chi_{\mu\nu}^{-1}(k)\langle A_\nu(k)|\ , \tag{9.45}$$

where the sums over $\mu$ and $\nu$ extend over all seven hydrodynamic variables $\vec{N}$,

$\vec{M}$, and $\epsilon$. Related discussion is given in connection with eq. (5.45b). The point is that the operator Q in (9.45) is a scalar under rotations, translations, and reflections, and that it indeed removes, in (9.44), all slow fluctuations from $\sigma_{\mu\nu}(kz)$. As a consequence, $\sigma_{\mu\nu}(kz)$ is a smooth function of z near z = 0. Moreover, if we use the fomal representation

$$\sigma_{\mu\nu}(kz) = \int \frac{d\omega}{2\pi i}\ \gamma_{\mu\nu}(k\omega)/(\omega - z) \tag{9.46}$$

and compare the eq. (9.44) with the equally fomal expression

$$C_{\mu\nu}(kz) = \langle A_\mu(k) | \frac{i}{z-L} | A_\nu(k)\rangle \tag{9.47}$$

and with (9.33), then it is clear that the damping functions $\gamma_{\mu\nu}(k\omega)$ must have precisely the same symmetry and positivity properties as those of $\chi''_{\mu\nu}(k\omega)/\omega$ which we discussed in section 9.2. For example, $\gamma_{11}(k\omega)$ and $\gamma_{22}(k\omega)$ are real, and in fact positive, even functions of $\omega$. $\gamma_{12}(k,\omega) = \gamma_{21}(-k,-\omega)$ is an imaginary, odd function of $\omega$. Consequently, at $z = i0$, $\sigma_{11}(k,i0)$ and $\sigma_{22}(k,i0)$ are positive while $\sigma_{12}(k,i0) = -\sigma_{21}(k,i0)$ is imaginary. The former contribute to the damping of the spin waves, the latter result in a correction to their energy $\hbar\omega_k = \hbar ck$.

For hydrodynamics the important question concerns the dependence of $\sigma_{\mu\nu}(k,z)$ on $\vec{k}$. Noting that $A_2 \equiv M_y$ is conserved but that $A_1 \equiv N_x$ is not it is clear from (9.44) that to lowest order in k and z

$$\sigma_{11}(k,z) = \gamma \quad \text{and} \quad \sigma_{22}(k,z) = k^2\sigma_\perp \ , \tag{9.48}$$

where $\gamma$ and $\sigma_\perp$ are positive transport coefficients.

$\sigma_{12}(k,z)$ and $\sigma_{21}(k,z)$ must be explicitly of order k at least since, by eq. (9.44), they both involve one factor $\dot{M}_y = i\, k_i j_{yi}$. Note, however, that $\sigma_{12}(\vec{k},z)$ and $\sigma_{21}(\vec{k},z)$ must be even functions of $\vec{k}$ on account of parity or rotational invariance in real space, and thus both functions must in fact be of order $k^2$. If we therefore insert these results into (9.43) to obtain

$$\Sigma_{\mu\nu}(k,i0) = \begin{pmatrix} (R'/N_o^2)\gamma k^2 & k^2 \ldots \\ k^4 \ldots & (\sigma_\perp/\chi_\perp)k^2 \end{pmatrix}, \qquad (9.49)$$

it is clear that the off-diagonal terms are of higher than hydrodynamic order in k and can be omitted. It is also clear that, because of the explicit powers $k^2$ in $\Sigma$, we can replace $\Sigma(k,z)$ by $\Sigma(k,i0)$ to hydrodynamic accuracy.

The correlation functions are thus determined by the matrix equation (9.34), or

$$\begin{pmatrix} z + ik^2(R'/N_o^2)\gamma & -i\,N_o/\chi_\perp \\ ik^2R'/N_o & z+ik^2(\sigma_\perp/\chi_\perp) \end{pmatrix}_{\mu\lambda} C_{\lambda\nu}(kz) = i\beta^{-1}\chi_{\mu\nu}(k), \qquad (9.50)$$

from which we can calculate the correlation functions. They are determined by spin wave modes whose dispersion relation, including damping, follows easily from (9.50) in the form

$$z = \pm\, ck - \frac{i}{2}\,\Gamma\,k^2 \qquad (9.51)$$

to second order in k. The speed c is given in (9.40), and the attenuation constant $\Gamma$ is given by

$$\Gamma = (\sigma_\perp/\chi_\perp) + \gamma\,R'/N_o^2. \qquad (9.52)$$

(9.51) verifies the assumption, made in section 9.3, that for very large wavelength the spin wave damping is negligible. (9.52) separates the damping constant $\Gamma$ into two physically distinct contributions of which the first is due to the diffusion of spins in a system in which the relative orientation of nearest neighbors is held fixed, while the second describes the reverse process. This separation is useful if one wants to do a microscopic calculation of $\Gamma$. The reader might find amusement in drawing pictures of the two distinct configurations.

Solving (9.50) explicitly we obtain, for example,

$$C_{M_y M_y}(kz) = i\beta^{-1}\chi_\perp \frac{z + ik^2(\Gamma - \sigma_\perp \chi_\perp^{-1})}{z^2 - c^2k^2 + izk^2\Gamma} \tag{9.53}$$

and corresponding expressions for the other elements of $C_{\mu\nu}(kz)$. Taking the real part of (9.53) we obtain the absorptive spin correlation function in the form

$$\frac{1}{\omega}\chi''_{M_y M_y}(k\omega) = \chi_\perp \cdot \frac{c^2k^4\,\Gamma + \chi_\perp^{-1}\sigma_\perp k^2(\omega^2 - c^2k^2)}{[\omega^2 - c^2k^2]^2 + [\omega k^2\Gamma]^2} . \tag{9.54}$$

The reader should take note of the similarity of this result to the Brillouin spectrum (4.44a) for sound waves in normal liquids. There is, of course, no analogue for the diffusive Rayleigh component. The other two correlation functions are given by

$$\frac{1}{\omega}\chi''_{N_x N_x}(k\omega) = \frac{N_o^2}{R'k^2} \cdot \frac{c^2k^4\Gamma + \gamma(R'/N_o^2)k^2(\omega^2 - c^2k^2)}{[\omega^2 - c^2k^2]^2 + [\omega k^2\Gamma]^2}, \tag{9.55}$$

$$\frac{1}{\omega}\chi''_{N_x M_y}(k\omega) = \frac{i\omega k^2\Gamma N_o}{[\omega^2 - c^2k^2]^2 + [\omega k^2\Gamma]^2} . \tag{9.56}$$

These results are in agreement, as they must be, with the thermodynamic sum rules

$$\lim_{k\to 0} \int \frac{d\omega}{\pi}\frac{1}{\omega}\,\chi''_{M_y M_y}(k\omega) = \chi_\perp , \tag{9.57a}$$

$$\int \frac{d\omega}{\pi}\frac{1}{\omega}\,\chi''_{N_x N_x}(k\omega) = \frac{N_o^2}{R'k^2}\,(1 + O(k^2)). \tag{9.57b}$$

They also agree with the statement (9.14) of the broken rotational symmetry,

$$\int \frac{d\omega}{\pi} \chi''_{N_x M_y}(k\omega) = i N_o \ . \tag{9.58}$$

However, the sum rule in (9.16), which provides a bound for the stiffness constant R′, diverges if (9.54) is inserted; the hydrodynamic result is rigorously correct for small $\omega$, but it falls off too slowly in the wings. To complete the usual catalogue note that we can extract the two Kubo-like relations

$$\sigma_\perp = \lim_{\omega\to 0} \lim_{k\to 0} \frac{\omega}{k^2} \chi''_{M_y M_y}(k\omega), \tag{9.59}$$

$$\gamma = \lim_{\omega\to 0} \lim_{k\to 0} \omega\, \chi''_{N_x N_x}(k\omega), \tag{9.60}$$

where the absence of a factor $k^{-2}$ in (9.60) is again due to the fact that $N_x(rt)$ undergoes hydrodynamic fluctuations not because it is conserved, which it is not, but because it displays long-ranged static fluctuations.

## 9.5 Para-, Ferro-, and Antiferromagnets

It is illustrative to compare the transverse hydrodynamic fluctuations in the isotropic antiferromagnet with those found in the corresponding ferromagnet, and in the paramagnetic temperature range for either system. For convenience we reproduce the magnetic correlation functions which we derived in chapters 2 and 8, namely

$$\frac{1}{\omega} \chi''_{M_y M_y}(k\omega) = \chi \frac{k^2 D}{\omega^2 + (k^2 D)^2} \qquad \text{(paramagnet)}, \tag{9.61}$$

$$\frac{1}{\omega} \chi''_{M_y M_y}(k\omega) = \frac{M_o^2}{2Rk^2} \frac{k^4 \gamma_\perp (\omega^2 + r^2 k^4)}{[\omega^2 - r^2 k^4]^2 + [\omega k^4 \gamma_\perp]^2} \qquad \text{(ferromagnet)}, \tag{9.62}$$

where (9.62) holds for an ideal ferromagnet which is oriented in the z-direction.

It is the magnetization whose fluctuations are probed by neutrons, and in a neutron scattering experiment in which neutrons are inelastically scattered at very small angles, i.e., for small k, one would measure either one of the three functions given in eqs. (9.54), (9.61), and (9.62). We are here not concerned with the relative orders of magnitude, in different systems and at various temperatures, of the constants $\chi$, D etc. which occur in these equations. Much more effort would be required to calculate these. We are concerned, instead, with the structure of the spectra that one should expect to find. Their qualitative features are summarized in Fig. 9.2 which is, I hope, self-explanatory.

These results, eqs. (9.54), (9.61) and (9.62), are rigorously correct for sufficiently small k and $\omega$ if the assumptions can be believed on the basis of which they have been derived. There is no doubt, in particular, that our results are correct for the isotropic Heisenberg magnets which can be described by the Hamiltonian (9.1). Indeed, they hold more generally. They will hold, for example, for models of ferromagnetism in which the spin-carrying electrons are itinerant rather than localized. The really crucial assumption, on the basis of which our results will stand or fall, is the assumption of spin isotropy: our derivation will only make sense for systems whose energy is invariant under simultaneous rotation of all the spins.

In real systems, this assumption is often not warranted. The interaction between the spins and phonons or impurities may be invariant under spin rotations, but it often is not. If it is, our theory holds; in this case, the only effect of phonons or impurities is in the numerical values of the thermodynamic and transport coefficients. For the many systems, on the other hand, for which anisotropic contributions to the spin interaction energy are large, terms which violate spin conservation, very little can be said about the spectrum of spin fluctuations even at long wavelength, except by explicit microscopic calculation (Lubensky 1971).

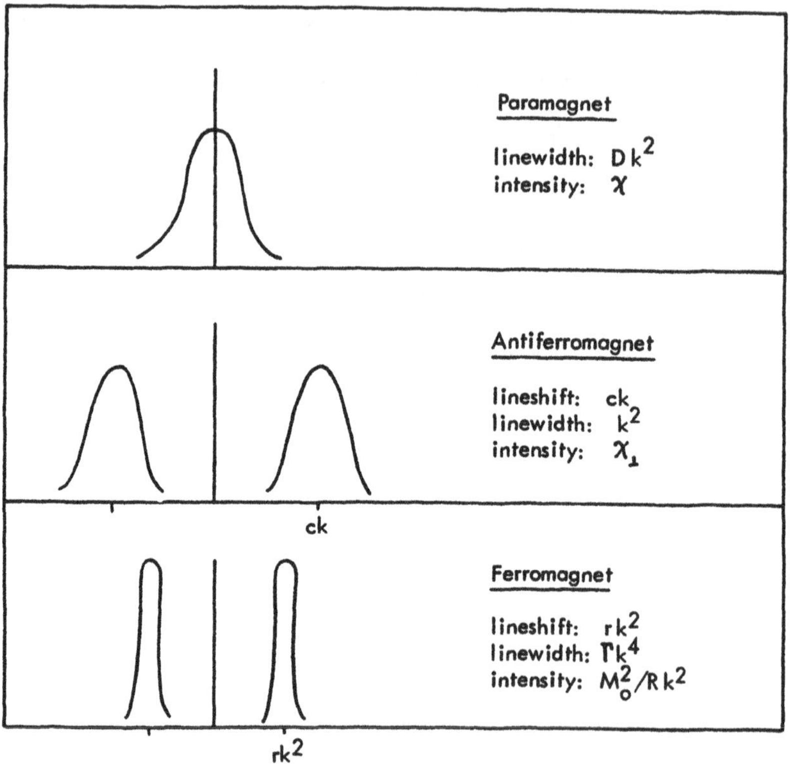


Fig. 9.2 A comparison of $\omega^{-1}\chi''_{M_yM_y}(k\omega)$ [vertical axis] versus $\omega$ [horizontal axis] for the isotropic paramagnet, antiferromagnet, and ferromagnet. The units are arbitrary. The wavevector k is fixed, and ka 1.

In order to further illustrate this point, let us consider the magnetic response function, $C_{MM}(kz)$, for a paramagnet in which there is strong coupling between the magnetic spins and the lattice, say, so that $M(r,t)$ is not conserved.

Let us set $k = 0$, $C_{MM}(0,z) \equiv C(z)$, and disregard all indices. Then eqs. (5.28) lead to the formula

$$C(z) = i\beta^{-1} \frac{\chi}{z + i\Sigma(z)} , \qquad (9.63)$$

where $\Sigma(z) \equiv \Sigma_{MM}(k=0,z)$ is finite, and analytic for $\mathrm{Im}\, z > 0$. We can find the resonances by setting

$$z_o + i\Sigma(z_o) = 0. \qquad (9.64)$$

If $\Sigma(z)$ varies slowly near $z_o = -i/T$, then we can replace (9.63) phenomenologically by

$$C(z) \approx i\beta^{-1} \frac{Z\chi}{z + i/T} , \qquad (9.65)$$

thus predicting a Lorentzian line of width $1/T$, and of strength

$$Z = [1 + i \frac{\partial\Sigma(z)}{\partial z}]^{-1}_{z=z_o} . \qquad (9.66)$$

This result is equivalent to a phenomenological Bloch equation of the form $\dot{M} = -M/T$. However, there is no reason in general for assuming that near $z_o$ $\Sigma(z)$ varies more slowly that the correlation function $C(z)$ itself. Thus, while for conserving magnets the results of fig. 9.2 are (asymptotically) rigorously Lorentzian, the Bloch equation for non-conserving systems is not. Indeed, in many cases the observed lineshape is more nearly Gaussian. Its precise form is interaction dependent and can only be obtained from a detailed microscopic theory. The reader will also note that for non-conserving systems the memory function formalism can be used, but that it is not nearly as informative as it is in isotropic magnets.

## CHAPTER 10

## SUPERFLUIDS

Liquid $Helium^4$ boils, under atmospheric pressure, at 4.2°K. Or, conversely, the gas liquefies at this temperature. The temperature can be further lowered by reducing the gas pressure above the liquid, i.e., by continuous pumping. In a demonstration experiment in which the Helium can be observed through a window in the Dewar one sees it bubbling violently as the temperature is lowered. Then, abruptly, there are no more bubbles. Below a temperature $T_\lambda = 2.18°K$ liquid $Helium^4$ looks clear and serene, despite continued pumping. At $T_\lambda$, Helium has become a superfluid.

As this experiment impressively demonstrates, Helium can exist in two physically distinct liquid phases, Helium I which is a normal liquid, and superfluid Helium II which has a number of spectacular properties. The phase diagram of $He^4$ is shown in fig. 10.1. The low-temperature phase is appropriately named a

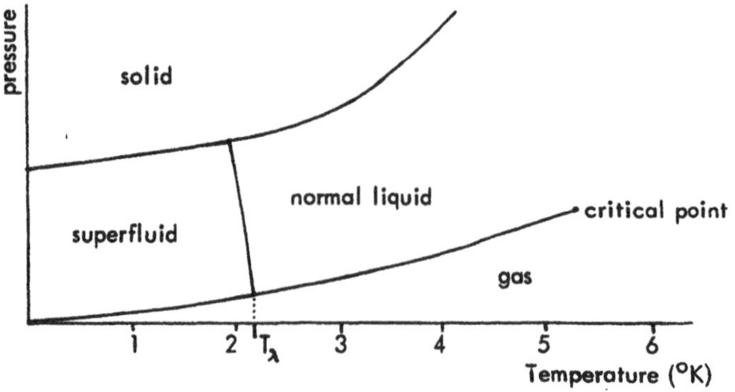


Fig. 10.1

"superfluid" because He II has been found (Kapitza 1938) to flow through narrow channels and cracks with no measureable viscosity.

The fundamental principles of superfluidity are described, e.g., in several excellent articles by Pines (1965) and Nozières (1966); see also Nozières and Pines (1964). In accord with the basic theme of this book I will discuss, in this chapter, the theory of hydrodynamic fluctuations in a superfluid. I will, in particular, give a derivation of the "two-fluid model" of superfluidity. This derivation, based on the memory function formalism of chapter 5, seems to be new in detail although in substance it is closely related to the discussion given by Hohenberg and Martin (1965). I will concentrate on those macroscopic features of superfluids which can be understood with little detailed calculation but which are in close formal analogy to our discussion of, say, ferromagnets. Only peripherally will I touch on more microscopic features of He II, such as the spectrum and nature of the elementary excitations. As a discussion of He II, this treatment is thus vastly inexhaustive. The curious reader is referred to the literature (e.g Wilks 1967).

## 10.1 Superflow and Long-Ranged Correlations

In order to apply the formalism of the last few chapters it would be most convenient to have an order parameter, say $\langle\psi\rangle$ , which vanishes in the normal liquid but is not zero in the superfluid phase as a result of some broken symmetry. We will introduce such a quantity below. However, the order parameter and the broken symmetry (see Anderson 1964a, 1966) appropriate for a superfluid are not as intuitive physically as the spontaneous magnetization and rotational symmetry are in a ferromagnet. We will therefore proceed first along a route which may be more tortuous but which is physically more transparent.

What do we mean by saying that a liquid is superfluid? That it does not bubble? Why not. But that is not an awfully useful characterization. (It is nonetheless interesting. Below $T_\lambda$ the bubbling stops because in He II temperature inhomogeneities can propagate in waves, namely as second sound. Second sound is a much more efficient way of spreading out local fluctuations of the temperature than the slow heat diffusion process which acts in normal liquids. Therefore, superfluid Helium does not overheat locally, and thus there are no bubbles; all evaporation takes place at the surface.) A more useful paradigmatic characterization of superfluidity is by what the name implies, the existence of persistent macroscopic flow — superflow (Pines 1965).

Consider, for example, a torus-shaped container filled with liquid $He^4$. Suppose that, initially, the container walls and the liquid co-rotate with angular velocity $\Omega$; then, at $t = 0$, the container is stopped. At first, even a normal liquid will continue to rotate, of course. However, because of the viscous interaction between liquid and walls, after a few minutes at most all (macroscopic) motion will cease. As $t \to \infty$, which means after a few minutes, the momentum density $\langle \vec{g} \rangle$ vanishes for a normal liquid. If the liquid is superfluid, on the other hand, all or at least a part of it will continue to rotate <u>indefinitely</u> (which means for days or indeed for years). Even as $t \to \infty$, the momentum density $\langle \vec{g} \rangle$ does not vanish, despite the proximity of stationary walls. The two-fluid model describes this peculiar behavior by saying that $He^4$ behaves as if it were a mixture of two liquids. One of these, the "normal component," has the normal viscous interaction with the container walls, and will therefore come to rest quickly; the normal part constitutes a temperature-dependent fraction $\rho_n/\rho$ of the liquid. The other, the "superfluid component," has no viscous interaction with either the walls or the normal part. This fraction $\rho_s/\rho = 1 - \rho_n/\rho$ of the liquid thus continues to move indefinitely. (This is, of course, only a description of what happens. To understand why Helium behaves this way one must consider, microscopically, all the processes by which the liquid might exchange momentum with the walls, and thus come to rest. A celebrated argument by Landau (see Khalatnikov, 1965) shows that the measured spectrum of elementary excitations in $He^4$ is such as to rule out such processes because they would violate energy and momentum conservation.)

In order to put this phenomenological characterization of superfluidity in mathematical terms it is a little easier to consider, instead of a torus, an infinitely long pipe filled with Helium. And instead of stopping the walls, let us start the system at complete rest, at $t = -\infty$, and then slowly accelerate the walls to a final velocity, at $t = 0$, of $\delta\vec{v}$. For a normal liquid, at $t = 0$ all of the liquid moves with velocity $\delta\vec{v}$ so that the momentum density $\delta\langle\vec{g}\rangle$ is given by

$$\delta\langle\vec{g}\rangle = \rho\,\delta\vec{v} \qquad \text{(normal liquid; } T > T_\lambda). \tag{10.1}$$

In superfluid Helium, however, the superfluid component remains behind. Only the normal component will be dragged along, and therefore the momentum density at $t = 0$ is given by

$$\delta\langle\vec{g}\rangle = \rho_n \cdot \delta\vec{v} \qquad \text{(superfluid; } T < T_\lambda), \tag{10.2}$$

where $\rho_n < \rho$.

We can compare this phenomenological statement with the results of linear response theory if we know the Hamiltonian that is appropriate in the presence of moving, rather than stationary, walls. Now clearly the motion of the walls adds to the Hamiltonian H of the liquid a term

$$\delta\mathcal{H} = -\vec{P}\cdot\delta\vec{v} = -\int d\vec{r}\;\vec{g}(r)\cdot\delta\vec{v} \tag{10.3}$$

so that, for $t < 0$, we can write the total Hamiltonian in the form

$$\mathcal{H}(t) = H - \int d\vec{r}\, \vec{g}(r) \cdot \delta\vec{v}(r) \cdot e^{\epsilon t} , \qquad (10.4)$$

which allows for a spatially varying impressed velocity $\delta\vec{v}(r)$ which is slowly ($\epsilon \to 0$) increased from zero. Linear response theory gives us immediately the momentum density which results at time t, namely

$$\delta\langle g_i(r,t)\rangle = \int_{-\infty}^{t} dt' \int d\vec{r}'\, 2i\, \chi''_{ij}(rt, r't')\, \delta v_j(r')\, e^{\epsilon t'} \qquad (10.5)$$

according to eq. (3.3), where

$$\chi''_{ij}(rt, r't') = \langle \frac{1}{2\hbar} [g_i(rt), g_j(r't')] \rangle \qquad (10.6)$$

is the momentum density correlation function, averaged in the equilibrium ensemble <u>at rest.</u> If the Helium system is sufficiently large, this function can be represented by

$$\chi''_{ij}(rt, r't') = \int \frac{d\omega}{2\pi}\, e^{-i\omega(t-t')} \int \frac{d^3k}{(2\pi)^3}\, e^{i\vec{k}\cdot(\vec{r}-\vec{r}')}\, \chi''_{ij}(\vec{k},\omega) . \qquad (10.7)$$

Therefore the induced momentum density at time $t = 0$ and, say, at $\vec{r} = 0$ can be written as

$$\delta\langle g_i\rangle = \int d^3r' \int \frac{d^3k}{(2\pi)^3}\, e^{-i\vec{k}\cdot\vec{r}'}\, \chi_{ij}(\vec{k}) \cdot \delta v_j , \qquad (10.8)$$

where

$$\chi_{ij}(\vec{k}) = \int \frac{d\omega}{\pi}\, \chi''_{ij}(\vec{k}\,\omega)/\omega \qquad (10.9)$$

as usual, and we have taken $\delta\vec{v}$ to be constant. The translationally invariant representation (10.7) can be used only for points $\vec{r}$ and $\vec{r}'$ deep in the bulk of the liquid, of course. This is fine since we are only interested in the bulk properties of Helium. However, when evaluating (10.8) it pays to be careful; we do want to retain enough information about the presence of walls to characterize an open pipe, as compared to a closed one, say.

Let us therefore consider a well-defined geometry (Baym 196 ). For simplicity, assume that the liquid is enclosed in a finite rectangular box, with dimensions $L_x$, $L_y$ and $L_z$. The spatial integrals in (10.8) extend over the volume occupied by the liquid, and using

$$\int_{-L_x/2}^{L_x/2} dx\, e^{-ik_x x} = \frac{\sin(k_x L_x/2)}{k_x/2} \equiv 2\pi\Delta(k_x) , \tag{10.10}$$

we obtain

$$\delta\langle g_x\rangle = \int d^3k\ \Delta(k_x)\,\Delta(k_y)\,\Delta(k_z)\ \chi_{xx}(\vec{k})\cdot\delta v_x , \tag{10.11}$$

where we have assumed that the velocity $\delta\vec{v}$ is in the x-direction.

It is a useful check on our procedure to first inquire what (10.11) predicts for a closed pipe. If we let the transverse dimensions $L_y$ and $L_z$ go to infinity first, and only then $L_x \to \infty$, we are describing a closed system in which the walls at $\pm L_x/2$ bodily push all of the mass along. Superfluid or not, in this case we expect that $\delta g_x = \rho\,\delta v_x$. Noting that

$$\Delta(k_x) \to \delta(k_x) \text{ as } L_x \to \infty, \tag{10.12}$$

we obtain

$$\rho = \lim_{k_x\to 0}\ [\lim_{\substack{k_y\to 0\\ k_z\to 0}} \chi_{xx}(\vec{k})] . \tag{10.13}$$

Eq. (10.13) is in fact a restatement of the f-sum rule (4.27d). It verifies that (10.3) is a proper Hamiltonian for the present purpose. The corresponding experiment with a torus would be one in which a rigid separation wall has been inserted into the ring, see fig. 10.2.

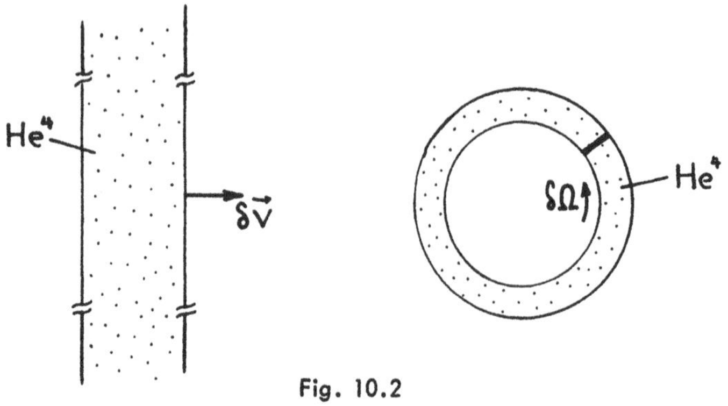


Fig. 10.2

In this case, if the walls of the torus are brought to the angular velocity $\delta\Omega = \delta v/R$, we would predict from (10.8) and (10.13) that the liquid acquires the angular momentum $\delta L = I\,\delta\Omega$ where $I = \rho R^2 V$ is the classical moment of inertia.

An open pipe is obtained if we let the longitudinal dimension, $L_x$, go to infinity first. In this case the interaction which drags the liquid along is the viscous one at the transverse walls, and by (10.2) we must have $\delta g_x = \rho_n \delta v_x$. Equation (10.11) thus requires that for a superfluid

$$\rho_n = \lim_{\substack{k_y \to 0 \\ k_z \to 0}} \left[ \lim_{k_x \to 0} \chi_{xx}(\vec{k}) \right]. \tag{10.14}$$

The corresponding experiment is with an open torus, fig. 10.3. Because of (10.14), its moment of inertia is given by $I_n = \rho_n R^2 V$, and it is smaller than the classical one: $\rho_n < \rho$.

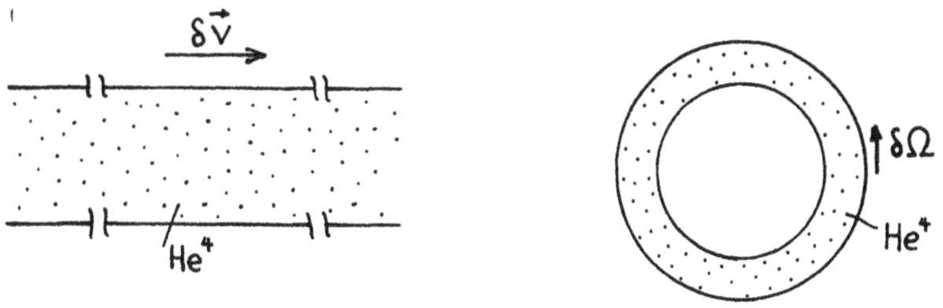


Fig. 10.3

What we have given here is not, of course, a theory of superfluidity — it is simply a description of it. Since superfluid $He^4$ at rest is isotropic, $\chi''_{ij}$ is of the form

$$\chi''_{ij}(\vec{k}\omega) = \hat{k}_i \hat{k}_j \chi''_\ell(k\omega) + (\delta_{ij} - \hat{k}_i\hat{k}_j)\, \chi''_t(k\omega), \qquad (10.15)$$

where $\hat{k} = \vec{k}/k$; thus also

$$\chi_{ij}(\vec{k}) = \hat{k}_i \hat{k}_j \chi_\ell(k) + (\delta_{ij} - \hat{k}_i\hat{k}_j)\, \chi_t(k) . \qquad (10.16)$$

The f-sum rule

$$\chi_\ell(k) = \int \frac{d\omega}{\pi} \chi''_\ell(k\omega)/\omega = \rho \qquad (10.17)$$

is equivalent to (10.13); it is a direct consequence of "gauge invariance" (i.e., number conservation plus the commutator (4.34)) and thus holds for any fluid. A system is superfluid if

$$\lim_{k\to 0} \int \frac{d\omega}{\pi} \chi''_t(k\omega)/\omega = \rho_n < \rho . \qquad (10.18)$$

If we define the superfluid density $\rho_s$ by

$$\rho_s = \rho - \rho_n , \qquad (10.19)$$

we can write the last three equations in the form

$$\lim_{k\to 0} \chi_{ij}(\vec{k}) = \rho_n \delta_{ij} + \rho_s \hat{k}_i \hat{k}_j \quad . \tag{10.20}$$

This is a rather curious result. $\chi_{ij}(\vec{k})$ is, of course, the Fourier transform of the static, spatial momentum density correlation function,

$$\chi_{ij}(\vec{k}) = \int d\vec{r}\, e^{-i\vec{k}\cdot\vec{r}} \chi_{g_i g_j}(\vec{r}) \tag{10.21}$$

So long as $\chi_{g_i g_j}(\vec{r})$ is of finite range, the limit $\lim_{k\to 0} \chi_{ij}(k)$ must equal $\chi_{ij}(k = 0)$, and thus cannot depend on the direction in $\vec{k}$-space from which $\vec{k} = 0$ is approached. Since in a superfluid we have found, by contrast, that the limit is not isotropic, we must conclude that

$$\int d\vec{r}\, \chi_{g_i g_j}(\vec{r}) = \infty \quad \text{in a superfluid.} \tag{10.22}$$

Thus we arrive at the most fundamental characterization of a superfluid: it is a system in which momentum correlations are of infinite range (P. C. Martin, quoted in Pines 1965).

How long is the infinitely long range? Well, simply do the Fourier transform of (10.20); for the non-mathematician, that is easy. Namely,

$$\frac{k_i k_j}{k^2} \to \frac{1}{4\pi} \nabla_i \nabla_j \frac{1}{r} = \frac{-1}{4\pi r^3} [\delta_{ij} - 3 r_i r_j / r^2] .$$

Thus for large $|\vec{r} - \vec{r}\,'|$, we have

$$\chi_{g_i g_j}(r - r') \approx \frac{-\rho_s}{4\pi} \frac{1}{|\vec{r} - \vec{r}\,'|^3} , \tag{10.23}$$

which agrees with (10.22).

For a classical system we have calculated the momentum density correlation function explicitly. This is possible since the momenta of two classical particles are always uncorrelated. From the result (4.37), or

$$\chi_{g_i g_j}(r-r') = \rho\,\delta(\vec{r}-\vec{r}\,')\,\delta_{ij} \qquad \text{(classically)}, \tag{10.24}$$

we see that a classical system can never be superfluid. The phenomenon of superfluidity is a quantum effect.

## 10.2 Broken Gauge Symmetry

We have seen that the persistent macroscopic flow which characterizes superfluidity is an aspect of long-ranged momentum correlations. What generates this peculiar order? In our discussion of magnetic systems we found that if a continuous symmetry of the Hamiltonian is broken, long-ranged order results. Can the superfluid phase of Helium be described in terms of a broken symmetry? The answer is yes. However, the symmetry which is broken is not one of the conventional ones, translations or rotations under which superfluid Helium is certainly invariant. Rather, the broken symmetry is gauge invariance, its generator the particle number operator N. In this section we will give a heuristic derivation of this connection.

In section 5.1 we pointed out that a static susceptibility like $\chi_{ij}(k)$ has the properties of a scalar product in a Hilbert space of operators. Suppose we choose a basis set of local operators $A_n(k)$, suitably orthogonalized, and expand $\chi_{ij}(k)$ in terms of these:

$$\langle g_i(k) | g_j(k) \rangle = \sum_n \langle g_i(k) | A_n(k) \rangle \langle A_n(k) | A_n(k) \rangle^{-1} \langle A_n(k) | g_j(k) \rangle \tag{10.25}$$

or equivalently

$$\chi_{g_i g_j}(k) = \Sigma\, \chi_{g_i A_n}(k)\, \chi^{-1}_{A_n A_n}(k)\, \chi_{A_n g_j}(k), \tag{10.26}$$

for either the longitudinal or the transverse susceptibility. The individual terms in this sum give the contribution of the basis operator $A_n$ to momentum correlations. Most of these will only generate local correlations, and thus contribute equally to $\chi_\ell$ and to $\chi_t$ at least as $k \to 0$. But there must be at least one operator, say $A_o$, which correlates more strongly with $g_\ell$ than with $g_t$. This variable $A_o(r)$ must therefore be more longitudinal than transverse, and with no loss of generality it can be taken as a purely longitudinal vector. Let us denote $A_o(r)$ by $g_s(r)$ so that (10.26) reads

$$\chi_\ell(k) = \chi_{g_\ell g_s}(k)\, \chi^{-1}_{g_s g_s}(k)\, \chi_{g_s g_\ell}(k) + \sum_{n \neq 0} \ldots ,$$

$$\chi_t(k) = 0 + \sum_{n \neq 0} \ldots , \tag{10.27}$$

where the rest sum $\sum_{n\neq 0} \ldots$ is that part which contributes equally to $\chi_\ell$ and $\chi_t$ as $k \to 0$. Then the longitudinal excess is given by

$$\rho_s = \lim_{k \to 0} \chi_{g_\ell g_s}(k)\, \chi^{-1}_{g_s g_s}(k)\, \chi_{g_s g_\ell}(k) . \tag{10.28}$$

We have a normalization of $g_s$ at our avail. We shall choose it so that

$$\lim_{k\to 0} \chi_{g_s g_s}(k) = \rho_s , \tag{10.29a}$$

which means also that

$$\lim_{k\to 0} \chi_{g_\ell g_s}(k) = \rho_s . \tag{10.29b}$$

If there are supercurrents, there must be a local observable $g_s$ with these properties.

Let us reformulate once more. Since $\vec{g}_s$ is a purely longitudinal vector it can be derived from a scalar potential (operator) $\varphi^{op}(\vec{r})$ by

$$\vec{g}_s(r) = \rho_s \cdot \frac{\hbar}{m} \vec{\nabla} \varphi^{op}(r), \tag{10.30}$$

where the constants are put there to make $\varphi^{op}(r)$ dimensionless, and m is the mass of a Helium atom. In terms of $\varphi$, eq. (10.29a) now reads

$$\chi_{\varphi\varphi}(k) = \int \frac{d\omega}{\pi} \omega^{-1} \chi''_{\varphi\varphi}(k\omega) = \frac{(m/\hbar)^2}{\rho_s k^2} \tag{10.31}$$

as $k \to 0$. We are, of course, still assuming that $\varphi^{op}(r)$, whatever it is, is a local operator, and we find that its correlations are on infinite range. Eq. (10.31) will be a starting point when we come to analyze the hydrodynamic properties of superfluid Helium. More interesting, for the moment, is eq. (10.29b). Invoking particle conservation and eq. (10.30) we have

$$\chi''_{g_\ell g_s}(k\omega) = -\frac{i\hbar}{m} \rho_s k \chi''_{g_\ell \varphi}(k\omega) = -i\hbar\omega \rho_s \chi''_{n\varphi}(k\omega)$$

and therefore

$$\lim_{k\to 0} \int \frac{d\omega}{\pi} \chi''_{n\varphi}(k\omega) = i/\hbar. \tag{10.32}$$

The hydrodynamic properties of superfluid $He^4$, i.e., the two fluid model, can be derived from the two equations (10.31) and (10.32) and standard assumptions. Of course, such a formal derivation would not help one's physical intuition very much. We do, for example, still have to identify what $\varphi^{op}(r)$ really means. But first of all, let us notice that (10.32) is identical to the statement

$$\lim_{k\to 0} \int d\vec{r}\, e^{-i\vec{k}\cdot\vec{r}} \langle [n(r), \varphi^{op}(r')] \rangle = i \ . \tag{10.33}$$

This statement presents us with a fascinating choice. $\int d^3r\, n(r) = N$ is of course the (operator for the) total number of particles. If the limit as $k \to 0$ could be freely taken, we would have

$$i = \mathrm{Tr}\, \rho[N, \varphi^{op}] = \mathrm{Tr}[\rho, N]\varphi^{op} \qquad (10.34)$$

because of the cyclic invariance of the trace. But the "usual" ensembles of equilibrium statistical mechanics, such as the canonical or grand canonical ensembles, are described by density matrices $\rho$ which commute with the number operator N. Thus either we use such a canonical ensemble; then we have to cope with the very inconvenient fact that (10.33) holds but that

$$\int d\vec{r}\, \langle [n(r), \varphi^{op}(r')] \rangle_{can.} = 0;$$

in such an ensemble the $k \to 0$ limit could not be uniform. Or, we insist that an ensemble be chosen in which the limit is uniform. Then we arrive at the conclusion that

$$[\rho_\eta, N] \neq 0 \qquad (10.35)$$

for that ensemble (which we shall call the "$\eta$-ensemble"). We shall choose the second, much more convenient, route. We shall describe the superfluid phase of Helium by an ensemble operator $\rho_\eta$ which breaks that symmetry of the Hamiltonian whose generator is N. This symmetry is called gauge symmetry.

## 10.3 Bose Condensation and the Order Parameter

When we discussed the ordered state of a ferromagnet we found it very convenient that this state could be characterized by the average of the magnetization operator, $\langle \vec{M}(r) \rangle$, which vanishes in the paramagnetic phase as a consequence of rotational invariance, but is different from zero in the ferromagnetic state in which rotational symmetry is broken. Can we describe the superfluid state of $He^4$ in an analogous fashion? Again, the answer is yes, and we will now jump right into it. The order parameter operator in question is $\Psi(r)$, the particle destruction operator, or its Hermitian adjoint, the creation operator $\Psi^+(r)$. The superfluid phase of $He^4$ can be characterized by the fact that

$$\langle \Psi(r) \rangle = \sqrt{n_o(r)}\; e^{+i\phi(r)} \neq 0,$$

$$\langle \Psi^+(r) \rangle = \sqrt{n_o(r)}\; e^{-i\phi(r)} \neq 0, \qquad (10.36)$$

which can be taken as the defining property of a superfluid in the same sense in which ferromagnetism was described by $\langle \vec{M}(r) \rangle \neq 0$.

Equations (10.36) need some explanation. The operators $\Psi(r)$ and $\Psi^+(r)$ are the field operators of second quantization. (The reader who is unfamiliar with this language should not despair; we will need very little of it. Any introductory book on quantum field theory will provide more than enough if the next few pages should still leave him or her unsatisfied. For instance, see Fetter and Walecka 1971). $\Psi(r)$ removes a particle at point $\vec{r}$, $\Psi^+(r)$ puts one there. For $He^4$, the particles in question are, of course, Helium atoms. $\Psi$ and $\Psi^+$ are operators in a Hilbert space, and they obey the commutation relations

$$[\Psi(\vec{r}),\ \Psi^+(\vec{r}\,')] = \delta(\vec{r} - \vec{r}\,'),$$

$$[\Psi(\vec{r}),\ \Psi(\vec{r}\,')] = [\Psi^+(\vec{r}),\ \Psi^+(\vec{r}\,')] = 0, \qquad (10.37)$$

which characterize a system of Bosons. The most important operators, namely the particle density n(r), momentum density $\vec{g}(r)$, and energy density $\varepsilon(r)$ can be written in the form

$$n(r) = \Psi^+(r)\,\Psi(r),$$
$$\vec{g}(r) = \frac{\hbar}{2i}[\Psi^+(r)\,\vec{\nabla}\Psi(r) - (\vec{\nabla}\Psi^+(r))\,\Psi(r)]\,,$$
$$\epsilon(r) = \frac{\hbar^2}{2m}\,\vec{\nabla}\Psi^+(r)\cdot\vec{\nabla}\Psi(r) + \frac{1}{2}\int d\vec{r}'\,\Psi^+(r)\Psi^+(r')\,v(r-r')\Psi(r')\Psi(r),$$
$$(10.38)$$

which are identical to eqs. (4.6) but expressed in a new language. The total number operator is, of course, given by

$$N = \int d\vec{r}\,n(r) = \int d\vec{r}\,\Psi^+(r)\Psi(r), \qquad (10.39)$$

and the total momentum $\vec{P}$ and Hamiltonian H are similarly the integrals of $\vec{g}$ and of $\epsilon$. The reader who is unfamiliar with the $\Psi$, $\Psi^+$ dialect may exercise a little by showing that indeed

$$[n(r), \vec{g}(r')] = -i\hbar\vec{\nabla}\,n(r)\,\delta(\vec{r}-\vec{r}') \qquad (10.40a)$$

and

$$\partial_t n(r,t) = (i/\hbar)\,[H, n(r,t)] = -\vec{\nabla}\cdot\vec{g}(r,t)/m, \qquad (10.40b)$$

as we had shown in eqs. (4.70) and (4.34).

Equations (10.36) are a mathematical statement of Bose condensation. The superfluid phase is described by the existence of a macroscopic wave function; $\langle\Psi(r)\rangle \equiv \underline{\overline{\Psi}}(r)$ is that wave function. The square of its amplitude, $n_o(r)$, is the density of particles in the "condensate". $\phi(r)$ is the macroscopic phase; its gradient, $\vec{v}_s = (\hbar/m)\vec{\nabla}\phi$, is the superfluid velocity. (Words can be helpful.) How can (10.36), $\langle\Psi\rangle \neq 0$, come about?

Let us assume that the Helium atoms sit in a box, with the usual periodic boundary conditions. We can then write

$$\Psi(r) = V^{-1/2} \sum_k e^{i\vec{k}\cdot\vec{r}} a_k,$$

$$\Psi^+(r) = V^{-1/2} \sum_k e^{-i\vec{k}\cdot\vec{r}} a_k^+ . \tag{10.41}$$

Since the total number and momentum operators are

$$N = \sum_k a_k^+ a_k ,$$

$$\vec{P} = \sum_k \hbar\vec{k}\, a_k^+ a_k , \tag{10.42}$$

it is clear that $a_k^+$ creates a particle with momentum $\hbar\vec{k}$. Now consider the state

$$|N_o\rangle = (N_o!)^{-1/2} (a_o^+)^{N_o} |0\rangle, \tag{10.43}$$

where $|0\rangle$ is the vacuum state so that $a_k|0\rangle = 0$. This state contains $N_o$ particles, all with zero momentum. First, you might check explicitly that

$$\langle N_o | \Psi(r) | N_o \rangle = 0. \tag{10.44}$$

However, the state

$$|\tilde{N}_o\rangle = (2m+1)^{-1/2} \sum_{\nu=-m}^{m} |N_o + \nu\rangle, \tag{10.45}$$

which for $1 \ll m \ll N_o$ has just a few particles more or less, and cannot be very different, physically, from $|N_o\rangle$, gives

$$\langle \tilde{N}_o | \Psi(r) | \tilde{N}_o \rangle = (N_o/V)^{1/2} = \sqrt{n_o}. \tag{10.46}$$

This result obtains because the states $|N_o + \nu\rangle$ in (10.45) are superimposed with coherent relative phases. If we had to deal with free Bosons, $|N_o\rangle$ would be the ground state since all particles would settle into the zero momentum state whose energy is the smallest possible. $N_o$ would be the total number of particles, so that $n_o \sim 10^{23}\ \text{cm}^{-3}$. If the particle number were to fluctuate a little, $|\tilde{N}_o\rangle$ would be just right. In $He^4$, where the atoms interact quite strongly, roughly that seems to happen. At zero temperature the ground state contains a macroscopic fraction of particles with zero momentum. However, not all particles are condensed into $|N_o\rangle$; indeed, for Helium $N_o$ is only a small fraction, maybe 8% (Penrose and Onsager 1956, Kerr et al. 1970), of all atoms at T = 0.

Now we are concerned with finite temperatures $0 < T < T_\lambda$ where $\langle\ \rangle$ is an average over some statistical ensemble. Which ensemble? First, there is an interesting item to notice. If you use the commutation relations (10.37), you will easily see that

$$[N, \Psi(r)] = -\Psi(r),$$
$$[N, \Psi^+(r)] = \Psi^+(r), \tag{10.47}$$

from which it follows that

$$e^{iaN}\Psi(r)\,e^{-iaN} = \Psi(r)\,e^{-ia},$$
$$e^{iaN}\Psi^+(r)\,e^{-iaN} = \Psi^+(r)\,e^{ia}. \tag{10.48}$$

The transformation $e^{iaN}$ which adds a phase to the field operators is called a gauge transformation. It is, in fact, closely related to the familiar gauge transformations of electrodynamics. Of course,

$$[N,H] = 0, \tag{10.49}$$

the Hamiltonian is invariant under the gauge transformation, as are all physical operators since they are built up from an equal number of $\Psi$'s and $\Psi^+$'s. What about the state of the system? Well, from (10.36) and (10.47) it is clear that

$$\langle\Psi(r)\rangle \equiv \mathrm{tr}\,\rho\,\Psi(r) = -\mathrm{tr}\,\rho[N,\Psi(r)]$$
$$= \mathrm{tr}[\rho,N]\,\Psi(r) \neq 0 \tag{10.50a}$$

only if

$$[\rho, N] \neq 0\,. \tag{10.50b}$$

If the condensate density $n_o$ is finite, the state of the system is not gauge invariant. In a Bose-condensed system, gauge symmetry is broken.

That, of course, means that the usual canonical ($\rho \sim e^{-\beta(H-\mu N)}$) ensembles cannot be used. These ensembles average over all phases of the order parameter $\langle \Psi \rangle$ while we have to select one particular, though arbitrary, phase. We have discussed an entirely analogous situation in section 7.2. Continuing the formal analogy, we would expect to have better luck with an ensemble in which, cf. eq. (7.31), an "external source field $\eta(r)$" gives a particular phase of $\Psi(r)$ slightly more weight. We will thus use the following "$\eta$-ensemble" (Hohenberg and Martin 1965):

$$\rho_\eta \sim \exp -\beta[H-\mu N - \int d\vec{r}\, \{\eta^*(r)\Psi(r) + \eta(r)\, \Psi^+(r)\}]\,. \tag{10.51}$$

where $\eta(r)$ is a complex function. Since N does not commute with $\Psi(r)$, it will not commute with $\rho_\eta$. Indeed, for the homogeneous superfluid at rest, with no vortices, a constant, real $\eta$ is sufficient. As in eqs. (7.32) and (7.33) the assertion behind (10.51) is that

$$\lim_{\eta\to 0}\ \mathrm{tr}\, \rho_\eta\, \Psi(r) = \sqrt{n_o}\ e^{i\phi} \neq 0 \tag{10.52}$$

even though

$$\mathrm{tr}\, \rho_{\eta=0}\, \Psi(r) = 0. \tag{10.53}$$

A microscopic theory of superfluidity in $He^4$ would have to prove this assertion, obviously. At least for weakly interacting Bosons that is not very difficult (Bogoliubov 1960) although for real Helium, whose atoms interact fairly strongly, there has not been so much progress. Here we will take the same point of view we adopted in magnetic systems: we will take it for granted that below $T_\lambda$ Helium is Bose-condensed, $\langle\Psi\rangle \neq 0$, and we will draw conclusions from that assertion.

## 10.4 Long-Ranged Phase Coherence

We are now ready to make contact with our introductory discussion in sections 10.1 and 10.2. Let us consider a stationary superfluid in which $\langle\Psi(r)\rangle$ is spatially constant. This is the case in a bucket of $He^4$ that has been left in peace for a while. The two equations

$$\langle[N,\Psi(r)]\rangle = -\langle\Psi(r)\rangle = -\sqrt{n_o}\, e^{i\phi} ,$$

$$\langle[N,\Psi^+(r)]\rangle = \langle\Psi^+(r)\rangle = \sqrt{n_o}\, e^{-i\phi} \qquad (10.54)$$

entail long-ranged order, as generally shown in section 7.4. However, the two equations (10.54) contain the same information since $\Psi$ and $\Psi^+$ are Hermitian adjoint. Let us therefore reshuffle the two operators, and define two Hermitian ones:

$$\varphi^{op}(r) = \frac{1}{2i\langle\Psi^+\rangle\langle\Psi\rangle}\left[\langle\Psi^+(r)\rangle\,\Psi(r) - \langle\Psi(r)\rangle\,\Psi^+(r)\right] , \qquad (10.55a)$$

$$n^{op}(r) = \langle\Psi^+(r)\rangle\,\Psi(r) + \langle\Psi(r)\rangle\,\Psi^+(r) . \qquad (10.55b)$$

If it were possible to write the operator $\Psi(r)$ in terms of an amplitude and a phase, both Hermitian operators, as

$$\Psi(r) = \sqrt{n_o + \delta n^{op}}\; e^{i[\phi + \delta\varphi^{op}]} , \qquad (10.56)$$

then for small fluctuations, these two operators would be formally given by (10.55). Actually, the decomposition (10.56) is problematic but the operators (10.55) are well-defined local operators at least so long as $n_o \neq 0$, i.e., so long as the superfluid is superfluid. From (10.54) we then obtain

$$\langle [N, n^{op}(r)] \rangle = -2i\, n_o \langle \varphi^{op}(r) \rangle = 0, \tag{10.57}$$

$$\langle [N, \varphi^{op}(r)] \rangle = \frac{i}{2n_o} \langle n^{op}(r) \rangle = i \ . \tag{10.58}$$

Now through all of this the reader has, hopefully, kept the lessons of chapter 7 in mind. It is immediately apparent that there is nothing special about $n^{op}(r)$. Because $\langle \varphi^{op} \rangle = 0$ there is no reason to expect that $n^{op}(r)$ has correlations of unusual range — the amplitude of the order parameter is a function of the temperature as $M_o$ is in ferromagnets; changing it would require finite forces. Thus $n^{op}(r)$ is not a hydrodynamic variable, it relaxes to equilibrium within a few collision times.

Things are different, however, with the phase operator $\varphi^{op}(r)$. The Bogoliubov inequality (7.60), applied to (10.58), requires that

$$\int \frac{d\omega}{\pi} \frac{\chi''_{\varphi\varphi}(k\omega)}{\omega} = \chi_{\varphi\varphi}(k) \geqq \frac{\left| \int \frac{d\omega}{\pi} \chi''_{n\varphi}(k\omega) \right|^2}{\int \frac{d\omega}{\pi}\, \omega\, \chi''_{nn}(k\omega)} \ . \tag{10.59}$$

Indeed, we can directly compute the commutator

$$\int \frac{d\omega}{\pi} \chi''_{n\varphi}(k\omega) = \hbar^{-1} \int d(r-r')\, e^{-i\vec{k}\cdot(\vec{r}-\vec{r}')} \langle [n(r), \varphi^{op}(r')] \rangle$$

$$= i/\hbar \tag{10.60}$$

even for all k, not just at k=0 as (10.58) would indicate. And since the denominator in (10.59) is of order $k^2$, $\chi_{\varphi\varphi}(k)$ must diverge at least as $1/k^2$. If it diverges no worse, we can write

$$\chi_{\varphi\varphi}(k) = \frac{(m/\hbar)^2}{\rho_s k^2} \tag{10.61}$$

and obtain the bound

$$\rho_s \leqq \rho = (m^2/k^2) \int \frac{d\omega}{\pi} \, \omega \, \chi''_{nn}(k\omega) \tag{10.62}$$

by the f-sum rule.

Now isn't that beautiful? From the single assumption of Bose condensation, eq. (10.54), we have been led to the eqs. (10.60) and (10.61). They are identical to the eqs. (10.31) and (10.32) which our earlier and more heuristic considerations had yielded. But now, in addition, we know what the operator $\varphi^{op}(r)$ is. Eq. (10.61), which in coordinate space says that

$$\langle \varphi^{op}(r)\, \varphi^{op}(r') \rangle - \langle \varphi^{op}(r) \rangle \langle \varphi^{op}(r') \rangle \sim \frac{1}{\rho_s |\vec{r}-\vec{r}'|} \tag{10.63}$$

as $|\vec{r}-\vec{r}'| \to \infty$ , indicates that superfluidity is a consequence of Bose condensation, the existence of a macroscopic wave function $\langle \Psi \rangle$ , and long-ranged coherence of its phase. On this basis, it is obvious now how to return to our original, phenomenological, statement of superfluidity, namely to eq. (10.20). Using particle conservation, we find from eq. (10.60) that

$$\chi_{g_i\varphi}(k) = \int \frac{d\omega}{\pi} \, \omega^{-1} \, \chi''_{g_i\varphi}(k\omega) = \frac{im}{\hbar k^2} \, k_i \ . \tag{10.64}$$

Therefore the phase fluctuations contribute by

$$\chi_{g_i\varphi}(k) \, \chi^{-1}_{\varphi\varphi}(k) \, \chi_{\varphi g_j}(k) = \rho_s \, \hat{k}_i \hat{k}_j \tag{10.65}$$

to the momentum density correlations. Unless there is a second and independent variable with long-ranged correlations, which we believe there is not, the remainder,

$$\tilde{\chi}_{g_i g_j}(k) = \chi_{g_i g_j}(k) - \chi_{g_i \varphi}(k)\, \chi^{-1}_{\varphi\varphi}(k)\, \chi_{\varphi g_j}(k), \qquad (10.66a)$$

is of short range, see eq. (7.100), and as $k \to 0$ it is therefore given by

$$\lim_{k\to 0} \tilde{\chi}_{g_i g_j}(k) = \rho_n \delta_{ij}, \qquad (10.66b)$$

where $\rho_n = \rho - \rho_s$. ( Are the two definitions (10.18) and (10.61) rigorously equivalent? For some discussion of this question see p. 254 below.)

## 10.5 Hydrodynamics without Dissipation

Let us now discuss hydrodynamic fluctuations in superfluid $He^4$. We will do so, once again, in terms of the general memory function formalism of chapters 5 and 7. The results which we will derive are not new, of course. They have been first proposed by Tisza (1938), and are discussed in terms of correlation functions by Hohenberg and Martin (1965).

Apart from the conserved densities $\rho$, $\vec{g}$, and $\varepsilon$ there is one additional hydrodynamic variable in superfluid Helium, namely the quantum phase $\varphi^{op}$. The latter is, of course, a scalar operator, and in the isotropic liquid it cannot couple to the transverse components of the momentum density. This being so, the transverse components of $\vec{g}$ will undergo the same diffusive motion as in a normal liquid. As for the longitudinal fluctuations it is convenient, and conventional, to introduce the superfluid velocity operator,

$$\vec{v}_s^{\,op}(rt) \equiv \frac{\hbar}{m} \vec{\nabla} \varphi^{op}(rt), \qquad (10.67)$$

which is a purely longitudinal vector. As we did in chapter 4 when discussing normal liquids, we shall also replace the energy density $\varepsilon$ by the variable

$$q(rt) = \varepsilon(rt) - \frac{\varepsilon + p}{\rho} \rho(rt), \qquad (10.68)$$

which represents an entropy density. The notation follows that of chapter 4 in detail except that, in keeping with tradition, we shall discuss the superfluid in terms of the mass density $\rho(rt) = mn(rt)$ instead of the particle density $n(rt)$. Thus we have the 4 longitudinal hydrodynamic variables

$$\{A_\mu\} \equiv \{\rho, q, g_\ell; v_s\} \ . \tag{10.69}$$

All we have to do is to apply the general recipes of eqs. (5.28) to their matrix of Kubo correlation functions

$$C_{\mu\nu}(kz) = \beta^{-1} \int \frac{d\omega}{\pi i} \chi''_{\mu\nu}(k\omega)/\omega(\omega-z). \tag{10.70}$$

We begin with the matrix $\chi_{\mu\nu}(k)$ of susceptibilities for which we have already done all the work. It is given by

$$\chi_{\mu\nu}(k) = \begin{pmatrix} \rho(\frac{\partial\rho}{\partial p})_T & T(\frac{\partial\rho}{\partial T})_p & 0 & 0 \\ T(\frac{\partial\rho}{\partial T})_p & \rho c_p T & 0 & 0 \\ 0 & 0 & \rho & 1 \\ 0 & 0 & 1 & \rho_s^{-1} \end{pmatrix} . \tag{10.71}$$

as $k \to 0$. (10.71) is correct to order $k^2$ since all matrix elements of $\chi_{\mu\nu}(k)$ are even functions of k. The elements in the lower right corner are simply restatements of eqs. (10.17), (10.61), and (10.64). Those in the upper left corner do not refer to superfluid properties in particular; they have been calculated in (4.27). Finally, the zeros in (10.71) are zeros to all orders in k because of time reversal symmetry.

The unnormalized memory matrix

$$\omega_{\mu\nu}(k) = \int \frac{d\omega}{\pi} \chi''_{\mu\nu}(k\omega) \tag{5.28}$$

is given by

$$\omega_{\mu\nu}(k) = \begin{bmatrix} 0 & 0 & k\rho & k \\ 0 & 0 & 0 & -Tsk \\ k\rho & 0 & 0 & 0 \\ k & -Tsk & 0 & 0 \end{bmatrix}, \qquad (10.72)$$

omitting only terms of order $k^3$. Again, time reversal symmetry eliminates all terms involving only $\{\rho, q\}$, or only $\{g_\ell, v_s\}$. $\omega_{\rho g_\ell} = k\rho$ follows easily from the commutator (4.34), and $\omega_{\rho v_s} = k$ is a restatement of eq. (10.60). $\omega_{q g_\ell} = 0$, up to terms of order $k^3$, is not a consequence of symmetry. Rather, it vanishes because this is how q has been defined; the reader will have no difficulty in showing that

$$\lim_{k\to 0} k^{-1}\, \omega_{q g_\ell}(k) = \lim_{k\to 0} \chi_{q\,\tau_{33}}(k) = T\frac{\partial\langle\tau_{33}\rangle}{\partial T}\Big|_p = 0 . \qquad (10.73)$$

The sum rule

$$\lim_{k\to 0} k^{-1} \int \frac{d\omega}{\pi}\, \chi''_{q v_s}(k\omega) = -Ts , \qquad (10.74)$$

where $ms = (S/N)$ is the entropy per particle, requires some explanation, however. Note that as $k\to 0$,

$$\begin{aligned} \lim_{k\to 0} k^{-1} \int \frac{d\omega}{\pi}\, \chi''_{q v_s}(k\omega) &= \frac{-i\hbar}{m} \lim_{k\to 0} \int \frac{d\omega}{\pi}\, \chi''_{q\varphi}(k\omega) \\ &= (1/im) \int dr\, \langle [q(r), \varphi^{op}(r')] \rangle \\ &= (1/im)\, \langle [H-\mu N - \Lambda, \varphi^{op}(r')] \rangle \qquad (10.75) \\ &\quad + (1/im)\,(\mu - \frac{\varepsilon+p}{n})\, \langle [N, \varphi^{op}(r')] \rangle + (1/im)\, \langle [\Lambda, \varphi^{op}(r')] \rangle , \end{aligned}$$

where $\Lambda = \int dr[\eta^*(r)\,\Psi(r) + \eta(r)\,\Psi^\dagger(r)]$ is the source term in the ensemble operator

(10.51). The averages in (10.75) are, of course, taken in the $\eta$-ensemble. Therefore, the first term vanishes since

$$\mathrm{tr}\, \rho_T [H - \mu N - \Lambda] \varphi^{op} = \mathrm{tr}[\rho_\eta, H - \mu N - \Lambda] \varphi^{op} = 0. \qquad (10.76)$$

Similarly, $\langle [\Lambda, \varphi^{op}] \rangle = i\, \mathrm{Re}\, \eta / \langle \Psi \rangle$ by explicit calculation, and vanishes as $\eta \to 0$. This leaves us with the second term in (10.75), and using (10.58) and the thermodynamic identity

$$\frac{\varepsilon + p}{\rho} = \frac{\mu}{m} + Ts \qquad (10.77)$$

we obtain the sum rule (10.74).

Let us harvest the fruits of what has been done. At least to lowest non-vanishing order in k, we have the matrices $\underset{\approx}{\chi}$ and $\underset{\approx}{\omega}$, and thus the normalized frequency matrix $\underset{\approx}{\Omega} = \underset{\approx}{\omega}\underset{\approx}{\chi}^{-1}$, and if the memory matrix $\underset{\approx}{\Sigma}(kz)$ can be omitted (it is of higher order in k) we can calculate $C_{\mu\nu}(kz)$ from the master formula (5.28), namely

$$[z\delta_{\mu,\lambda} - \Omega_{\mu\lambda}(k)]\, C_{\lambda\nu}(kz) = i\beta^{-1} \chi_{\mu\nu}(k). \qquad (10.78)$$

This is a 4 by 4 matrix equation. We first examine the poles by solving $\det[z - \underset{\approx}{\Omega}(k)] = 0$. We obtain two pairs of propagating solutions, namely

$$z = \pm\, c_1 k\,,$$
$$z = \pm\, c_2 k \qquad (10.79a)$$

where the constants are given by

$$c_1^2 + c_2^2 = \left(\frac{\partial p}{\partial \rho}\right)_S + \frac{\rho_s}{\rho_n} \frac{Ts^2}{c_v}\,,$$

$$c_1^2\, c_2^2 = \left(\frac{\partial p}{\partial \rho}\right)_T \cdot \frac{\rho_s}{\rho_n} \frac{Ts^2}{c_v}\,. \qquad (10.79b)$$

For the normal liquid we would, at this stage, have found one sound pole, $z = \pm\, ck$, and the diffusive heat pole $z = 0$, to order k. (The results for the normal liquid are obtained by setting $\rho_s \equiv 0$). In the superfluid, by contrast, there are two sound-like modes. They are called first and second sound, respectively. Since $He^4$ is only superfluid at very low temperatures where $c_p \approx c_v$ and thus $(\frac{\partial p}{\partial \rho})_S \approx (\frac{\partial p}{\partial \rho})_T$, we obtain

$$c_1^2 = \frac{dp}{d\rho} \qquad \text{(first sound)} \qquad (10.80a)$$

and

$$c_2^2 = \frac{\rho_s}{\rho_n} \cdot \frac{Ts^2}{c_v} = -\frac{\rho_s}{\rho_n} \frac{\partial T}{\partial(1/s)} \qquad \text{(second sound)}. \qquad (10.80b)$$

As these expressions indicate, $c_1$ is the speed of compressional waves --normal sound. What is the nature of the mode which propagates with the speed $c_2$? This question is answered if we diagonalize the matrix $C_{\mu\nu}(kz)$, and determine which variable, or combination of variables, exhibits the second sound pole. After some algebra one finds that (if $c_p \approx c_v$)

$$\beta C_{qq}(kz) = i\rho c_p T \cdot \frac{z}{z^2 - c_2^2 k^2}, \qquad (10.81)$$

as one would have expected from eq. (10.80b), or from the eq. (4.42b) for normal fluids. Therefore, while first sound is a compressional wave, second sound is an entropy wave. In a first sound mode, the normal and the superfluid components oscillate in phase. In a second sound mode, the two fluids, normal and super, move against each other, in such a fashion that the total density remains uniform, $\delta\rho = \delta\rho_s + \delta\rho_n = 0$. This also answers the question how one would go about exciting, and observing, the second sound mode. By eq. (4.20b), at $\delta\rho = 0$, fluctuations of the local temperature and the variable q are equivalent, namely

$$\delta T(r,t) = (1/\rho c_v)\, \delta q(r,t). \qquad (10.82)$$

If one therefore creates a heat pulse at one end of a superfluid sample, the temperature pulse does not slowly diffuse away as it does in any normal liquid.

Rather, the pulse excites a wave which propagates rapidly through the sample with velocity $c_2$ -- a temperature wave. Experimentally, all one needs to observe second sound is therefore a resonant cavity filled with $He^4$, with a heater on one wall and a thermometer on the other (Snyder 1963). Indeed, the fact that pumped He below the λ-point does not bubble is evidence for the existence of temperature waves, as mentioned above.

## 10.6 Dissipation

We include now the contributions coming from the damping matrix $\underset{\approx}{\Sigma}(kz) = \underset{\approx}{\sigma}(kz)\,\underset{\approx}{\chi}^{-1}(k)$ in (5.28). As shown there, $\sigma_{\mu\nu}(kz)$ is generally given by the formal expression

$$\sigma_{\mu\nu}(\vec{k}z) = \beta\langle \mathring{A}_\mu(k)\,|Q\frac{i}{z-QLQ}Q\,|\,\mathring{A}_\nu(k)\rangle\,, \qquad (10.83)$$

where the projector Q removes fluctuations of all the six hydrodynamic variables $\rho$, $q$, $\vec{g}$, and $\varphi^{op}$ or $v_s$. Indeed, while the first five obey the usual continuity equations (4.7), $v_s(rt)$ obeys the "quasi-conservation law"

$$\partial_t \vec{v}_s(r,t) - \vec{\nabla}\frac{\hbar}{m}\,\dot{\varphi}^{op}(r,t) = 0$$

or

$$\dot{v}_s(k) = -ik\frac{\hbar}{m}\,\dot{\varphi}^{op}(k) \qquad (10.84)$$

by its definition. Obviously, the phase operator $\varphi^{op}(r,t)$ is not conserved; if we were to carry through the discussion in terms of $\varphi^{op}$ rather than $v_s$, we would see that its contributions to $\Sigma_{\mu\nu}(kz)$ are of explicit order $k^2$ because of the diverging susceptibility (10.61). Eq. (10.84) is a convenient representation of this fact. It is clear, then, that all elements of $\sigma_{\mu\nu}(kz)$ are explicitly of order $k^2$. For $z = i0$ they can be written in the form

$$\sigma_{\mu\nu}(k,i0) = k^2 \begin{pmatrix} \kappa T & 0 & 0 \\ 0 & \zeta_2 + \frac{4}{3}\eta & \zeta_1 \\ 0 & \zeta_4 & \zeta_3 \end{pmatrix} \tag{10.85}$$

to lowest order in k, where the 1st, 2nd, and 3rd row and column refer to q, $g_\ell$, and $v_s$, respectively. All matrix elements involving the density $\rho$ vanish identically. The coefficients in (10.85) are all real, and are constrained by

$$\kappa \geqq 0; \ \zeta_2 \geqq 0; \ \eta \geqq 0; \ \zeta_3 \geqq 0;$$
$$\zeta_2 \zeta_3 \geqq \zeta_1 \zeta_4 \tag{10.86}$$

and the Onsager relation

$$\zeta_1 = \zeta_4 \ . \tag{10.87}$$

Since this is not the first time we have played with memory matrices, some brief comments will suffice to explain (10.85). (1) Since $\sigma_{\mu\nu}(kz)$, and therefore also $\Sigma_{\mu\nu}(kz)$, is explicitly of order $k^2$, the z-dependence can only introduce contributions of higher order, and thus hydrodynamically $\sigma_{\mu\nu}(kz)$ can be replaced by $\sigma_{\mu\nu}(k,0)$. Because of the projector Q in (10.83) $\sigma_{\mu\nu}(kz)$ is expected to be a smooth function of z, for small z. (2) Because $\dot{\rho} = ikg_\ell$, and $g_\ell$ is itself conserved and thus removed by the projector Q in (10.83), $Q|\dot{\rho}(k)\rangle = 0$. The reader will do well to recall the rather more laborious argument given in section 4.6, see eqs. (4.61) and (4.62), to prove that there are no transport coefficients that refer to the particle density. The reason is, here and there, momentum conservation. (3) The matrix elements $\sigma_{qg_\ell}$ and $\sigma_{qv_s}$ must be odd functions of $\vec{k}$ from parity, and at least of order $k^2$ because of the conservation laws. Thus, they must be at least of order $k^3$, in fact, and do not contribute to hydrodynamic ($\sim k^2$) order. (4) The eqs. (10.86) and (10.87) are a consequence of the general symmetry and positivity properties of

$\sigma_{\mu\nu}(k\omega)$, eqs. (5.31) and (5.32). Noting that $\vec{v}_s(rt)$ is odd under both time reversal and parity, the reader will have no difficulty verifying that. The reader may take note that, contrary to popular belief, a superfluid does not have zero viscosity. Indeed, instead of 2 viscosities $\eta$ and $\zeta$ as in the normal liquid, we need 4 viscosity coefficients, plus a heat conductivity as usual, for the complete description of the transport properties of superfluid $He^4$. The coefficient $\eta$ in (10.85) is the usual shear viscosity. In terms of it the transverse memory function is given by

$$\sigma_{g_t g_t}(k, i0) = k^2\eta, \tag{10.88}$$

to order $k^2$. A symmetry argument, based on the fact that the stress tensor is symmetric, shows that not only $\zeta_2 + \frac{4}{3}\eta$, but $\zeta_2$ itself is positive.

We can now insert $\sigma_{\mu\nu}$ in the master formula (5.28), and derive the hydrodynamic expressions for all correlation functions of interest. Obviously, this is laborious. We shall therefore leave this exercise in inverting matrices to the patience of the reader, and quote only the most important results (Hohenberg and Martin 1965). The correlation functions have poles when

$$\Delta(kz) = (z^2 - c_1^2k^2 + ik^2z\,D_1)\,(z^2 - c_2^2k^2 + ik^2z\,D_2) \tag{10.89}$$

vanishes, where the attenuation constants are given by

$$D_1 + D_2 = \frac{\varkappa}{\rho c_V} + (\zeta_2 + \tfrac{4}{3}\eta)/\rho_n + (\rho_s/\rho_n)(\rho\zeta_3 - 2\zeta_1), \tag{10.90a}$$

$$c_2^2D_1 + c_1^2D_2 = \frac{\varkappa}{\rho c_p}\,c^2 + \frac{\rho_s}{\rho_n}\,\frac{\zeta_2 + \frac{4}{3}\eta}{\rho}\left[c^2 + \frac{Ts^2}{c_V} - \frac{2Ts}{\rho c_V}\left(\frac{\partial p}{\partial T}\right)_\rho\right]$$

$$+ \rho\zeta_3\frac{\rho_s}{\rho_n}\,c^2 - 2\zeta_1\frac{\rho_s}{\rho_n}\left[c^2 - \frac{Ts}{\rho c_V}\left(\frac{\partial p}{\partial T}\right)_\rho\right], \tag{10.90b}$$

where $c_1^2$ and $c_2^2$ are given by (10.79b), and we have used the abbreviation

$c^2 = (\partial p/\partial \rho)_S$. At Helium temperatures, where $c_p \approx c_v$ to high accuracy, these formulae simplify to

$$D_1 = (\zeta_2 + \tfrac{4}{3}\eta)/\rho, \tag{10.91a}$$

$$D_2 = \frac{\kappa}{\rho c_v} + \frac{\rho_s}{\rho_n}[\zeta_3 \rho - 2\zeta_1 + (\zeta_2 + \tfrac{4}{3}\eta)/\rho]. \tag{10.91b}$$

Readers who find these equations too cloudy for their taste should turn the next few pages. For the others, here are a few correlation functions. The longitudinal momentum density function is of the form

$$C_{g_\ell g_\ell}(kz) = i\beta^{-1}\rho\frac{z}{\Delta(kz)}[z^2 - k^2\frac{Ts^2}{c_v}\frac{\rho_s}{\rho_n} + ik^2 z(D_1 + D_2 - \tilde{\zeta}_2/\rho)]. \tag{10.92}$$

Let us, before we proceed, add its transverse brother without further ado. Since the superfluid order is purely longitudinal in $He^4$, the transverse correlation function is the same as in normal liquids, namely

$$C_{g_t g_t}(kz) = i\beta^{-1}\rho_n \cdot \frac{1}{z + ik^2\eta/\rho_n}. \tag{10.93}$$

Yes, not quite the same. Because of the sum rule (10. 18) the normal fluid density $\rho_n$ appears in (10.93) where we had $\rho$ in (4.39); a small change but one from which we have, in a sense, derived all the other features of macroscopic superfluidity.

The density-density correlation function, $C_{\rho\rho}(kz)$, can be derived from (10.92) since

$$z^2 C_{\rho\rho}(kz) - k^2 C_{g_\ell g_\ell}(kz) = i\beta^{-1} z\, \chi_{\rho\rho}(k) \simeq i\beta^{-1} z\, \rho(\frac{\partial\rho}{\partial p})_T, \tag{10.94}$$

where the last equation holds for small k. Equation (10.94) is simply mass conservation. The autocorrelation function of the superfluid velocity is given by

$$C_{v_s v_s}(kz) = \frac{i\beta^{-1}}{\rho_s \Delta(kz)} [z^3 - k^2 z(\frac{\rho_n}{\rho} c^2 + \frac{\rho_s}{\rho} \frac{Ts^2}{c_v} \frac{\rho_s}{\rho_n} - \frac{2Ts\rho_s}{\rho^2 c_v} (\frac{\partial p}{\partial T})_\rho)$$

$$+ ik^2 z^2 (D_1 + D_2 - \rho_s \zeta_3) - ik^4 \frac{\rho_n c^2}{\rho} \cdot \frac{\varkappa}{\rho c_p} ], \quad (10.95)$$

while $C_{g_\ell v_s}$ is given by

$$C_{g_\ell v_s}(kz) = i\beta^{-1} \frac{z}{\Delta(kz)} [ z^2 - k^2 (\frac{Ts^2}{c_v} \frac{\rho_s}{\rho_n} + \frac{Ts}{\rho c_v} (\frac{\partial p}{\partial T})_\rho)$$

$$+ ik^2 (D_1 + D_2 - \zeta_1)]. \quad (10.96)$$

Note that

$$zC_{\rho v_s}(kz) = kC_{g_\ell v_s}(kz). \quad (10.97)$$

Finally, the autocorrelation function of the entropy density q, which describes the propagation of temperature waves, is of the form

$$C_{qq}(kz) = i\beta^{-1} \frac{\rho c_p T}{\Delta(kz)} [z^3 - k^2 z(c^2 + \frac{Ts^2}{c_v} \frac{\rho_s}{\rho_n} (1 - \frac{c_v}{c_p}))$$

$$+ ik^2 z^2 (D_1 + D_2 - \frac{\varkappa}{\rho c_p}) - ik^4 \gamma \frac{\rho_s}{\rho_n} ], \quad (10.98)$$

where $\gamma$ is a linear combination of $\zeta_1$, $\zeta_3$, and $\zeta_2 + \frac{4}{3}\eta$, with rather cumbersome thermodynamic derivatives as coefficients.

All of these expressions, which hold as usual for Im$z > 0$, are asymptotically rigorous, in the sense which was indicated in eq. (4.43). As a practical matter, if one uses the $C_{\mu\nu}(kz)$ to find, for example,

$$\omega^{-1} \chi''_{\rho\rho}(k\omega) = \beta \,\mathrm{Re}\, C_{\rho\rho}(k, \omega + i\epsilon), \quad (10.99)$$

the hydrodynamic expressions are valid for $k\ell \ll 1$ and $\omega\tau \ll 1$. It is the latter restriction which is quite severe in Helium. The collision time $\tau$ is determined by the interactions of elementary excitations (rotons and phonons), and at 1°K the density of phonons and rotons is so small that $\tau$ becomes rather large, $\tau \approx 10^{-9}$ sec. Note also, in this connection, that $\rho_n = \rho$, of course, at $T_\lambda \simeq 2.18$°K but that $\rho_n/\rho$ decreases rapidly as the temperature is lowered. At 1.5°K, $\rho_n/\rho \approx 0.11$ while at 1.1°K, $\rho_n/\rho \approx 0.015$ (Donnelly, 1967, gives some useful tables). At $T = 0$, $\rho_n = 0$.

Even with this caveat, the hydrodynamic formulae are interesting. The reader will note that above $T_\lambda$, $\rho_s \equiv 0$, and all the expressions which we have derived agree with those obtained for the normal liquid in chapter 4 where we employed an alternative technique. In superfluid Helium the function $C_{v_s v_s}$ is of particular interest. Because the phase operator $\varphi^{op}(r)$, and therefore the superfluid velocity $\vec{v}_s(r)$, is a linear function of the fundamental fields $\Psi(r)$ and $\Psi^+(r)$, we can obtain an expression for the field correlation function $G(kz)$. It is customary to define the latter by (Kadanoff and Baym 1962)

$$\langle [\Psi(rt), \Psi^+(rt')] \rangle = \int \frac{d\omega}{2\pi} \int \frac{d^3k}{(2\pi)^3} e^{-i\omega(t-t') + i\vec{k}\cdot(\vec{r}-\vec{r}')} A(k\omega) , \tag{10.100a}$$

$$G(kz) = \int \frac{d\omega}{2\pi} \frac{A(k\omega)}{z-\omega} . \tag{10.100b}$$

Its importance is that all thermodynamic, and much other, information can be obtained from it. Using eqs. (10.55), and remembering that $n^{op}(r)$ is of short range and can be disregarded in comparison with $\varphi^{op}(r)$, one obtains

$$G(kz) = -\frac{m^2 n_o/\hbar}{k^2} [i\beta z C_{v_s v_s}(kz) + 1/\rho_s] , \tag{10.101}$$

which omits only terms of relative order $k^2$. Microscopically, these terms are important; note that (10.101) fails to satisfy the fundamental sum rule $\int d\omega A(k\omega) = 2\pi$ because it yields an approximation for $A(k\omega)$ which is odd in $\omega$. A result similar to (10.101), but more cautious in this respect, has been given by Hohenberg and Martin (1965). Note, however, that at $z = 0$, where (10.101) should be appropriate, it states that

$$\int \frac{d\omega}{2\pi\hbar} \frac{A(k\omega)}{\omega} = \frac{(m/\hbar)^2}{\rho_s k^2} n_o . \tag{10.101a}$$

This relation between the condensate density, $n_o$, the superfluid density $\rho_s$, and the fluctuation spectrum $A(k\omega)$ is due to Josephson (1966). It will be reconsidered in section 10.9 below where we will discuss one of its interesting consequences: there can be no Bose condensation in 2 dimensions.

## 10.7 Kubo Relations

Apart from thermodynamic derivatives and the superfluid mass density $\rho_s$, the hydrodynamic equations which we have derived are specified by five dissipative coefficients: the heat conductivity $\kappa$, the shear viscosity $\eta$, and three longitudinal viscosities $\zeta_2$, $\zeta_3$, and $\zeta_1 = \zeta_4$. As usual, these coefficients can be rigorously expressed as limits on the correlation functions. We list the resulting Kubo formulae below.

$$\kappa T = \lim_{\omega\to 0} \lim_{k\to 0} \frac{\omega}{k^2} \chi''_{qq}(k\omega), \tag{10.102a}$$

$$\eta = \lim_{\omega\to 0} \lim_{k\to 0} \frac{\omega}{k^2} \chi''_{t}(k\omega), \tag{10.102b}$$

$$\zeta_2 + \frac{4}{3}\eta = \lim_{\omega\to 0} \lim_{k\to 0} \frac{\omega}{k^2} \chi''_{\ell}(k\omega), \tag{10.102c}$$

$$\zeta_3 = \lim_{\omega\to 0} \lim_{k\to 0} \frac{\omega}{k^2} \chi''_{v_s v_s}(k\omega), \tag{10.102d}$$

$$\zeta_1 = \zeta_4 = \lim_{\omega\to 0} \lim_{k\to 0} \frac{\omega}{k^2} \chi''_{g_\ell v_s}(k\omega) . \tag{10.102e}$$

The reader should have no difficulty in deriving these expressions from the results of the last section. Alternatively he can use the definitions in terms of the memory functions $\sigma_{\mu\nu}$, for example

$$\kappa T = \lim_{\omega\to 0} \lim_{k\to 0} \frac{1}{k^2} \operatorname{Re} \sigma_{qq}(k,\omega+i0). \tag{10.103}$$

The last equation still contains the projector Q in its dynamics, see eq. (10.83). Q can be removed, however, by use of the general formula (5.58) which has an obvious matrix equivalent. As in normal liquids the formulae for $\zeta_2$ and $\eta$ can be combined in the form (4.59b) which makes it clear that not only $\eta$ and $\zeta_2 + \frac{4}{3}\eta$, but the bulk viscosity $\zeta_2$ itself is a positive coefficient. Moreover, since

$$\lim_{\omega\to 0} \lim_{k\to 0} \frac{\omega}{k^2} \chi''_{\rho\rho}(k\omega) = \lim_{\omega\to 0} \lim_{k\to 0} \frac{\omega}{k^2} \chi''_{\rho q}(k\omega) = 0, \tag{10.104}$$

the entropy density q in (10.102a) can be replaced by the energy density $\epsilon$. Finally, the equations

$$\lim_{\omega\to 0} \lim_{k\to 0} \frac{\omega}{k^2} \chi''_{qg_\ell}(k\omega) = \lim_{\omega\to 0} \lim_{k\to 0} \frac{\omega}{k^2} \chi''_{qv_s}(k\omega) = 0 \tag{10.105}$$

are an expression of the general Onsager–Casimir property that there are no dissipative coefficients which link quantities with opposite time reversal signatures.

The reason, of course, why all of these formulae are of the same formal structure as they are in normal systems is because we have employed the pseudo-conserved superfluid velocity $\vec{v}_s^{op}$, rather than the quantum phase operator

$\varphi^{op}$ which is the microscopically more meaningful quantity. In terms of the latter, we obtain

$$\zeta_3 = (\hbar/m)^2 \lim_{\omega\to 0} \lim_{k\to 0} \omega\, \chi''_{\varphi\varphi}(k\omega), \tag{10.106}$$

for example, which displays the now familiar absence of the factor $1/k^2$ because $\varphi^{op}(r,t)$ is a symmetry-restoring variable but not a conserved one.

We have listed a series of thermodynamic sum rules in previous sections, and shall not repeat them here. Suffice it to say that the expressions (10.92) through (10.98) are, of course, in accord with all the sum rules which are implicit in eqs. (10.71) and !0.72). Additional sum rules could be derived and employed in interpolation schemes (see, e.g., Puff (1965) and Kerr et al. (1970)), in a manner as suggested in section 4.8.

## 10.8 The Phenomenological Two-Fluid Equations

Hydrodynamics is conventionally discussed in terms of equations of motion for the non-equilibrium densities $\delta\langle\rho(rt)\rangle = \langle\rho(rt)\rangle_{\text{non-eq.}} - \langle\rho\rangle_{\text{eq}}$, as we did for normal fluids in section 4.1. While there are obvious advantages to the direct route to the hydrodynamic correlation functions which we have employed here, it is worthwhile to make contact with the conventional formulation (Enz 1974, Khalatnikov 1965). Of course, our results yield only the linearized equations of hydrodynamics although for Helium there is little difficulty in deriving, microscopically, those nonlinear contributions which are usually employed, as Hohenberg and Martin (1965) have shown.

We begin by returning to linear response theory. If we consider the Hamiltonian

$$\delta \mathcal{H}(t) = -\sum_{\mu} \int d^3r\, A_{\mu}(r)\, \delta a_{\mu}^{ext}(r)\, e^{\epsilon t}\, \eta(-t), \qquad (10.107)$$

where now $A_{\mu}(r)$ is the full set of hydrodynamic variables,

$$\{A_{\mu}\} = \{\rho, q, \vec{g}; \vec{v}_s\} , \qquad (10.108)$$

and $\delta a_{\mu}^{ext}$ are as yet unknown external forces, then at $t = 0$ the linear response is

$$\delta\langle A_{\mu}(r)\rangle = \chi_{\mu\nu}(k=0)\, \delta a_{\nu}^{ext}(r) \qquad (10.109)$$

by eq. (3.9), if we only assume that the external forces vary slowly in space. Since we know how to interpret $\delta\langle A_{\mu}\rangle$, and since $\chi_{\mu\nu}(0)$ are given by thermodynamic derivatives, eq. (10.109) can be used to infer the meaning of the external forces. For example, $\delta a_{\rho}^{ext} = \delta p^{ext}/\rho$ is an externally produced pressure excess, in agreement with (10.109), namely

$$\delta\langle\rho\rangle = \rho\left(\frac{\partial\rho}{\partial p}\right)_T \delta a_{\rho}^{ext} = \left(\frac{\partial\rho}{\partial p}\right)_T \delta p^{ext}, \qquad (10.110)$$

when all other forces vanish.

It is therefore suggestive to define the internal, space and time dependent, dynamical forces $\delta a_{\mu}(rt)$ by

$$\delta a_{\mu}(rt) = \chi_{\mu\nu}^{-1}(k=0)\, \delta\langle A_{\nu}(rt)\rangle . \qquad (10.111)$$

If we write

$$\delta a_{\rho}(rt) = \rho^{-1}\, \delta p(rt),$$

$$\delta a_{q}(rt) = T^{-1}\, \delta T(rt), \qquad (10.112)$$

then (10.111) guarantees that these symbols have their usual thermodynamic significance when variations in space and time are slow. Further, if we define

$\delta\vec{v}_n$ by

$$\delta\langle\vec{g}(rt)\rangle = \rho_n \delta\vec{v}_n(rt) + \rho_s \delta\vec{v}_s(rt), \tag{10.113}$$

where $\delta\vec{v}_s = \delta\langle\vec{v}_s^{op}\rangle$ is purely longitudinal, then we obtain

$$\begin{aligned} \delta\vec{a}_g(rt) &= \delta\vec{v}_n(rt) \;, \\ \delta\vec{a}_{v_s}(rt) &= \rho_s[\delta\vec{v}_s(rt) - \delta\vec{v}_{n\ell}(rt)] \,. \end{aligned} \tag{10.114}$$

At time $t > 0$, the response to the external Hamiltonian (10.107) is given by

$$\delta\langle A_\mu(kz)\rangle = \beta\, C_{\mu\nu}(kz)\, \delta a_\nu^{ext}(k), \tag{10.115}$$

which is eq. (3.14), with Fourier-Laplace transforms defined as in eq. (3.12). For small k and z, these amplitudes therefore obey the equations

$$z\delta\langle A_\mu(kz)\rangle - [\omega_{\mu\nu}(k) - i\sigma_{\mu\nu}(k,i0)]\,\delta a_\nu(kz) = i\delta\,\langle A_\mu(k,t{=}0)\rangle \,. \tag{10.116}$$

We will now write these equations in their conventional form, beginning with the continuity equations

$$\partial_t\, \rho + \vec{\nabla}\vec{g} = 0, \tag{10.117a}$$

$$\partial_t\, g_i + \nabla_j\, \tau_{ij} = 0, \tag{10.117b}$$

$$\partial_t\, \epsilon + \vec{\nabla}\,\vec{j}^{\epsilon} = 0, \tag{10.117c}$$

$$\partial_t\langle\vec{v}_s\rangle + m^{-1}\,\vec{\nabla}\mu^{\ell oc} = 0. \tag{10.117d}$$

The last equation reflects only the definition of $\vec{v}_s$ as a gradient. Equations (10.116) then contain constitutive relations between the currents and the internal forces. The $\omega_{\mu\nu}$ - term gives rise to reactive, reversible terms while $\sigma_{\mu\nu}$ contributes irreversible terms. Using (10.72) and (10.85), we find

$$\delta\langle \vec{j}^{\,\epsilon}\rangle = \mu\rho_s(\vec{v}_s - \vec{v}_n) + (\mu + Ts)\,\rho\,\vec{v}_n - \kappa\,\vec{\nabla}T, \tag{10.118a}$$

$$\delta\langle \tau_{ij}\rangle = \delta p\,\delta_{ij} - \delta_{ij}[\zeta_2\vec{\nabla}\cdot\vec{v}_n + \zeta_1\,\vec{\nabla}\cdot\rho_s(\vec{v}_s - \vec{v}_n)]$$
$$-\eta[\nabla_i v_{nj} + \nabla_j v_{ni} - \frac{2}{3}\,\delta_{ij}\,\vec{\nabla}\cdot\vec{v}_n], \tag{10.118b}$$

$$\frac{1}{m}\,\delta\mu^{\ell oc} = \frac{1}{m}\,\delta\mu - \zeta_3\,\vec{\nabla}\rho_s(\vec{v}_s - \vec{v}_n) - \zeta_4\,\vec{\nabla}\cdot\vec{v}_n\ . \tag{10.118c}$$

The local pressure and chemical potential which appear here deviate from their equilibrium values by $\delta p(rt)$ of (10.112) and by $(\rho/m)\delta\mu \equiv \delta p - \rho s\,\delta T$, of course. (We have not followed the custom to denote by $\mu$ the chemical potential per unit mass; the explicit factors of m above are the penalty.) Equations (10.117) and (10.118) are the standard (linearized) two-fluid equations of superfluid hydrodynamics. They are applied in the book by Khalatnikov (1965) to a variety of situations.

## 10.9 Absence of Bose Condensation in 2 Dimensions

Clearly, even this already long chapter does scant justice to the rich variety of phenomena observed in superfluid Helium. We have not even referred to the nature of the low-lying elementary excitations, longitudinal phonons and rotons, in terms of which the Landau theory is formulated. The spectrum of these excitations can be inferred, even without detailed microscopic calculation, from the sum rules implicit in (10.71) (Pines 1965, Puff 1965), and once known, it determines the thermodynamic properties of $He^4$ at low temperature. For many other topics -- vortex lines and rings with quantized circulation, to mention one of the really fascinating ones -- the reader will have to go to other books. An excellent but, alas, unpublished treatment of many aspects of the superfluid Bose liquid has been given by Nozières and Pines (1964); see also Pines (1965).

There is, however, one result which fits in directly with the theme of this chapter. As Hohenberg (1967) has shown, a Bose condensate may exist in 3 dimensions but it would by dynamically unstable in a two-dimensional world. The proof is based on the Bogoliubov inequality from which we concluded that $\chi_{\varphi\varphi}(k)$ diverges at least as $1/k^2$ as $k \to 0$, if $\langle \Psi \rangle \neq 0$. If we can show that $n_o \cdot \int d\vec{k}\ \chi_{\varphi\varphi}(k)$ must be finite in a thermodynamic system, we have ruled out $\langle \Psi \rangle \neq 0$ in 2 dimensions where the integrand has divergent contributions at small k. For a careful proof it is useful to define the quantities

$$\tilde{A}(r-r', t-t') = \langle [\tilde{\Psi}(rt), \tilde{\Psi}^+(r't')] \rangle, \tag{10.119}$$

$$\tilde{\varphi}(r-r', t-t') = \frac{1}{2} \langle \{ \tilde{\Psi}(rt), \tilde{\Psi}^+(r't') \} \rangle - n_o, \tag{10.120}$$

and their Fourier transforms, $\tilde{A}(k\omega)$ and $\tilde{\varphi}(k\omega)$. In these equations, we are using the modified Hamiltonian

$$\mathcal{H} = H - \mu N - \int d\vec{r} [\eta^*(r)\ \Psi(r) + \eta(r)\ \Psi^+(r)] \tag{10.121}$$

to generate the time dependence, according to

$$\tilde{\Psi}(rt) = e^{\frac{i}{\hbar}\mathcal{H}t}\ \Psi(r)\ e^{-\frac{i}{\hbar}\mathcal{H}t}. \tag{10.122}$$

The reason is largely a matter of formal convenience. With this definition, the $\eta$-ensemble average $\langle \tilde{\Psi}(rt) \rangle = n_o^{1/2}$ is time independent while $\langle \Psi(rt) \rangle = n_o^{1/2}\ e^{-i\mu t/\hbar}$ is not, even as $\eta \to 0$, as the alert reader will have noticed. Moreover, with (10.122) we do not need to change tack to prove, exactly as in section 3.5, that the fluctuation-dissipation theorem,

$$\tilde{\varphi}(k\omega) = \frac{1}{2} \coth(\hbar\omega\beta/2)\ \tilde{A}(k\omega), \tag{10.123}$$

holds even in the $\eta$-ensemble. Since $\tilde{\varphi}(k\omega)$ is obviously a positive quantity, it follows that $\omega\tilde{A}(k\omega) \geqq 0$, too. Indeed, just as in section 7.3 we find Bogoliubov inequalities, among them

$$\int\frac{d\omega}{2\pi\hbar}\,\omega^{-1}\tilde{A}(k\omega) \geqq \frac{\left|\int\frac{d\omega}{\pi}\,\tilde{\chi}''_{n\Psi}(k\omega)\right|^2}{\int\frac{d\omega}{\pi}\,\omega\,\tilde{\chi}''_{nn}(k\omega)}\,, \tag{10.124}$$

where the tilde indicates that the particle density $\tilde{n}(rt)$ in these terms is also given the time dependence (10.122). The reader will easily convince himself that this does not change the sum rules which appear on the right in (10.124), and he will therefore find that

$$\int\frac{d\omega}{2\pi\hbar}\,\omega^{-1}\tilde{A}(k\omega) \geqq \frac{(m/\hbar)^2}{k^2\rho}\,|\langle\Psi\rangle|^2\,. \tag{10.125}$$

This is essentially the same statement as (10.61) or (10.101a). If we further define

$$n(k) = \int d(\vec{r}-\vec{r}\,')\,e^{-i\vec{k}\cdot(\vec{r}-\vec{r}\,')}[\langle\Psi^+(r')\Psi(r)\rangle - n_o] \tag{10.126}$$

then it is clear that

$$n(k) + \frac{1}{2} = \int\frac{d\omega}{2\pi}\,\tilde{\varphi}(k\omega) \geqq (k_BT)\int\frac{d\omega}{2\pi\hbar}\,\omega^{-1}\tilde{A}(k\omega), \tag{10.127}$$

which uses the F-D theorem (10.123) and the fact that $x \coth x \geqq 1$ for all $x$. Now the average particle density $n = \langle\Psi^+(r)\Psi(r)\rangle$ is obtained by inverting (10.126),

$$n - n_o = \int\frac{d^dk}{(2\pi)^d}\,n(k) \geqq \int\frac{d^dk}{(2\pi)^d}\left[\frac{k_BT(m/\hbar)^2}{\rho k^2}\,|\langle\Psi\rangle|^2 - \frac{1}{2}\right], \tag{10.128}$$

in d dimensions. In 3 dimensions, the integral in (10.128) is finite. In 2 dimensions, however, it is proportional to $\int dk/k$ which diverges logarithmically, at small k. (Since $n(k) \geqq 0$ for all k, the subtracted term $-1/2$ cannot make up for the divergence.) However, $n-n_o$ is finite. The only way out is $\langle \Upsilon \rangle = 0$: Fluctuations of long wavelength are so strong, in two dimensions, that they destroy the condensate.

One final remark: we have encountered two definitions of the superfluid density $\rho_s$. The first of these, by eq. (10.18), is unquestionably the one to which measurements by rotation experiments refer, or by determining the velocity of second sound. The Josephson definition (10.101a), on the other hand, extracts $\rho_s$ from the one-particle spectral function $A(k\omega)$, essentially the spectrum of elementary excitations. Are the two definitions rigorously equivalent? The answer is by no means obvious; it has not, to my knowledge, been spelled out in the published literature. It may be well to summarize the "proof" of their equivalence which we have given in eqs. (10.58) through (10.66). It contains three elements: (a) the assertion that there is a gauge-restoring scalar variable $\varphi^{op}(r)$ whose fluctuations are of long range as in eq. (10.61), (b) the assertion that all other correlations are of short range, thus specifically that $\langle g_i(r)\, Q\, g_i(r')\rangle$ is short-ranged where Q is the projector which removes fluctuations of $\varphi^{op}$ as in (10.66a), and (c) the assertion that $\varphi^{op}(r)$ is linear in the fundamental field operator $\Psi(r)$, namely that $\varphi^{op}$ is microscopically given by eq. (10.55a). The difficulty is in combining (b) and (c), and while there has been related work and partial success ( e.g. Gavoret and Nozières 1964, Balian and DeDominicis 1964), I do not think the question is settled, particularly for finite temperatures. To indicate a possible source of difficulty: it may be that the operator $\varphi^{op}$ which fulfils the assertions (a) and (b) above is not strictly linear in $\Psi$ but contains a $\Psi^3$ admixture as well. That would not affect the definition of $\rho_s$ in terms of currents correlations, eq. (10.18), but it would affect eq. (10.101a). Note, however, that our hydrodynamic results are independent of the microscopic expression for $\varphi^{op}$.

# CHAPTER 11

# NEMATIC LIQUID CRYSTALS

Water is a gas at high temperatures, a liquid between 100°C and 0°C, and solid ice when it is cold. In the liquid and gas phases the $H_2O$ molecules are free to translate and rotate while in the solid both the translational and rotational degrees of freedom are frozen except for small oscillations. Not all substances fall into this simple scheme, however. The molecules of methane, for example, are so nearly spherical that they continue to rotate freely even in the solid, until at a very low temperature ($\sim$ 20°K) a second, orientational, freezing transition takes place.

Liquid crystals are systems, composed of elongated organic molecules, in which this sequence is reversed. A classical example is para-azoxyanisole (PAA) whose chemical composition is given in fig. 11.1. Above 135°C, at atmospheric

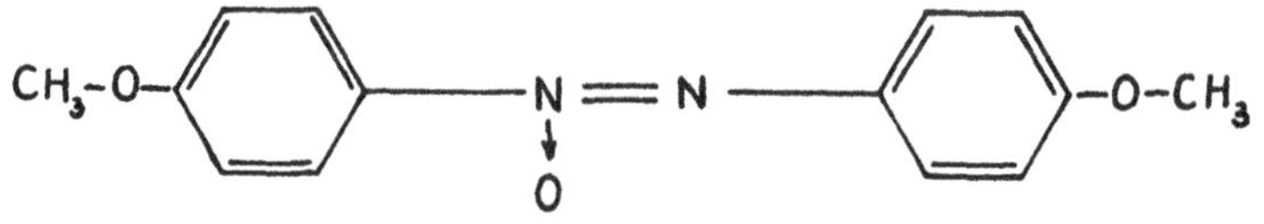


Fig. 11.1

pressure, PAA is a normal, isotropic liquid, while below 116°C PAA is a solid. In between PAA forms a "mesomorphic phase" -- a nematic liquid crystal. Phenomenologically, the liquid crystalline phases are liquid; for example, nematics flow like liquids. However, they are not isotropic like water. Rather, the electrical, magnetic, and mechanical properties of the nematic mesophase are quite strongly uniaxial. Our present understanding of the structure of a nematic is shown in fig. 11.2. As seen above, the PAA molecules are rod-like, with a length and width of roughly 20 Å and 5 Å, respectively. In the isotropic liquid these rods are positioned and oriented at random (fig. 11.2a), except for correlations of short range, and they translate and rotate chaotically.

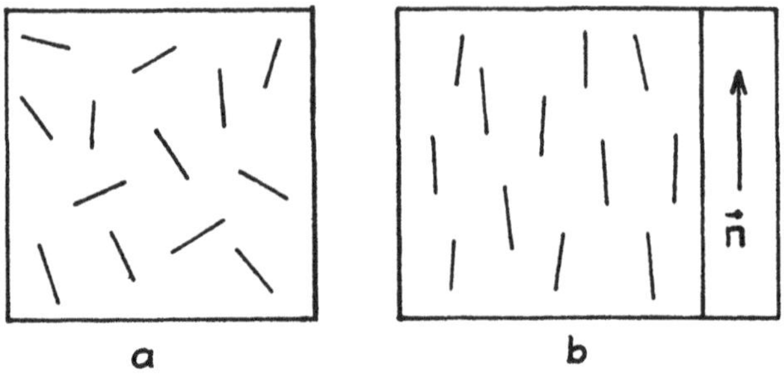


Fig. 11.2

In the nematic phase (fig. 11.2b) the centers of mass are still arranged at random, and the molecules move translationally as in a normal liquid. As in a conventional crystal, however, the molecular axes are aligned along some common direction $\vec{n}$, at least on the average. In contrast to a solid there is no translational order in a nematic. In contrast to a conventional liquid, however, there is orientational order.

The nematic structure is not the only one found in liquid crystalline systems. There are cholesterics, in which the molecular axes are arranged in a helical pattern, and there are several classes of smectics in which there is also at least some degree of translational order. I will not discuss these here, but refer the interested reader to the excellent book by de Gennes (1974) and the recent review by Stephen and Straley (1974). Even the discussion of nematics will be quite inexhaustive. As, for example, I did not consider vortex lines and rings in Helium, I will not consider here the threadlike macroscopic faults, "disclination lines", from which nematics derive their name (νημα = thread). What I will discuss are the hydrodynamic properties of nematics, and the microscopic principles from which they follow. The approach is that of Forster (1974a) see also Forster et al. (1971). Cholesterics, smectics, and solids have been considered by Martin et al. (1972).

## 11.1 The Free Energy of Distortion

In equilibrium all molecules are aligned, on the average, along some common direction $\vec{n}$. What determines this direction? Obviously, the interaction energy between the molecules is rotationally invariant. (There is a difference to the isotropic ferromagnet of eq. (7.19). The interaction between two PAA molecules changes if we just rotate their axes. It is only invariant if we rotate their center of mass positions as well.) Thus $\vec{n}$ must be determined externally, by boundary conditions or by external fields. In particular, for a uniaxial arrangement, as shown in fig. 11.2b, the magnetic susceptibility tensor $\chi_{ij}$ must be of the form

$$\chi_{ij} = \chi_{\parallel} n_i n_j + \chi_{\perp}(\delta_{ij} - n_i n_j), \tag{11.1}$$

and the magnetic free energy per unit volume is therefore

$$F_m/V = -\frac{1}{2} H_i \chi_{ij} H_j = -\frac{1}{2}\chi_{\perp} \vec{H}^2 - \frac{1}{2}\chi_a(\vec{n}\cdot\vec{H})^2, \tag{11.2}$$

where $\chi_a = \chi_{\parallel} - \chi_{\perp}$ is the anisotropic part of $\chi_{ij}$, and is usually positive. In a constant magnetic field $\vec{H}$, $F_m$ is minimized if the unit vector $\vec{n}$ is parallel to $\vec{H}$.

This also means that a slowly varying magnetic field $\vec{H}(r)$ distorts the uniform alignment. Such a distortion will, of course, produce internal thermodynamic forces which would drive the system back to uniformity were they not balanced by the magnetic forces which are contained in (11.2). Phenomenologically, we can describe the distortion by a spatially varying unit vector $\vec{n}(\vec{r})$, the "director", and if $\vec{n}(\vec{r})$ varies slowly on a molecular scale we can obtain an expression for the free energy of distortion by collecting all terms which are at most quadratic in $\nabla_i n_j$, and compatible with the uniaxial symmetry. The possible terms are further restricted by two symmetries. First, nematics are invariant under parity, $\vec{r} \to -\vec{r}$. Second, the state $\vec{n}$ and the state $-\vec{n}$ are physically identical. For PAA this must be so since the molecule itself does not know head from tail, see fig. 11.1. Even systems like MBBA, with the awful chemical name N-(p-methoxybenzylidene)-p-butylaniline, and whose molecules are not quite symmetric, have this up-down symmetry. If so, there are only three scalars that can be assembled from $\nabla_i n_j$ to quadratic order. They are given by the first three terms in

$$\delta F = \frac{1}{2} \int d^3r [K_1(\mathrm{div}\, \vec{n})^2 + K_2(\vec{n} \cdot \mathrm{curl}\, \vec{n})^2 + K_3(\vec{n} \times \mathrm{curl}\, \vec{n})^2] - \frac{1}{2} \int d^3r \, \chi_a(\vec{H} \cdot \vec{n})^2 \ . \qquad (11.3)$$

The three constants $K_i$, $i = 1,2,3$, must be positive so that the free energy increase upon distortion. We have also written the contribution of a magnetic field $\vec{H}(r)$, omitting however the first term in (11.2) which is independent of the direction of the molecules. Equation (11.3) is the fundamental formula of the continuum theory of nematics (Frank 1958). By pursuing its consequences (see de Gennes 1974) one can measure the three elastic constants $K_i$, conventionally called the Frank coefficients. They are found to be roughly equal, and of magnitude $\sim 10^{-6}$ dynes.

What does eq. (11.3) say in terms of correlation functions? Let us argue loosely first. Suppose we can find some local, microscopic vector, $\vec{n}^{op}(\vec{r})$, whose average value is the macroscopic director $\vec{n}(r)$. If we write, for small deviations from uniformity, $\vec{n}(r) = \vec{n}^o + \delta\vec{n}(r)$, then $\delta\vec{n}$ must be orthogonal to $\vec{n}^o$ in order to preserve $\vec{n}^2 = 1$. Thus $\delta\vec{n}$ has only two transverse (to $\vec{n}^o$) components. Now arguing as we did after eq. (7.73) we can easily see that, to 2nd order in $\delta n(r)$, the free energy of distortion is generally given by

$$\delta F = \frac{1}{2}\int d^3r \int d^3r' \, \delta n_i(r)\, K_{ij}(r-r')\, \delta n_j(r'), \tag{11.4}$$

where the suppressed sums over i and j go only over 1,2 if $\vec{n}^o$ is in z-direction, and where $K_{ij}(r-r')$ is the matrix inverse of the static susceptibility matrix $\chi_{n_i n_j}(\vec{r}-\vec{r}')$ of the operators $\vec{n}^{op}(r)$, defined as usual by eq. (3.10). The simplest coordinate system is one with $\vec{n}^o$ in the 3-direction and the wave vector $\vec{k}$ in the 1-3 plane. By comparing (11.4) to (11.3) one finds, in the absence of magnetic fields, that

$$\chi_{n_1 n_1}(\vec{k}) = \frac{1}{K_1 k_1^2 + K_3 k_3^2}\quad , \tag{11.5a}$$

$$\chi_{n_2 n_2}(\vec{k}) = \frac{1}{K_2 k_1^2 + K_3 k_3^2}\quad . \tag{11.5b}$$

In other words, one finds that for small k in the orientationally ordered system the director susceptibility diverges as $1/k^2$. This behavior will be the starting point for our theory of hydrodynamic fluctuations in nematics.

The $1/k^2$ singularity has, of course, been built in to the theory by insisting that the free energy be independent of the direction $\vec{n}^o$ along which the molecules happen to be aligned. In the presence of a finite external magnetic field $\vec{H}$, parallel to $\vec{n}^o$, we would have found instead that

$$\chi_{n_1 n_1}(\vec{k}) = [K_1 k_1^2 + K_3 k_3^2 + \chi_a \vec{H}^2]^{-1}, \tag{11.6}$$

because now it does cost energy to rotate $\vec{n}^o$. Equation (11.6) defines a coherence length $\xi(H)$ which is given by

$$\xi(H) = \frac{1}{H}\sqrt{K/\chi_a} \quad , \tag{11.7}$$

where K is some average of the $K_i$'s. In real space, director correlations now fall off as

$$\langle n_1(r)\, n_1(o)\rangle \sim \frac{k_B T/K}{4\pi r}\, e^{-r/\xi} \tag{11.8}$$

for large distances, but as $1/r$ in the field-free case.

[The Frank free energy (11.3) contains an interesting alternative. The equilibrium director $\vec{n}^o$ can be fixed by suitable treatment of the glass plates, say, which hold the nematic. Suppose again that $\vec{n}^o$ is in the 3-direction but that the external field $\vec{H}$ is in the 1-direction. In this case we find from (11.4) and (11.3) that

$$\chi_{n_1 n_1}(\vec{k}) = [\, K_1 k_1{}^2 + K_3 k_3{}^2 - \chi_a H^2 ]. \tag{11.9}$$

The susceptibility, $\chi_{n_1 n_1}(\vec{k})$, must of course be positive in a stable system but (11.9) violates this condition when $k > k_c = \xi^{-1}(H)$, assuming $K_1 \sim K_3$. The result is a transition to a state in which $\vec{n}^o$ is not spatially uniform. This is called the Frederiks transition (see de Gennes 1974) which has many of the properties of a phase transition. It can be used, for example, to measure the Frank coefficients.]

## 11.2 The Order Parameter

In all of our previous examples we characterized an ordered state by an order parameter whose nonvanishing value, in the ordered system, singles out the particular state into which the system has condensed. One's first idea is that in nematics there must be a molecular vector observable, say $\vec{n}^{op}(r)$, whose average $\langle \vec{n}^{op}(r) \rangle$ vanishes in the isotropic liquid but is proportional to $\vec{n}^{o}$ in the mesophase. However, since molecules like PAA have a center of symmetry (which accounts for the equivalence of $\vec{n}^{o}$ and $-\vec{n}^{o}$) no molecular vector can have a non-zero average. The next possibility is to choose a tensorial order parameter. If $R_{ij}(\vec{r})$ is a symmetric and traceless second rank tensor, its average must vanish in an isotropic system but in a uniaxial system it is given by

$$\langle R_{ij}(r) \rangle = \rho S(n_i^o n_j^o - \frac{1}{3} \delta_{ij}), \tag{11.10}$$

where $S = S(T) \neq 0$ and $\rho$ is the mass density. (A more general state of this kind would be biaxial: this would correspond to a system in which not only the long molecular axes are parallel on the average, but in which the planes of the benzene rings in fig. 11.1 are parallel to one another as well. Biaxial nematics have never been found.) Equation (11.10) has the $\vec{n}^{o} \to -\vec{n}^{o}$ symmetry built in. If we assume the molecules to be rigid we can write down a molecular expression for $R_{ij}(r)$, namely the moment of inertia tensor (Lubensky 1970 )

$$R_{ij}(\vec{r}) = \sum_{\alpha} m^{\alpha}(\xi_i^{\alpha} \xi_j^{\alpha} - \frac{1}{3} \delta_{ij}) \, \delta(r - r^{\alpha}), \tag{11.11}$$

where $\vec{\xi}^{\alpha}$ is the unit vector along the long axis of the $\alpha$-th molecule whose center of mass position is $\vec{r}^{\alpha}$ and whose mass is $m^{\alpha}$. In this case we find that

$$S = \frac{3}{2} \langle \cos^2 \theta^{\alpha} - 1/3 \rangle \;, \tag{11.12}$$

where $\theta^{\alpha}$ is the angle between $\vec{\xi}^{\alpha}$ and the preferred direction $\vec{n}^{o}$. It is easy to extend (11.11) to non-rigid molecules of arbitrary shape (de Gennes 1969, Forster 1975). In this case, S of (11.10) will still be given by (11.12) except for a multiplicative constant.

Which symmetry is broken in a state in which $\langle R_{ij} \rangle \neq 0$? The answer is, obviously, rotational symmetry, generated by the total angular momentum operator $\vec{L}$. Except for gradient terms whose average vanishes in a uniform system, the tensorial character of $R_{ij}$ is expressed by

$$\frac{1}{i\hbar}[L_i, R_{jk}] = \epsilon_{ij\ell} R_{\ell k} + \epsilon_{ik\ell} R_{j\ell}. \qquad (11.13)$$

It is much more convenient to define local vector components by

$$n_1(rt) = R_{13}(rt)/\rho S \text{ and } n_2(rt) = R_{23}(rt)/\rho S, \qquad (11.14)$$

if $\vec{n}^{\,o}$ is in the 3-direction, or generally by $n_i = (\delta_{ik} - n_i^o n_k^o) R_{k\ell} n_\ell^o/\rho S$. Then (11.13) leads to

$$\langle [L_i, n_j(r)] \rangle = i\hbar \epsilon_{ijk} n_k^o. \qquad (11.15)$$

This is of the form of eq. (7.53). The two broken symmetries are rotations about the 1- and 2-axes, generated by $L_1$ and $L_2$. The symmetry-restoring operators are $n_1 \sim R_{13}$ and $n_2 \sim R_{23}$, respectively, which play the role of director fluctuations. The reader will easily convince himself that (11.13) averages to zero for all other components of $R_{ij}$.

It would be tempting now to proceed as in section 7.4, i.e., to prove on the basis of a Bogoliubov inequality that $\chi_{n_i n_i}(\vec{k})$ must diverge, for small k, at least as $1/k^2$. This cannot be done (Straley 1972). The difficulty is that the formally defined angular momentum density, $\vec{\ell}(r)$, is not a local operator but contains reference to an origin, so that e.g., its autocorrelation function increases as $V^{5/3}$ with V. It is clear from (11.15) and (7.60) that

$$\chi_{n_i n_i}(k=0) \geq \frac{1}{\langle \dot{L}_2 \dot{L}_2 \rangle} = \infty \qquad (11.16)$$

at $k = 0$, but we cannot prove more since

$$\langle \dot{\ell}_2(-k)\, \dot{\ell}_2(k) \rangle = k_B T \int \frac{d\omega}{\pi}\, \omega \chi''_{\ell_2 \ell_2}(k\omega) \qquad (11.17)$$

is not of order $k^2$ as $k \to 0$. To obtain the form (11.15) we have to use (11.16), and rely on the intuitive assumption that the molecular variables $R_{ij}(r)$ are so local that $\chi^{-1}_{n_1 n_1}(r)$ is a short-ranged function. Related discussion has been given at the end of section 7.5. The reader will now appreciate why I started this discussion by considering the macroscopic, plausible and experimentally verified expression (11.3) for the Frank free energy, rather than by considering formal Bogoliubov inequalities.

Apart from providing an argument for why, even for a tensorial order parameter $R_{ij}$, there are only two variables $R_{13}$ and $R_{23}$ whose susceptibilities diverge as $k \to 0$, the present discussion does yield one other bonus. The locally meaningful quantities are, of course, $R_{13}$ and $R_{23}$, rather than $n_1$ and $n_2$ which are formally infinite in the isotropic phase. Let us therefore write (11.5) in the form

$$\chi_{R_{13}R_{13}}(\vec{k}) = \frac{\rho^2}{(K_1/S^2)k_1^{\,2} + (K_3/S^2)k_3^{\,2}} \quad . \qquad (11.18)$$

In the isotropic phase $\chi^{-1}_{R_{13}R_{13}}(k=0)$ is a finite constant, which is added to the denominator of (11.18). We therefore expect that the ratios $(K_i/S^2)$ are determined by the local, short-ranged, structure which is not very temperature dependent. It is indeed found that in their temperature dependence the Frank coefficients $K_i$ behave roughly like the square of the order parameter S. Since the $K_i$ have dimensions of an energy divided by a length, one further expects to get a reasonable estimate for $K_i/S^2$ by dividing a binding energy of $\sim 400°K \sim 6 \times 10^{-14}$ erg (the ordering temperature) by an average molecular length of 10 Å. This gives $K_i \sim 10^{-6}$ dynes which is about right.

## 11.3 The Light Scattering Intensity

The most striking observation one makes when cooling PAA below the isotropic to nematic transition temperature $T_c \approx 135°C$ is that the liquid, which is pretty clear above $T_c$, becomes turbid more or less suddenly below $T_c$. For this reason $T_c$ is called the clearing point. It was originally thought that the turbidity is due to strong Tyndall scattering by quasi-crystalline "swarms" of molecules suspended in a fluid phase. This is not so. We can now understand the strong light scattering intensity from nematics on the following basis.

We have seen in chapter 4 and in the Appendix that light is scattered by fluctuations $\delta\epsilon_{ij}(rt)$ of the local dielectric constant. The basic formula for the scattered intensity is eq. (A.28), namely

$$I_{i\to j}(\vec{k},\omega) = \int d^3r \int_{-\infty}^{\infty} dt\, e^{i\omega t - i\vec{k}\cdot\vec{r}} \langle \delta\epsilon_{ij}(rt)\,\delta\epsilon_{ij}(00)\rangle \qquad (11.19)$$

for light that is scattered from the initial frequency $\Omega$, wave vector $\vec{K}$, and polarization $\hat{i}$ to the final frequency $\Omega-\omega$, wave $\vec{K}-\vec{k}$, and polarization $\hat{j}$. Some kinematic factors are omitted from (11.19). In a uniaxial nematic $\delta\epsilon_{ij}$ will be uniaxial as well, and can be written in the form

$$\delta\epsilon_{ij}(rt) = \delta\epsilon(rt)\,\delta_{ij} + a\,\delta R_{ij}(rt) \quad , \qquad (11.20)$$

where a is roughly of the order of a molecular polarizeability. Omitting the isotropic term in (11.20) and integrating (11.19) over all frequencies we obtain the total scattering intensity in direction $\vec{k}$,

$$I_{i\to j}(\vec{k}) = \int \frac{d\omega}{2\pi}\, I_{i\to j}(\vec{k},\omega) = a^2 \langle \delta R_{ij}(r)\,\delta R_{ij}(0)\rangle(\vec{k}). \qquad (11.21a)$$

And using the classical fluctuation-dissipation theorem (at 130°C, PAA is certainly a classical system),

$$I_{i\to j}(\vec{k}) = a^2 k_B T \; \chi_{R_{ij}R_{ij}}(\vec{k}) \quad . \qquad (11.21b)$$

In particular, for depolarized light scattering from $\hat{i}$ in 1-direction to $\hat{j}$ in 3-direction, chosen as the direction of alignment, we find

$$I_{depol.}(\vec{k}) = \frac{(a\rho)^2 k_B T \cdot S^2}{K_1 k_1^{\,2} + K_3 k_3^{\,2}} \quad . \qquad (11.22)$$

By contrast, in an isotropic liquid we only have isotropic scattering, $\hat{i} \to \hat{i}$, generated by the first term in (11.20). The major contribution to $\delta\epsilon(rt)$ are fluctuations of the particle density $\delta n(rt)$, and we obtain

$$I_{isotr.}(\vec{k}) = a'^2 k_B T \; \chi_{nn}(\vec{k}) \approx a'^2 \, k_B T \cdot n \, \frac{\partial n}{\partial p})_T \, , \qquad (11.23)$$

where $a' = \partial\epsilon/\partial n$ is again of the order of a molecular polarizeability. Note the important difference: $I_{depol.}(\vec{k})$ diverges as $k^{-2}$ as $k \to 0$, $I_{isotr.}(k)$ is finite. Their ratio is of the order

$$I_{depol.}(\vec{k})/I_{isotr.}(\vec{k}) \sim 1/(ka)^2, \qquad (11.24)$$

where a must be a length by dimension. In a molecular liquid there is only one such length: the size of molecules, or the next neighbor distance, or the force range, all of the order of 10 Å, say. (de Gennes, 1974, arrives at (11.24) by a somewhat more careful estimate. In a sufficiently chaotic system, nothing is quite as powerful as dimensional analysis.) Since $k \sim 10^{-4}$ Å$^{-1}$ for optical wavelengths, the scattering in the nematic phase is about a million times larger than it is in the isotropic phase. That is the reason for the turpidity of nematics. Physically, as noted by de Gennes, a long wavelength modulation of $\epsilon$ in a conventional liquid requires a uniform dilation which requires a finite elastic energy. In a nematic liquid, on the other hand, fluctuations of $\epsilon$ can be induced by a rotation of the optical axis which, for nearly uniform rotation, $k \to 0$, costs very little energy.

It should be clear from this discussion that the turpidity of liquid crystals is not due to "swarms" but is an intrinsic property of orientational order. The main results of this section are first, that the scattering is unusually intense only for crossed polarizations $\hat{i}$ and $\hat{j}$, and second, that the scattering increases with decreasing k, i.e., towards the forward direction. These results were already obtained experimentally by Chatelain (1948), and verified in much detail by the Orsay group (1969,1971).

## 11.4 Transverse Hydrodynamic Fluctuations

We will now proceed to work out the equations of motion for slow, long-wavelength fluctuations in nematics. The procedure used is the canonical one (if it be permitted to use this highfalutin word for the recipes worked out in the course of the last ten chapters): we use the master formula (5.28) and work out the form which the matrices $\chi_{\mu\nu}$, $\omega_{\mu\nu}$, and $\sigma_{\mu\nu}$ take for small k and z.

The hydrodynamic variables are, of course, the conserved densities $\rho$, $\vec{g}$, and $\epsilon$ of mass, momentum and energy, plus the two transverse director components $n_1$ and $n_2$, the latter because they are symmetry-breaking local variables with large static fluctuations at small k. $n_1$ and $n_2$ are more convenient to write than the tensor components $R_{13}$ and $R_{23}$. We will again use a coordinate system in which $\vec{n}^o$ is in the 3-direction, and $\vec{k}$ in the 1-3 plane. Note, first, the symmetry signatures of these variables: $n_1$ and $n_2$ behave like vector components under rotations about the 3-axis (the only symmetry axis in the problem), but they are even under time reversal and under parity, and odd under the additional symmetry $\vec{n}^o \to -\vec{n}^o$. $\rho$ and $\epsilon$ are even under T and P as always, $\vec{g}$ is odd under both, and all five variables are even under $\vec{n}^o \rightarrow -\vec{n}^o$. The reader will then easily verify that there is no cross coupling between the two groups of variables

$$\{A_\mu ; \mu = 1 \,..\, 5\} = \{\rho, \epsilon, g_3, g_1, n_1\} \tag{11.25a}$$

and

$$\{A_\mu ; \mu = 6,7\} = \{g_2, n_2\}\,. \tag{11.25b}$$

In this section we will consider the fully transverse variables $g_2$ and $n_2$, and their correlation matrix

$$C_{\mu\nu}(kz) = \langle A_\mu(k) | \frac{i}{z-L} | A_\nu(k)\rangle\,, \tag{11.26}$$

($\mu$ =6,7) defined as always by, e.g., eqs. (3.44).

The two by two matrix of susceptibilities is diagonal because of time reversal, and given by

$$\chi_{\mu\nu}(\vec{k}) = \begin{pmatrix} \rho & 0 \\ 0 & [K_2 k_1^2 + K_3 k_3^2]^{-1} \end{pmatrix} \tag{11.27}$$

to lowest relative order in k. Nematics are not superfluid, of course, so that $\chi_{g_2 g_2}(k) = \rho$. Again from time reversal symmetry, the unnormalized frequency matrix $\omega_{\mu\nu}(k)$ has no diagonal components, and its off-diagonal elements are given by the Poisson bracket

$$\omega_{g_2 n_2}(\vec{k}) = \omega_{n_2 g_2}(\vec{k}) = i\langle [g_2(\vec{r}), n_2(\vec{r}\,')]_{P.B.}\rangle(k). \tag{11.28a}$$

Given an explicit mechanical expression for the order parameter $n_2(\vec{r}) \sim R_{23}(\vec{r})$, such as (11.11), this Poisson bracket can be evaluated (Forster 1974b). It is clear, in any case, that $\omega_{g_2 n_2}(k)$ vanishes as $k \rightarrow 0$ because a nematic is translationally invariant, and the only expression linear in $\vec{k}$ which is compatible with the uniaxial symmetry ($\omega_{g_2 n_2}$ is the 223 element of a third rank tensor) and with parity is

$$\omega_{g_2 n_2}(\vec{k}) = -\mu_2 k_3, \tag{11.28b}$$

where $\mu_2$ is real but need not be positive.

In all of our previous examples we could omit the memory matrix $\sigma_{\mu\nu}$, systematically to lowest order in k, and obtain results which were reversible and asymptotically correct for small k. Let us try that here (it is not permitted as we will see). From (11.28) and (11.27) we obtain the normalized frequency matrix in the form

$$\underset{\approx}{\Omega}(k) = \underset{\approx}{\omega}(k)\,\underset{\approx}{\chi}^{-1}(k) = -\mu_2 k_3 \begin{pmatrix} 0 & K_2 k_1^2 + K_3 k_3^2 \\ \rho^{-1} & 0 \end{pmatrix}. \tag{11.29}$$

From $\det[z - \underset{\approx}{\Omega}(k)] = 0$ we would then obtain two modes with the interesting dispersion relation

$$z = \pm k^2 \cdot D(\varphi), \tag{11.30a}$$

$$D(\varphi) = \mu_2[(K_3 \cos^2\varphi + K_2 \sin^2\varphi)/\rho]^{1/2}, \tag{11.30b}$$

where $\varphi = (\hat{k}, \vec{n}^{\circ})$ is the propagation angle. We would, in other words, find propagating waves of the type of ferromagnetic spin waves, with an anisotropic effective mass to boot. A spectacular result.

Alas, things are otherwise. Using only momentum conservation, director non-conservation, and symmetry we find that $\underset{\approx}{\sigma}(kz)$ is of the form

$$\sigma_{\mu\nu}(k,z) = \begin{pmatrix} \nu_2 k_1^2 + \nu_3 k_3^2 & -ik_3\tilde{\mu} \\ -ik_3\tilde{\mu} & \xi \end{pmatrix} \tag{11.31}$$

at $z = i0$, and omitting only terms of relative order $k^2$. Because of the general positivity of $\sigma_{\mu\nu}$, eq. (5.32), the viscosities $\nu_2, \nu_3$ and the rotational relaxation coefficient $\xi$ are positive while $\tilde{\mu}$ is real but can have either sign. The reader will first notice that $\tilde{\mu}$ contributes to the reactive, rather than dissipative, behavior of nematics. (It is easy to show that the matrix $\underset{\approx}{\Omega}(k)$ in (5.28) never contributes to dissipation. However, $\underset{\approx}{\Sigma}(kz)$ may contain reversible contributions. It always does at high frequencies, and for nematics it does at hydrodynamic frequencies as well.) Indeed, since only the sum $\underset{\approx}{\omega}(k) - i\underset{\approx}{\sigma}(kz)$ occurs in the master formula (5.28), only the sum

$$\mu_2 + \tilde{\mu} \equiv \frac{1}{2}(\lambda + 1) \tag{11.32}$$

enters the full hydrodynamic theory. If the reader now uses (11.31) and (11.27) to obtain the matrix $\underset{\approx}{\Sigma}(kz) \approx \underset{\approx}{\Sigma}(k, i0)$, she or he can compute the corrected mode dispersion relations from $\det[z - \underset{\approx}{\Omega}(k) + i\underset{\approx}{\Sigma}(kz)] = 0$. This leads to either one of two possibilities, namely

$$z = \pm D'k^2 - \frac{i}{2}\Gamma' k^2 \tag{11.33}$$

where $D'$ and $\Gamma'$ are real and $\varphi$-dependent, or

$$z = -ik^2[\Gamma_s(\varphi) \text{ or } \Gamma_f(\varphi)] , \tag{11.34}$$

depending on whether (omitting angle dependence for simplicity)

$$(\lambda + 1)^2 K/\rho - (\xi K - \nu/\rho)^2 > 0 \text{ or } < 0. \tag{11.35}$$

In fact, in PAA and all other nematics examined to date, $\nu^2 \gg K\rho$ so that the latter case is experimentally realized. In terms of the angle-dependent coefficients,

$$\begin{aligned} K(\varphi) &= K_2 \sin^2\varphi + K_3 \cos^2\varphi , \\ \nu_t(\varphi) &= \nu_2 \sin^2\varphi + \nu_3 \cos^2\varphi , \end{aligned} \tag{11.36}$$

the damping constants are explicitly given by

$$\Gamma_s + \Gamma_f = \nu_t/\rho + \xi K \quad ,$$

$$\Gamma_s \cdot \Gamma_f = (\nu_t/\rho)\xi K + \frac{1}{4}(\lambda+1)^2 (K/\rho) \cos^2\varphi \quad . \qquad (11.37)$$

In PAA the shear viscosities $\nu_2$ and $\nu_3$ are of the order of 0.1 poise (an order of magnitude larger than the viscosity of water). $\xi^{-1}$ has also the dimensions of a viscosity, and is of the same order of magnitude. For typical wavenumbers $k \approx 10^4$ to $10^5$ cm$^{-1}$, we find a rapidly relaxing process with a relaxation time $\tau_f \approx (k^2 \nu/\rho)^{-1} \approx 10^{-7}$ to $10^{-9}$ sec, and a slowly relaxing process with a relation time $\tau_s \approx (k^2 \xi K)^{-1} \approx 10^{-3}$ to $10^{-5}$ sec. The former is, essentially, conventional shear diffusion, the latter describes the orientational diffusion of director fluctuations.

Following (5.28) one can now also calculate the hydrodynamically rigorous expressions for the fully transverse correlation functions. They emerge in the form

$$C_{\mu\nu}(kz) = \frac{i\beta^{-1}\rho}{(z+ik^2\Gamma_s)(z+ik^2\Gamma_f)} \cdot r_{\mu\nu}(kz), \qquad (11.37a)$$

where

$$r_{g_2 g_2}(kz) = z + ik^2 \xi K ,$$

$$r_{n_2 n_2}(kz) = (z+ik^2\nu_t/\rho)/(\rho K k^2),$$

$$r_{g_2 n_2}(kz) = -\frac{1}{2} k(\lambda+1) \cos\varphi/\rho \quad . \qquad (11.37b)$$

Because $\Gamma_s \ll \Gamma_f$, these results could be considerably simplified for practical applications (Orsay Group 1969); but we will not do that. Since the three correlation functions $\chi''_{g_2 g_2}$, $\chi''_{n_2 n_2}$, and $\chi''_{g_2 n_2}$ are real, they can be obtained by taking the real part of $\beta C_{\mu\nu}(k,\omega)$. In particular, the director correlation function $\omega^{-1}\chi''_{n_2n_2}(k\omega)$, which is measured in elastic, depolarized light scattering, emerges in the form

$$\frac{1}{\omega}\chi''_{n_2n_2}(\vec{k}\omega) = \frac{\omega^2\xi + k^4(\nu/\rho K)\,\Gamma_s\cdot\Gamma_f}{[\omega^2 + (k^2\Gamma_s)^2][\omega^2 + (k^2\Gamma_f)^2]} \ . \tag{11.38}$$

It exhausts, of course, the "thermodynamic" sum rules

$$\lim_{k_1\to 0}\ \lim_{k_3\to 0} k_1^2 \int\frac{d\omega}{\pi}\,\omega^{-1}\chi''_{n_2n_2}(\vec{k}\omega) = 1/K_2 \ , \tag{11.39a}$$

$$\lim_{k_3\to 0}\ \lim_{k_1\to 0} k_3^2 \int\frac{d\omega}{\pi}\,\omega^{-1}\chi''_{n_2n_2}(\vec{k}\omega) = 1/K_3 \ . \tag{11.39b}$$

It also leads to a Kubo relation for the relaxation coefficient $\xi$ which is of the expected form,

$$\xi = \lim_{\omega\to 0}\ \lim_{k\to 0} \omega\chi''_{n_2n_2}(k\omega) \ . \tag{11.40}$$

Note, however, that if we use (11.37) to compute the cross correlation function $\chi''_{g_2n_2}(k\omega)$, we find that the hydrodynamically correct result fails to fulfil the sum rule

$$\lim_{k\to 0} k_3^{-1} \int\frac{d\omega}{\pi}\,\chi''_{g_2n_2}(\vec{k}\omega) = -\mu_2, \tag{11.41}$$

which expresses the eq. (11.28). Instead, (11.37) replaces $\mu_2$ in (11.41) by $\mu_2 + \tilde{\mu} = \frac{1}{2}(\lambda+1)$. The difference has to do with the fact that the matrix element $\sigma_{g_2n_2}(kz)$ vanishes as $z\to\infty$, and therefore does not contribute to the sum rule; but it need not vanish as $z\to i0$, and it therefore does contribute to hydrodynamics. Or in different words, only the part $\mu_2$ of the reactive coupling between orientations and shear stress is instantaneous, provided as it were by a thermodynamic mean field. The part $\tilde{\mu}$ appears to be instantaneous only if the system is probed at low frequency; $\tilde{\mu}$ is in fact due to collisions which require a finite if short time to become effective. $\mu_2$ can be calculated from equilibrium statistical mechanics (Forster 1974b). To compute $\tilde{\mu}$ (which no one has done yet) microscopically one would have to solve a Boltzmann-type kinetic equation.

Another remark is in place here. We have pointed out that both transverse modes in nematics are diffusive, see (11.34); there are no shear waves, in the strict sense of the word. It is instructive, however, to separate eqs. (11.37) in the form

$$C_{g_2g_2} = i\beta^{-1}\rho[-\frac{(\Gamma_s-\xi K)}{z+ik^2\Gamma_s} + \frac{\Gamma_f-\xi K}{z+ik^2\Gamma_f}](\Gamma_f-\Gamma_s)^{-1}, \qquad (11.42a)$$

$$C_{n_2n_2} = \frac{i\beta^{-1}}{Kk^2}[\frac{\Gamma_f-\xi K}{z+ik^2\Gamma_s} - \frac{\Gamma_s-\xi K}{z+ik^2\Gamma_f}](\Gamma_f-\Gamma_s)^{-1}, \qquad (11.42b)$$

which we have written so that all numerators are positive. Both $\omega^{-1}\chi''_{n_2n_2}$ and $\omega^{-1}\chi''_{g_2g_2}$ are superpositions of two Lorentzians, one narrow ($\Gamma_s$), the other broad ($\Gamma_f$). In $\omega^{-1}\chi''_{n_2n_2}$ the broad line appears as a background (of negative weight) which is hard to notice. In $\omega^{-1}\chi''_{g_2g_2}$, however, the main weight is in the positive broad component into which the narrow line (which has negative weight) burns a hole. This is shown in fig. 11.3, which is exaggerated for clarity. Unfortunately, one does not measure $\omega^{-1}\chi''_{gg}$ by light scattering. An ultrasonic reflection experiment has been performed by Martinoty and Candau (1971); it is carefully discussed by Stephen and Straley (1974).

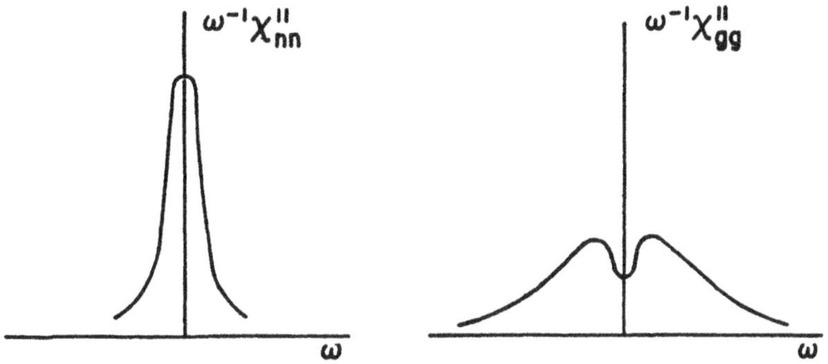


Fig. 11.3

Consider, however, what would happen if the inequality $\xi K < \nu_t/\rho$ were reversed -- to be more to the point, if the rotational relaxation time were shorter than the shear diffusion time. Then the hole would be burned out of a basically broad line for the correlation function of the orientation density $\vec{n} \sim R_{ij}$, and could be measured by light scattering. That is what happens in a large number of organic, not nematogen, liquids like anisaldehyde or quinoline (Stegeman and Stoicheff 1973, Andersen and Pecora 1971, Kristaponis and Forster 1975). One may quibble with whether one should call the observed doublet evidence for "shear waves" (since $\chi''_{gg}$ does not show it) but that is what it has been called.

Above the clearpoint in a nematic, one might expect to see the same doublet in the depolarized scattering spectrum. However, the isotropic to nematic transition is almost of second order, and just above $T_c$ the rotational relaxation time $\tau_{rot}$ [which replaces $(k^2\xi K)^{-1}$ in eqs. (11.42)] is still much larger than the shear diffusion time $(k^2\nu/\varrho)^{-1}$. Consequently pretransitional light scattering in nematics shows a spectrum of the structure of fig. 11.3, for all accessible temperatures above $T_c$.

## 11.5 Longitudinal Hydrodynamic Fluctuations

The analysis of those fluctuations which involve the longitudinal variables $\rho$, $\epsilon$, $g_3$, $g_1$, and $n_1$ is entirely analogous if more laborious in detail, and we will therefore only give some pertinent formulae. For more detail the reader is referred to the literature in which the present (Forster et al. 1971, Forster 1974a, Martin et al. 1972) or alternative (de Gennes 1974, Stephen and Straley 1974) formulations are employed. The simplest way of summarizing matters seems to be to write down, to hydrodynamic accuracy, the three matrices $\underset{\approx}{\chi}$, $\underset{\approx}{\omega}$, and $\underset{\approx}{\sigma}$. So here goes.

The matrix of susceptibilities has the elements

$$\chi_{\rho\rho} = \rho\left(\frac{\partial\rho}{\partial p}\right)_T \ ; \ \chi_{qq} = \rho c_p T ;$$

$$\chi_{\rho q} = T\left(\frac{\partial\rho}{\partial T}\right)_p \ ; \ \chi_{g_i g_j} = \rho\,\delta_{ij} ;$$

$$\chi_{n_1 n_1}(\vec{k}) = [K_1 k_1^2 + K_3 k_3^2]^{-1}, \tag{11.43}$$

where again $q(rt) = \epsilon(rt) - (\epsilon + p/\rho)\,\rho(rt)$ describes fluctuations of the entropy. Except for $\chi_{n_1 n_1}$, of course, these thermodynamic expressions are the same as those in an isotropic liquid. Note, in particular, that at equilibrium the pressure in a nematic is hydrostatic, so that the average stress tensor $\langle\tau_{ij}\rangle = p\,\delta_{ij}$ is isotropic; a nematic liquid crystal is liquid, not solid, its free energy independent of the shape of the container in which it is placed. This also leads to the observed fact that low frequency sound waves propagate with the same speed in all directions. Related is the fact that the only elements of $\underset{\approx}{\chi}(\vec{k})$ which do not vanish from time reversal symmetry, namely $\chi_{\rho n_1}(\vec{k})$ and $\chi_{q n_1}(\vec{k})$, vanish at k=0 because changes of pressure and temperature do not affect the direction of nematic alignment.

The frequency matrix $\underset{\approx}{\omega}(\vec{k})$ has the nonvanishing elements

$$\omega_{\rho g_i} = k_i \rho ,$$

$$\omega_{g_3 n_1} = -\mu_3 k_1 \; ; \; \omega_{g_1 n_1}(\vec{k}) = -\mu_2 k_3 \quad , \qquad (11.44)$$

to lowest order in k. It is clear that the matrix element $\omega_{g_1 n_1} \sim \omega_{g_1 R_{13}}$ must be proportional to $k_3 = (\vec{k}\,\vec{n}^o)$, and since, for $\vec{k}$ parallel to $\vec{n}^o$, $\omega_{g_1 n_1} = \omega_{g_2 n_2}$, the coefficient $\mu_2$ is identical to the one which occurs in (11.28b). It is somewhat more laborious to show that in fact

$$\mu_2 - \mu_3 = 1 \; . \qquad (11.45)$$

This equation is a consequence of the fact that in equilibrium a liquid rotates like a solid body. The only other elements which do not vanish from time reversal symmetry are $\omega_{qg_i}(\vec{k})$; they are of order $k^3$ rather than k by definition of the entropy variable q (and reflect Galilei invariance).

Finally, the <u>memory matrix</u>, $\sigma_{\mu\nu}(\vec{k},z) \approx \sigma_{\mu\nu}(\vec{k}, i0)$. Its nonvanishing elements, to lowest order in k, are of the form

$$\sigma_{qq} = T(\varkappa_\perp k_1^2 + \varkappa_\parallel k_3^2),$$

$$\sigma_{g_3 g_3} = \nu_3 k_1^2 + (2\nu_1 + 2\nu_5 + \nu_2 - \nu_4) k_3^2 \; ,$$

$$\sigma_{g_1 g_1} = (\nu_2 + \nu_4)\, k_1^2 + \nu_3 k_3^2 ,$$

$$\sigma_{g_3 g_1} = (\nu_3 + \nu_5) k_1 k_3 \; ,$$

$$\sigma_{n_1 n_1} = \xi ,$$

$$\sigma_{g_3 n_1} = -ik_1 \tilde{\mu} ,\; \sigma_{g_1 n_1} = -ik_3 \tilde{\mu} \; . \qquad (11.46)$$

The coefficients $\varkappa_\parallel$, $\varkappa_\perp$, $\nu_2$, $\nu_3$, $2(\nu_1 + \nu_5) + \nu_2 - \nu_4$, $\nu_4(2\nu_1 + \nu_2) - (\nu_5 - \nu_4)^2$, and $\xi$ are positive, and $\tilde{\mu}$ is real. $\tilde{\mu}$ and $\xi$ are the same coefficients which occur in the transverse fluctuations, see section 11.4.

I will leave things there. The reader who wants to insert the eqs. (11.43) through (11.46) into the general formula (5.28) to compute the correlation functions $C_{\mu\nu}$ or $\chi''_{\mu\nu}$ in the hydrodynamic limit is in for some labor but there is no difficulty in principle. The main results are as follows: Those correlation functions which link the director and the second shear momentum density component $\tilde{g}_t(\vec{k}t) = [k_3 g_1(\vec{k}t) - k_1 g_3(\vec{k}t)]/k$, namely $C_{n_1 n_1}$, $C_{\tilde{g}_t \tilde{g}_t}$, and $C_{\tilde{g}_t n_1}$, are of the same general form as those found in section 11.4. In particular, they contain two purely diffusive processes, one describing slow rotational diffusion, the other more rapid but still hydrodynamic shear diffusion. Conversely, those correlation functions which link the three longitudinal conserved densities, $\rho$, $q$, and $g_\ell = (\vec{k}\cdot\vec{g})/k$, contain two Brillouin modes and the Rayleigh heat diffusion mode. Indeed, these functions are of the same form as they are in normal liquids, namely of the form of (4.42) or (4.44). The speed of sound is the conventional one, $c = (\partial p/\partial \rho)_S^{1/2}$. The attenuation constants $\Gamma$ and $D_T$, however, depend on the angle of propagation $\varphi$ in nematics. Finally, the correlation functions which link these two sets of variables contain all five hydrodynamic poles. An unusual feature is explained in Forster (1974a): the strengths with which the poles enter into the correlation functions cannot fully be determined by hydrodynamic considerations, even to lowest order in k.

I do, however, want to explain the relation (11.45). Because of it, there is only one coefficient which couples director motion and shear non-dissipatively, namely $\lambda$ as given by (11.32), rather than two as one might expect in a system of uniaxial symmetry. The matter is quite simple, really. If we write

$$\langle [g_i(\vec{r}), n_j(\vec{r}\,')]_{P.B.} \rangle = \mu_{ijk} \nabla_k \delta(\vec{r}-\vec{r}\,') + \nabla\nabla\ldots, \quad (11.47)$$

then $\mu_{113} = \mu_{223} = \mu_2$ and $\mu_{311} = \mu_1$. Now $\vec{g}(\vec{r})$ is the total microscopic momentum density. We can write it in the form

$$\vec{g}(r) = \sum_{\alpha}\sum_{\nu} \vec{p}^{\alpha\nu}\, \delta(\vec{r} - \vec{r}^{\alpha\nu}), \tag{11.48}$$

where $\vec{r}^{\alpha\nu}$ and $\vec{p}^{\alpha\nu}$ are, respectively, the position and momentum of the $\nu$th particle in the $\alpha$th molecule. Defined in this fashion, it is apparent that $\vec{g}(r)$ describes both translational and rotational motion of the molecules, and that the total angular momentum $\vec{L}$ is given by

$$\vec{L} = \int d^3r\, \vec{r} \times \vec{g}(r) \quad . \tag{11.49}$$

This operator generates uniform rotations, whose effect on $\vec{n}(r)$ has been stated in eq. (11.15). If we therefore multiply (11.47) by $\epsilon_{mni}r_n$ and integrate, we find $\mu_{113}-\mu_{311} = 1$, that is $\mu_2-\mu_3 = 1$.

This relation has an interesting implication. Namely, because of the fluctuation-dissipation theorem (3.39) and momentum conservation, we can identify

$$\mu_{ijk}k_k = -\beta k_k \langle \tau_{ik}(\vec{k}) | n_i(\vec{k}) \rangle = -k_k \chi_{\tau_{ik}, n_i}(\vec{k}), \tag{11.50}$$

as $k \to 0$. $\tau_{ij}(\vec{r})$ is the fully microscopic stress tensor, resolved into the contributions from each particle within each molecule. $\tau_{ij}(\vec{r})$ is therefore of a form akin to that given in (4.6), and must be symmetric. (Related and detailed discussion is given in Martin et al., 1972.) If, thus, $\chi_{\tau_{ik}, n_i}(\vec{k})$ were uniform as $\vec{k} \to 0$, $\mu_{113}$ would have to equal $\mu_{311}$, contrary to what we just found. Consequently, the limit of $\chi_{\tau_{ik}, n_i}(\vec{k})$ as $\vec{k} \to 0$ depends on the direction from which the limit is taken, indicating that in a nematic the correlations between the stress tensor $\tau_{ij}$ and the molecular orientation density $R_{ij}$ are of infinite range. This can be seen as the fundamental property of liquid crystals. We will briefly come back to this property in section 11.7.

## 11.6 The Phenomenological Equations of Motion

It is useful to rephrase the results which we have obtained in the more conventional language of equations of motion for small macroscopic fluctuations $\delta\langle\rho(r,t)\rangle$ etc., not the least because this allows easy comparison with the various alternative, but equivalent, formulations of nematic hydrodynamics (see de Gennes 1974, and references quoted there). The procedure has already been explained in section 10.8, and so we can be brief. (Many authors prefer to interpret the hydrodynamic equations as Langevin equations of the type briefly considered in section 6.2. The present author does not.)

The microscopically conserved densities obey, of course, the continuity equations

$$\dot{\rho} + \vec{\nabla}\cdot\vec{g} = 0, \tag{11.51a}$$

$$\dot{g}_i + \nabla_j \tau_{ij} = 0, \tag{11.51b}$$

$$\dot{\varepsilon} + \vec{\nabla}\cdot\vec{j}^{\,\varepsilon} = 0 . \tag{11.51c}$$

In addition, there is an equation of motion for the director, $\vec{n}(r,t)$ which is constrained to be orthogonal to $\vec{n}^{\,o}$. If we write

$$\dot{n}_i + X_i = 0 \tag{11.51d}$$

we have simply invented a new name, $X_i(r,t)$, for the microscopic quantity $-\dot{n}_i(r,t)$. Now conventional hydrodynamics expands the small, macroscopic, non-equilibrium fluctuations $\delta\langle\vec{g}\rangle$, $\delta\langle\underset{\approx}{\tau}\rangle$, $\delta\langle\vec{j}^{\,\varepsilon}\rangle$, and $\delta\langle\vec{X}\rangle$ in terms of the gradients of the "thermodynamic forces", $\delta a_\mu$. In (10.109) we saw that a thermodynamically consistent definition of the latter is by

$$\delta a_\mu(\vec{k},t) = \chi^{-1}_{\mu\nu}(\vec{k})\, \delta\langle A_\nu(\vec{k},t)\rangle \,, \tag{11.52}$$

where it is sufficient to take $\chi^{-1}_{\mu\nu}(\vec{k})$ to the lowest nonvanishing order in $\vec{k}$. As in normal fluids, we find that $\delta a_\rho = \delta p/\rho$, $\delta a_q = \delta T/T$, and $\delta a_{g_i} = \delta v_i$. The forces conjugate to the director are given by

$$-\delta a_{n_1}(rt) \equiv \delta h_1(rt) = [K_1\nabla_1^2 + K_2\nabla_3^2]\,\delta n_1(rt),$$

$$-\delta a_{n_2}(rt) \equiv \delta h_2(rt) = [K_2\nabla_1^2 + K_3\nabla_3^2]\,\delta n_2(rt). \tag{11.53}$$

As we have obtained the director susceptibilities $\chi_{n_i n_j}(k)$ from the Frank free energy, these local stresses $\delta\vec{h}(rt)$ can be so derived (Orsay Group 1969). As (11.53) makes clear, these local elastic stresses are set up in a nematic only when there is curvature in the direction of alignment.

The expansion of fluxes in terms of forces can now be read off from our results. We first give the <u>reversible</u> contributions which come from the matrices $\underset{\approx}{\omega}(\vec{k})$ and from the real part of $i\underset{\approx}{\sigma}(\vec{k}, i0)$:

$$\delta\langle\vec{g}\rangle^R = \rho\,\delta\vec{v}\,, \tag{11.54a}$$

$$\delta\langle\vec{j}^\epsilon\rangle^R = (\epsilon + p)\,\delta\vec{v}\,, \tag{11.54b}$$

$$\delta\langle\tau_{ij}\rangle^R = \delta_{ij}\delta p - \frac{1}{2}\lambda(\delta h_i n_j^0 + \delta h_j n_i^0) - \frac{1}{2}\epsilon_{ijk}(\delta h \times n^0)_k\,, \tag{11.54c}$$

$$\delta\langle X_i\rangle^R = (n^0 \times \omega)_i - \lambda A_{ij} n_j^0 \;. \tag{11.54d}$$

In the last equation we have introduced the customary symbols $\vec{\omega}(r) = \frac{1}{2}\vec{\nabla} \times v(\vec{r})$ for the local vorticity of the fluid, and $A_{ij} = \frac{1}{2}(\nabla_i v_j + \nabla_j v_i)$ for the shear flow tensor. The first two equations follow from Galilei invariance which is automatically incorporated in our procedure. The coefficient in front of $\vec{n}^0 \times \vec{\omega}$ in the last equation, unity, is in accord with the requirement that $\partial_t\vec{n} + (\vec{n} \times \vec{\omega}) = 0$ for pure solid body rotation. About the stress tensor, $\delta\langle\tau_{ij}\rangle^R$, we will talk some more below. First, let us also write down, for completeness, the dissipative contributions to the fluxes:

$$\delta\langle\vec{g}\rangle^D = 0, \tag{11.55a}$$

$$\delta\langle\vec{j}^{\varepsilon}\rangle^D = -\kappa_\perp\vec{\nabla}\delta T - (\kappa_\parallel - \kappa_\perp)\vec{n}^o(\vec{n}^o\cdot\vec{\nabla})\delta T, \tag{11.55b}$$

$$\begin{aligned}\delta\langle\tau_{ij}\rangle^D &= -2\nu_2 A_{ij} - 2(\nu_3-\nu_2)[A_{ik}n_k^o n_j^o + A_{jk}n_k^o n_i^o] \\ &\quad -(\nu_4-\nu_2)\delta_{ij}A_{kk} - 2(\nu_1+\nu_2-2\nu_3)n_i^o n_j^o n_k^o n_\ell^o A_{k\ell} \\ &\quad -(\nu_5-\nu_4+\nu_2)[\delta_{ij}n_k^o n_\ell^o A_{k\ell} + n_i^o n_j^o A_{kk}],\end{aligned} \tag{11.55c}$$

$$\delta\langle\vec{X}\rangle^D = \xi\,\delta\vec{h}. \tag{11.55d}$$

That's it. In a conventional isotropic liquid, $\kappa_\parallel = \kappa_\perp = \kappa$, $\nu_1 = \nu_2 = \nu_3 = (\nu_4 - \nu_5) = \eta$ and $\frac{1}{3}(2\nu_4 + \nu_5) = \zeta$, and the equations above reduce to those given in section 4.1. Note that we have only derived the linearized equations of hydrodynamics, although there is no difficulty in obtaining, by similar methods (Kawasaki 1970), the usual reversible non-linear terms as well.

The reader will notice that the dissipative part of the stress tensor, as given above, is complicated, but at least it is symmetric in line with what we suggested earlier. However, the reactive part in (11.54c) is not. What has gone wrong? The answer is, of course, nothing, a point which has received considerable attention in the literature. (11.54) is certainly not what it seems to be, the average of the microscopic stress tensor, because the latter is symmetric, but the only role which $\delta\tau_{ij}$ plays in hydrodynamics is as $\nabla_j\delta\tau_{ij}$, and that leaves many choices. We can replace (11.54c) by a symmetric expression of the same content. If we insist that that expression still give $\delta\tau$ in terms of the forces $\delta\vec{h}$, then it must involve very non-local objects like $\nabla_i\nabla_j/\nabla^2 \sim k_ik_j/k^2$. This simply reflects the long-ranged static correlations between the director and the stress tensor mentioned earlier. A more satisfactory choice, which we write for simplicity in our earlier coordinate system ($\vec{n}^o = \hat{3}$, i and j = 1 or 2), is

$$\delta\langle\tau_{ij}\rangle^R = [p - K_1\nabla_3(\vec{\nabla}\cdot\vec{n})]\,\delta_{ij} - K_2\nabla_3[\nabla_i n_j + \nabla_j n_i - 2(\vec{\nabla}\cdot\vec{n})\,\delta_{ij}],$$

$$\delta\langle\tau_{i3}\rangle^R = \delta\langle\tau_{3i}\rangle^R = -\frac{1}{2}(\lambda+1)\,K_3\nabla_3^2 n_i$$

$$-\frac{1}{2}(\lambda-1)[K_1\nabla_i(\vec{\nabla}\cdot\vec{n}) + K_2(\vec{n}^o\times\vec{\nabla})_i(\vec{\nabla}\times\vec{n})_3]$$

$$\delta\langle\tau_{33}\rangle^R = p + K_3\nabla_3(\vec{\nabla}\cdot\vec{n}) \ , \qquad (11.56)$$

where δ's have been omitted on the right hand side. The point of symmetrizing the stress tensor is that, for $\tau_{ij} = \tau_{ji}$, angular momentum conservation is a consequence of momentum conservation, for any finite sample with no external fields and with free surfaces. Consequently, volume torques can only be introduced by external fields or at the boundaries, and that is good.

## 11.7 What is a Solid?

No, I do not mean to solicit enumerations of lattice structures, atomic arrangements and the like. These are interesting, but macroscopically one might not necessarily be aware of them. Macroscopically, a solid is a system which resists, quasistatically, external stresses that are applied at the surface, even if these stresses preserve the volume. (All physical systems respond elastically to external forces which change the volume, of course.) A liquid does not; it flows away. The question is being asked here because it elucidates the intermediate role which liquid crystals play between the "traditional" phases, liquids and solids.

What must happen in a solid as an external force $\delta\vec{F}(r)$ is applied at the surface is that balancing stresses are set up in the bulk. Now the contribution to the Hamiltonian due to $\delta\vec{F}$ will be of the general form

$$\delta\mathcal{H} = -\oint d\vec{r}\ \vec{A}(r)\cdot\delta F(\vec{r}), \tag{11.57}$$

where $\vec{A}(r)$ is some local variable (a displacement variable), and the integral extends only over the surface region. By linear response theory the resulting stress $\delta\langle\tau\rangle$ in the bulk, say at $\vec{r}$, is then

$$\delta\langle\tau_{ij}(\vec{r})\rangle = \oint\chi_{\tau_{ij},A_k}(\vec{r},\vec{r}\,')\ \delta F_k(\vec{r}\,')d\vec{r}\,' . \tag{11.58}$$

This is, again, a surface integral, and consequently $\delta\langle\tau\rangle$ in the bulk will vanish unless the equilibrium correlation function, $\chi_{\tau A}(r,r')$, falls off with distance no faster than as

$$\chi_{\tau_{ij}A_k}(r,r') \sim 1/|\vec{r}-\vec{r}\,'|^2 . \quad \text{(solid)} \tag{11.59}$$

Let us, first of all, see what we have in a nematic liquid crystal (which we already know not to be a solid). From the commutator (11.47) between the momentum density and the order parameter we know that (using momentum conservation $\dot{g}_i + \nabla_j\tau_{ij} = 0$)

$$k_j\chi_{\tau_{ij},n_k}(k) = i\,\chi_{\dot{g}_i,n_k}(k)$$

$$= \langle\frac{1}{\hbar}[g_i(r),n_k(r')]\rangle(k) = -\mu_{ij}^{(k)}\,k_j\ , \tag{11.60}$$

where we have rearranged the indices for easy legibility, and omitted terms of order $k^2$. The commutator between $g_i(r)$ and $n_j(r')$ <u>must</u> vanish when $\vec{r}$ and $\vec{r}\,'$ are far apart, and therefore $\mu_{ij}^{(k)}$ has the uniaxial symmetry of the unstrained nematic. Since we know that $\mu_{13}^{(1)}-\mu_{31}^{(1)}$ $(= \mu_{113}-\mu_{311}) = 1$, by the argument following eq. (11.47), we can write

$$\mu_{ij}^{(k)} = \mu_{ij}^{o(k)} + \epsilon_{ij\ell}\, m_\ell^{(k)} , \qquad (11.61)$$

where

$$\mu_{ij}^{o(k)} = \mu[n_i^o(\delta_{jk} - n_j^o n_k^o) + n_j^o(\delta_{ik} - n_i^o n_k^o)] , \qquad (11.61a)$$

$$m_\ell^{(k)} = \frac{1}{2}\epsilon_{\ell kn}\, n_n^o . \qquad (11.61b)$$

$\mu$ is a constant, namely $\mu = \mu_2 - 1/2$. Now we have pointed out that $\chi_{\tau_{ij}, n_k}$ must be symmetric in i and j, and since $\mu_{ij}^{(k)}$ is not, the eq. (11.60) leads to

$$\chi_{\tau_{ij}, n_k}(\vec{k}) = -\mu_{ij}^{o(k)} + (\vec{m}^{(k)} \times \hat{k})_i \hat{k}_j + (\vec{m}^{(k)} \times \hat{k})_j \hat{k}_i , \qquad (11.62)$$

as $\vec{k} \to 0$, where $\hat{k} = \vec{k}/k$. This shows more explicitly the statement made at the end of section 11.5, that the stress tensor correlation function $\chi_{\tau n}(\vec{k})$ is not uniform in the limit as $\vec{k} \to 0$. Since $\hat{k}_i\hat{k}_j \to \nabla_i \nabla_j (-4\pi r)^{-1}$ in space, we find that

$$\chi_{\tau_{ij} n_k}(\vec{r}) = -\frac{3}{4\pi} \frac{(m^{(k)} \times \hat{r})_i \hat{r}_j + (m^{(k)} \times \hat{r})_j \hat{r}_i}{r^3} , \qquad (11.63)$$

except for terms of finite range, where $\hat{r} = \vec{r}/r$. Comparing to (11.59) it is apparent that nematics are liquid: they cannot transmit static shear stresses. (For this reason the macroscopic expression (11.56) for the elastic stress tensor contains only second derivatives of $\delta\vec{n}$, not first derivatives.)

What can it transmit from the surface to the bulk? If we define the angular momentum density $\vec{\ell}(r,t)$ by

$$\vec{\ell}(r,t) \equiv \vec{r} \times \vec{g}(r,t), \qquad (11.64)$$

it obeys the continuity equation

$$\frac{\partial}{\partial t} \ell_i(r,t) + \nabla_j M_{ij}(r,t) = 0, \tag{11.65a}$$

where

$$M_{ij}(r,t) = \epsilon_{ik\ell} r_k \tau_{\ell j}(r,t) \tag{11.65b}$$

is a torque density. It is clear from (11.63) that

$$\chi_{M_{ij} n_k}(r) = -\frac{3}{4\pi} \frac{m_i^{(k)} \hat{r}_j - (\hat{r} \cdot \vec{m}^{(k)}) \hat{r}_i \hat{r}_j}{r^2} \tag{11.66}$$

falls off only as $1/r^2$. Thus, while a nematic cannot transmit static shear stresses, in contrast to a solid, it can transmit static torques, in contrast to a normal liquid which cannot do either one of these things because for normal liquids the stress tensor correlations are of finite and usually quite short range. The reader may want to take note that these statements have been derived here, ultimately, from equation (11.15) and thus the existence of an orientational order parameter, but that they make no explicit reference to the more phenomenological Frank free energy.

Let us now return to eq. (11.59) and the question of solidity. It is clear that to obtain $\chi_{\tau,A}(r) \sim r^{-2}$ there must be one factor of k less, in the commutator between g(r) and some local variable A(r), than we had in (11.60). In other words, $k_i \chi_{\tau_{ij}, A_k}(\vec{k})$ must be finite as $k \to 0$ which is only possible if the Fourier transform of

$$\langle \frac{1}{i\hbar} [g_i(r), A_j(r')] \rangle = a_{ij} \, \delta(r-r') + \nabla \ldots \tag{11.67}$$

is finite as $k \to 0$. But this means that

$$\langle \frac{1}{i\hbar} [P_i, A_j(r')] \rangle = a_{ij} \neq 0, \tag{11.68}$$

and since the total momentum operator $\vec{P}$ generates infinitesimal displacements, we find, to nobody's surprise, that in a solid translational invariance must be broken.

Let us, for simplicity, consider a glass, by which we mean a solid which is nevertheless translationally invariant (in the sense that correlation between local variables at $\vec{r}$ and $\vec{r}'$ depend only on the vector distance $\vec{r}-\vec{r}'$ so that we can define Fourier transforms as usual -- how this is possible we shall see below) and isotropic. In this case, $a_{ij} = \delta_{ij}$ if we remove a normalization constant. The Bogoliubov inequality argument, applied as in eq. (7.60), immediately shows that $\chi_{AA}(k)$ must diverge at least as $1/k^2$ as $k\to 0$, and if we separate $\chi_{A_iA_j}$ into a longitudinal and two transverse components, we have

$$\chi_{A_iA_j}(k) = \hat{k}_i\hat{k}_j[K_\ell k^2]^{-1} + (\delta_{ij}-\hat{k}_i\hat{k}_j)[K_t k^2]^{-1} \qquad (11.69)$$

for small k, where $K_\ell$ and $K_t$ are a longitudinal and a transverse elastic constant. Consequently, the three vector components $\vec{A}(rt)$ are hydrodynamic, and by applying our now standard routine the reader can derive the spectrum of hydrodynamic fluctuations. He will find that there are 8 hydrodynamic modes in all, of which 6 are propagating longitudinal and transverse sound waves ($z = \pm c_{\ell,t}k - \frac{i}{2}k^2\Gamma_{\ell,t}$) and 2 are diffusive ($z = -ik^2D_T$ and $z = -ik^2D'$). There are 5 dissipative coefficients: 2 viscosities $\eta_\ell$ and $\eta_t$, one heat conductivity $\kappa$, one order parameter relaxation constant $\xi = \sigma_{\dot{A}\dot{A}}(0,i0)$, and a constant $\zeta = \sigma_{i_q\dot{A}_\ell}(0,i0)$ which couples the order parameter to heat flow, and which may have either sign but is constrained by $\kappa T\xi \geqq \zeta^2$. In a (translationally invariant, but not rotationally isotropic) crystal the number and general structure of these modes is the same but the number of coefficients needed to describe them is much larger. Cholesteric and the various classes of smectic liquid crystals, finally, are layered systems which are translationally invariant at least in some directions though not in others. Formally this means that the matrix $a_{ij}$ in (11.67) is only of rank 1 (in cholesterics and smectic A and C), and thus some gradient terms in (11.67) have to be kept as well. For an extensive discussion of all these systems the reader is referred to the paper by Martin et al. (1972).

What is the variable $\vec{A}(r)$ physically? In terms of its fluctuations, the reactive part of the phenomenological stress tensor can be written in the form

$$\delta\langle\tau_{ij}\rangle^R = [\delta p^\circ + K_\ell(\vec{\nabla}\delta\vec{A})]\,\delta_{ij} + K_t[\nabla_i\delta A_j + \nabla_j\delta A_i - \frac{2}{3}\,\delta_{ij}\vec{\nabla}\cdot\delta\vec{A}\,] \,. \tag{11.70}$$

Its longitudinal part can clearly be absorbed in the definition of the pressure, $\delta p = \delta p^\circ + K_\ell \vec{\nabla}\delta\vec{A}$. The transverse part shows that $\delta\vec{A}$ is identical with the displacement vector, usually called $\vec{u}$, which occurs in the continuum theory of solids. It is usually said that $\vec{\nabla}\cdot\vec{u} = \delta\rho$ is the mass density. If this were true microscopically, $\delta A_\ell$ would not be an independent hydrodynamic variable, and there would only be one diffusive mode in solids, namely heat diffusion. The extra mode, with $z = -ik^2D'$, thus accounts for displacements which are not accompanied by mass motion, and can be identified as describing vacancy diffusion.

We have, in this hurried discussion, assumed that it is possible to write $\chi_{AB}(\vec{r},\vec{r}\,') = \chi_{AB}(\vec{r}-\vec{r}\,')$ and thus to define Fourier transforms as usual, even though eq. (11.68) indicates that translational invariance is broken. What does that mean, anyway? Let us consider a glass. The present microscopic picture of, say, amorphous $S_i$ is that of a fairly rigid structure, netted by roughly tetrahedral bonds. The system is disordered, in the sense that there is no macroscopic difference between the situation at $\vec{r}$ and at $\vec{r}+\vec{\xi}$ if $|\vec{\xi}\,|$ is many bond lengths. However, local displacements, and in particular infinitesimal displacements, <u>are</u> connected with large energies and are noticeable because of the rigid local structure. This means that we can assume that for some A(r),

$$\langle A(\vec{r}+\vec{\xi})\rangle = \langle e^{-i\vec{P}\cdot\vec{\xi}/\hbar} A(\vec{r})\, e^{i\vec{P}\cdot\vec{\xi}/\hbar}\rangle \approx \langle A(\vec{r})\rangle , \qquad (11.71)$$

if $\xi \gg \xi_c$, say, even though for infinitesimal displacements $\delta\vec{\xi}$,

$$\delta\langle A(r)\rangle = \langle \frac{1}{i\hbar}[\vec{P},A(r)]\rangle\cdot\delta\vec{\xi} \neq 0. \qquad (11.72)$$

In other words, the density matrix $\rho$ can be assumed to be invariant under translations by large distances even if $[\rho,\vec{P}] \neq 0$. The hydrodynamic theory, of course, is only concerned with fluctuations of wavelength $\lambda \gg \xi_c$. The reader can try to make these notions more precise by building a theory on the basis of "gross" variables which he may define by

$$\bar{A}(\vec{r}) = \xi_c^{-3} \int d^3\xi\, A(\vec{r}+\vec{\xi}), \qquad (11.73)$$

where the integral extends over a volume $\xi_c^3$ which surrounds the point $\vec{r}$. In terms of such variables, the system is translationally invariant but $\langle[\vec{P},\bar{A}(r)]\rangle \neq 0$. In a crystalline solid the situation is entirely similar.

# CHAPTER 12

# SUPERCONDUCTORS

Perhaps the most spectacular performance of which many-particle systems are capable is superconductivity: the ability of many metals to conduct an electric current without any resistance when they are cooled below a characteristic temperature $T_c$. Both the experimental development of the many fascinating aspects of superconductivity, and the successful microscopic theory of the effect by Bardeen, Cooper and Schrieffer, constitute undoubtedly one of the really great triumphs of twentieth century physics. Alas, an even cursory treatment of the phenomenon of superconductivity would be beyond the limitations of a brief chapter, and probably beyond those of its author. The goal of this final chapter is thus a modest one. I want to show the close analogy between superconductivity and superfluidity, between electric supercurrents in Lead and macroscopic superflow in Helium, between the Meissner effect and the reduction of the moment of inertia. Since superconductivity is, of course, intrinsically connected with electric charge I also want at least to indicate what can happen in systems in which there are forces of long range, although the discussion will be quite incomplete. And in so doing, I want, one more time, to demonstrate the versatility of the formalism which we have used throughout, based on response functions and the associated memory functions.

## 12.1 Bose Condensation in a Fermi System

In a system of Bose particles we found that the macroscopic aspects of

289

superfluidity could be explained if we assumed that $\langle \Psi^+ \rangle \neq 0$ where $\Psi^+(r)$ is the operator which creates one particle at point r. More precisely, the assumption, which has been microscopically verified at least for weak repulsive interparticle forces, is that the "one-particle density matrix",

$$G(r-r') \equiv \langle \Psi^+(r')\Psi(r)\rangle \rightarrow \langle \Psi^+(r')\rangle\langle \Psi(r)\rangle, \qquad (12.1)$$

does not vanish when $\vec{r}$ and $\vec{r}\,'$ are far apart. It is tempting to assume that the properties of superconductors can be similarly explained. However, the electrons in a metal are Fermions whose field operators $\Psi_s(r)$ and $\Psi_s^+(r)$ (where the spin index s takes on the two values ↑ or ↓ ) obey not commutation relations like (10.37) but the anticommutation relations

$$[\Psi_s(r), \Psi_{s'}^+(r')]_+ = \delta_{ss'}\,\delta(\vec{r}-\vec{r}\,') \ ,$$

$$[\Psi_s(\vec{r}), \Psi_{s'}(\vec{r}\,')]_+ = [\Psi_s^+(\vec{r}), \Psi_{s'}^+(\vec{r}\,')]_+ = 0 \ , \qquad (12.2)$$

where $[A,B]_+ \equiv AB + BA$. At $\vec{r} \neq \vec{r}\,'$ the equivalent one-particle function $G_{ss'}(\vec{r}-\vec{r}\,') = \langle \Psi_{s'}^+(\vec{r}\,')\Psi_s(\vec{r})\rangle$ must therefore be antisymmetric under interchange of $\vec{r}$,s with $\vec{r}\,'$,s', and so it plainly cannot factorize at large distances as in (12.1), (Yang 1962). Physically, the Pauli principle forbids more than one electron from condensing into the same macroscopic state.

The next simplest possibility is for the electrons to form pairs -- Cooper pairs. For this to happen we need an attractive force between the electrons which is provided by the exchange of phonons. These two-electron "molecules" have Boson properties, and thus there is no physical contradiction in assuming that the pairs in turn condense in the same macroscopic state, or a mathematical contradiction in assuming that $\langle \Psi_s^+(r)\Psi_{s'}^+(\vec{r}\,')\rangle \neq 0$ where $\Psi_s^+(\vec{r})\Psi_{s'}^+(\vec{r}\,')$ is the operator which creates a pair of electrons at $\vec{r}$ with spin s and at $\vec{r}\,'$ with spin s', respectively.

Since $\psi_\uparrow^+(r)\psi_\uparrow^+(r) = 0$, it is favorable to form pairs with opposite spin. We shall therefore assume that in a superconductor below $T_c$

$$\langle\psi_\uparrow(\vec{r})\psi_\downarrow(\vec{r}')\rangle = \Delta(\vec{r}-\vec{r}') \neq 0, \tag{12.3}$$

and see what follows.

The reader should accept these heuristic arguments with a strong "as it were". The BCS theory shows that (12.3) does indeed characterize the stable equilibrium state. However, the phonon-mediated interaction is very weak, and thus the size of the pairs, which is measured by the spatial range of the function $\Delta(\vec{r}-\vec{r}')$, is very large, of the order of many hundred interparticle distances, which makes the picture of two-electron molecules a tenuous one.

To proceed, note that as in a condensed Bose fluid the state in which $\Delta \neq 0$ cannot be gauge invariant. The total number of electrons is given by

$$N = \int d\vec{r}\; n(r) = \int d\vec{r}\,[\psi_\uparrow^+(r)\psi_\uparrow(r) + \psi_\downarrow^+(r)\psi_\downarrow(r)]\,. \tag{12.4}$$

Using the anticommutation relations (12.2) it is easy to see that

$$\langle[N, \psi_\uparrow(\vec{r})\psi_\downarrow(\vec{r}')]\rangle = -2\Delta(\vec{r}-\vec{r}') \neq 0, \tag{12.5}$$

so that the density matrix cannot commute with $N$; $[\rho, N] \neq 0$. In contrast to the Bose fluid, we have not a single order parameter $n^o$, but a function $\Delta(\vec{r}-\vec{r}')$. That is inconvenient for our purposes. However, we can simplify matters by the following device. In the BCS theory the gap function $\Delta(\vec{r})$ is s-like, and falls off at large r exponentially. Let us assume that we can construct a complete and orthogonal set of functions, $\Delta_\mu(r)$, such that $\Delta(r) \equiv \Delta_o(r)$ and

$$\int d\vec{r}\; \Delta_\mu^*(r)\Delta_o(r) = n_o\delta_{\mu,o}\,, \tag{12.6}$$

where $n_o$ is a normalization constant. If we then define local operators $a_\mu(r)$ by

$$a_\mu(r) = \frac{1}{\sqrt{n_o}} \int d\vec{r}'\, \Delta^*_\mu(\vec{r}')\, \Psi_\uparrow(\vec{r}+\tfrac{1}{2}\vec{r}')\, \Psi_\downarrow(\vec{r}-\tfrac{1}{2}\vec{r}'), \qquad (12.7)$$

we find that

$$[N, a_o(r)] = -2\, n_o^{1/2} \qquad (12.8a)$$

but

$$[N, a_\mu(r)] = 0, \quad \mu \neq 0. \qquad (12.8b)$$

Thus only the operator $a_o(\vec{r})$ and its Hermitian adjoint $a_o^+(\vec{r})$ break gauge invariance, and if, in analogy to eq. (10.55), we define the Hermitian phase $\varphi^{op}$ by

$$2\,\varphi^{op}(\vec{r}) = \frac{1}{2i\langle a_o^+\rangle\langle a_o\rangle}\,[\langle a_o^+\rangle a_o(r) - \langle a_o\rangle a_o^+(r)], \qquad (12.9)$$

we find

$$[N, \varphi^{op}(r)] = i, \qquad (12.10)$$

in analogy to (12.58). The factor of two in (12.9) is due to the fact that in a Fermi system the condensed entities are pairs of particles.

Equation (12.10) entails long-ranged phase coherence. The argument is completely parallel to that given in section 10.4. The momentum density operator for the electrons is given by

$$\vec{g}(r) = \frac{\hbar}{2i}[\Psi_\uparrow^+(r)\vec{\nabla}\Psi_\uparrow(r) - (\nabla\Psi_\uparrow^+(r))\,\Psi_\uparrow(r)] + [\uparrow \to \downarrow], \qquad (12.11)$$

and since

$$\partial_t n(rt) + \vec{\nabla}\cdot\vec{g}(rt)/m = 0 \qquad (12.12a)$$

and

$$[n(r), \vec{g}(r')] = -i\hbar\,\vec{\nabla}n(r)\,\delta(\vec{r}-\vec{r}'), \qquad (12.12b)$$

where m is the electron mass, we obtain the f-sum rule

$$\int \frac{d\omega}{\pi}\, \omega \chi''_{nn}(k\omega) = \frac{n}{m} k^2 \tag{12.13}$$

for any k, where n is the average electron mass density. The reader will also easily verify that

$$\int \frac{d\omega}{\pi}\, \chi''_{n\varphi}(k\omega) = \frac{i}{\hbar} \cdot n_o^{-1} \int d\vec{r}\; |\Delta(r)|^2 \, e^{-2i\vec{k}\cdot\vec{r}}, \tag{12.14}$$

which approaches $i/\hbar$ smoothly as $k \to 0$ since the gap function $\Delta(r)$ is of finite range. We thus find, using the Bogoliubov inequality (10.59), that

$$\chi_{\varphi\varphi}(k) = \frac{m/\hbar^2}{n_s k^2} \quad \text{as } k \to 0, \tag{12.15}$$

where $n_s \leqq n$. Thus again we find long-ranged fluctuations in the quantum phase. The parameter $n_s$ can be thought of as the number density of "superconducting electrons." Proceeding as in section 10.4 we also find that the phase fluctuations contribute by

$$\chi_{g_i\varphi}(k)\, \chi^{-1}_{\varphi\varphi}(k)\, \chi_{\varphi g_j}(k) = m n_s \hat{k}_i \hat{k}_j \tag{12.16}$$

to the momentum density correlations as $k \to 0$. This contribution is purely longitudinal. Assuming that all other fluctuations are short-ranged and thus contribute equally to longitudinal and transverse momentum density correlations, we find that as $k \to 0$,

$$\chi_{g_i g_j}(\vec{k}) = mn\, \hat{k}_i \hat{k}_j + m n_n (\delta_{ij} - \hat{k}_i \hat{k}_j) \tag{12.17}$$

where $n_n \equiv n - n_s$ is the number density of "normal electrons". Thus below $T_c$ where $n_s \neq 0$, the electronic momentum density has correlations of infinite range. The superfluidity of the electrons in metals, expressed by (12.17), has its most drastic manifestations because of their electromagnetic properties.

## 12.2 Superconductivity and Meissner Effect

To investigate these properties we shall assume that a small external electromagnetic field is coupled to the system of electrons. We choose a gauge in which the external scalar potential $\varphi^{ext}$ vanishes so that $\vec{E}^{ext} = -\frac{1}{c}\dot{\vec{A}}^{ext}$ and $\vec{B}^{ext} = \vec{\nabla} \times \vec{A}^{ext}$. Then the interaction Hamiltonian is, for $t < 0$,

$$\delta\mathcal{H} = -\frac{1}{c}\int d^3r\, \vec{J}(\vec{r})\, \delta\vec{A}^{ext}(r)\, e^{\epsilon t} \quad , \tag{12.18}$$

if a small static external field is slowly turned on. We have chosen rationalized Gaussian units, i.e., the units for which Maxwell's equations contain only the speed of light c, and no factors $4\pi$. Because the velocity of a particle of momentum $\vec{p}$ in the presence of $\vec{A}$ is not $\vec{p}/m$ but $(\vec{p} - \frac{e}{c}\vec{A}(rt))/m$, the current operator $\vec{J}(rt)$ is given by

$$\vec{J}(rt) \;=\; \vec{j}(rt) - \frac{e}{mc}\vec{A}(rt)\,\rho(rt) \tag{12.19}$$

where (omitting an inconsequential paramagnetic spin current)

$$\vec{j}(rt) \;=\; \frac{e}{m}\vec{g}(rt) \tag{12.20a}$$

$$\rho(rt) \;=\; e\, n(rt) \tag{12.20b}$$

are the translational current and the charge density, respectively, given by (12.4) and (12.11). In this chapter $\rho$ will always be used for the charge density, never for the mass density.

In this section we will assume that $\vec{A}$ in (12.19) can be replaced by $\vec{A}^{ext}$. This assumption, which we will reconsider in section 12.4, amounts to treating even the internally generated electromagnetic field as an external perturbation, and simplifies matters considerably (Pines and Noziéres 1966, Baym 1967). $\vec{J}$ in (12.18) can then be replaced by $\vec{j}$, and by a straightforward application of the linear response formalism we obtain

$$\delta\langle J_i^{ind}(\vec{k})\rangle = [\chi_{j_i j_j}(\vec{k}) - \omega_p^2\delta_{ij}]\frac{1}{c}\delta A_j^{ext}(\vec{k}) , \qquad (12.21)$$

where $\omega_p^2 = e^2 n/m$ is the square of the electronic plasma frequency. The first term is analogous to eq. (3.9); the second term is simply the average of the explicit contribution to the current in (12.19).

The reader will now want to recall the argument presented at length in section 10.1. If the external field $\delta\vec{A}^{ext}(r)$ in (12.21) varies slowly in space so that only small $\vec{k}$ are important, then the limit on $\chi_{ij}(\vec{k}) - \omega_p^2\delta_{ij}$ which is appropriate for an infinitely long wire, or a metallic torus, is the one in which the longitudinal component of $\vec{k}$ goes to zero first. In this limit we find from (12.17) that

$$\lim_{k_t\to 0}\lim_{k_\ell\to 0} \chi_{j_i j_j}(\vec{k}) = \frac{e^2}{m^2} mn_n\delta_{ij} , \qquad (12.22)$$

so that

$$\delta\langle\vec{J}^{ind}\rangle = -\omega_p^2(n_s/n)\frac{1}{c}\delta\vec{A}^{ext}. \qquad (12.23)$$

Thus a constant external vector potential, which is not accompanied by an electric field since $\vec{E} = -\dot{\vec{A}}/c$, will nevertheless induce a finite current. Conversely, a finite current in a stationary wire will flow forever. This is superconductivity.

Now the moving electrons will in turn induce a magnetic field or vector potential. What they react to is in fact the total field,

$$\vec{A}^{tot} = \langle\vec{A}^{ind}\rangle + \vec{A}^{ext}, \qquad (12.24)$$

so that we should actually write

$$\delta\langle J_i^{ind}(k)\rangle = [\chi_{j_i j_i}(\vec{k}) - \omega_p^2 \delta_{ij}]\frac{1}{c}\,\delta\,\vec{A}^{tot}(k)\,. \qquad (12.25)$$

Evidently this equation accounts only for the averaged effect of the internal field, neglecting field fluctuations. ( They are discussed extensively by Pines and Nozières (1966) who show that their effects are usually small. See also Martin (1967) and section 12.4 below.) $\langle A\rangle^{ind}$ can be calculated from Maxwell's equations since

$$[\frac{1}{c^2}\partial_t^2 - \nabla^2]\langle\vec{A}^{ind}(rt)\rangle = \frac{1}{c}\langle\vec{J}^{ind}(rt)\rangle \qquad (12.26)$$

for the transverse components. We are analyzing stationary fields so that $\partial_t \equiv 0$. From the last three equations we then obtain the total field in terms of the externally applied one, namely

$$\vec{A}^{tot}(k) = [1 - \frac{\chi_{jj}^T(k) - \omega_p^2}{c^2k^2}]^{-1}\ \vec{A}^{ext}(k)\,. \qquad (12.27)$$

For small k we therefore have

$$\vec{A}^{tot}(k) = \frac{k^2}{k^2 + \lambda^{-2}}\ \vec{A}^{ext}(k)\,, \qquad (12.28)$$

where

$$\lambda = (mc^2/n_s e^2)^{1/2}\,. \qquad (12.29)$$

If the superconductor occupies the x > 0 half-space and the external magnetic field is constant (12.28) predicts that

$$\vec{B}^{tot}(x) = \vec{B}^{ext}\ e^{-x/\lambda}. \qquad (12.30)$$

The magnetic field is thus expelled from the superconductor, penetrating only to a depth λ. This is the Meissner effect. The result (12.29) for the penetration depth holds only in the London limit when $\lambda$ is larger than all microscopic lengths, i.e., close to $T_c$ where $n_s \to 0$, or in typical strong coupling type I superconductors. If the coherence length $\xi$ (that is, the spatial size of the gap function $\Delta(\vec{r})$) is larger than $\lambda$, then the terms in $\chi^T(k)$ which are of order $k^2$ can not be neglected, and the actual magnetic penetration depth lies between $\lambda_L$ and $\xi$. This is, of course, an involved topic; see for example de Gennes (1966). Still, the reader may want to note that the obvious generalization of the above to fields that vary slowly in both space and time leads to the London equation,

$$\langle \vec{J}(rt) \rangle = -\frac{n_s}{n}\omega_p^2 \frac{1}{c}\vec{A}^{tot}(rt), \qquad (12.31)$$

which implies zero resistance.

## 12.3 Coulomb Forces

It is tempting now to go ahead and search for collective, hydrodynamic modes in the system of superconducting electrons. We shall not do so. The reason is that in the presence of long-ranged Coulomb forces some of the assumptions about the k→0 limit, which we have previously taken for granted, become invalid. A full discussion of the peculiar difficulties (and simplifications) in charged systems is given in the excellent book by Pines and Nozières (1966).

To see what can happen, however, we note that Poisson's equation, $\nabla^2\varphi = -\rho(r)$, implies that the static charge correlation function

$$\chi_{\rho\rho}(k) = \int \frac{d\omega}{\pi}\omega^{-1}\chi''_{\rho\rho}(k\omega) = k^2 \left(= \frac{"\partial\rho"}{\partial\varphi}\right) \qquad (12.32)$$

vanishes as $k \to 0$. If we therefore use the general equation (5.28) for the two variables $\rho$ and the longitudinal current $j_\ell$, omit the memory matrix $\Sigma$, and use the commutator (12.12b) to calculate the frequency matrix

$$\omega_{\rho j}(k) = \int \frac{d\omega}{\pi} \chi''_{\rho j}(k\omega) = \omega_p^{\,2} k , \tag{12.33}$$

then from $\det[z \chi(k) - \omega(k)] = 0$ we obtain a mode with the dispersion relation

$$z = \pm \omega_p \tag{12.34}$$

as $k \to 0$. This is the plasmon. Its energy is finite as $k \to 0$, despite the fact that charge is conserved, because the long-ranged Coulomb force leads to a finite restoring force on a charge imbalance, even for infinitely long wavelength. Since the charge current is not in general conserved, due to collisions of the electrons with phonons and impurities, the lifetime of the plasmon will be finite as well as $k \to 0$, $z = \pm \omega_p - i\tau_{eff}^{-1}$. In a neutral system where $\chi_{nn}(k \to 0) = \partial n/\partial\mu$ is finite, the corresponding mode would be the hydrodynamic mode for concentration fluctuations, $z = -i Dk^2$ where D is the diffusion coefficient. Because of the Coulomb force there are thus no new hydrodynamic modes in a superconductor, despite the fact that a continuous symmetry is broken. The modes which exist, longitudinal phonons and plasma oscillations, have nothing to do with superconductivity. An excellent account of this whole affair has been given by Martin (1969).

Things would be different in a "neutral superconductor," that is in a condensed system of electrically neutral Fermions. It appears that at very low temperatures $He^3$ does in fact become superfluid. Moreover, because of the strong repulsion of $He^3$ atoms at close range, $He^3$ "Cooper pairs" condense not in an s-state but in a p-state: The most probable gap function for the A-phase of $He^3$ is of the form (Anderson and Morel 1961, Ambegaokar et al. 1974)

$$\langle \Psi(\vec{r})\, \Psi(\vec{r}\,')\rangle = \Delta(\vec{r}-\vec{r}\,'), \tag{12.35}$$

where $\Delta(\vec{r}) = (x + iy)\Delta'(|r|)$ is an $\ell = 1$, $m = 1$ function. (The spins, which are omitted, consequently must be in a triplet state.) Such a state breaks not only gauge symmetry, but it is anisotropic and breaks rotational symmetry about the x- and y-axes as well. Thus one has a very exciting combination of the properties of a normal isotropic superfluid with those of a nematic liquid crystal. The hydrodynamics of such a system has been worked out by Graham (1974), and the reader might have some fun reconstructing his results within the formalism presented here. It should be mentioned that the experimental relevance of hydrodynamic modes in $He^3$ is not clear. The A-phase is stable between 0.002 and 0.0027° K; at such low temperatures collisions are infrequent, and the hydrodynamic restriction $\omega\tau_c < 1$ is hard to fulfil.

## 12.4 Linear Response of Charged Systems

Let us now return to systems of charged particles. In section 12.2 we obtained the linear electromagnetic response of such systems under an essential and simplifying assumption: that field fluctuations could be neglected. In a very elegant paper Martin (1967 and 1968) has considered the problem if this assumption is not made, and proved a number of fundamental sum rules. In this final section we will reconstruct these results on the basis of the memory function formalism which provides some additional physical insight into the nature of the electromagnetic response.

In a general system of particles of several species with mass $m_i$, charge $e_i$ and magnetic moment $\mu_i$, the electric current is given by

$$\vec{J}(r,t) = \sum_i \sum_\alpha \frac{e_i}{2m_i} \{\vec{p}_i^{\,\alpha}(t) - \frac{e_i}{c} \vec{A}(r,t),\ \delta(\vec{r}-\vec{r}_i^{\,\alpha}(t)\}$$

$$+ c\,\vec{\nabla} \times \sum_i \mu_i \sum_\alpha (\vec{s}_i^{\,\alpha}/s_i)\ \delta(\vec{r}-\vec{r}_i^{\,\alpha}(t)) \qquad (12.36)$$

where the last term, omitted from eq. (12.19), is the contribution from the particle spins. (12.36) is the most general expression for a system of particles whose dynamics can be treated nonrelativistically. $\vec{J}(rt)$ is the operator which occurs in the microscopic Maxwell equations

$$\vec{\nabla} \times \vec{E} = -\frac{1}{c}\dot{\vec{B}} ,$$
$$\nabla \times \vec{B} = \frac{1}{c}(\vec{J} + \dot{\vec{E}}) . \qquad (12.37)$$

The electric and magnetic field operators are given by

$$\vec{E} = -\frac{1}{c}\dot{\vec{A}} - \vec{\nabla}\varphi ,$$
$$\vec{B} = \vec{\nabla} \times \vec{A} , \qquad (12.38)$$

in terms of the scalar potential $\varphi$ and the vector potential $\vec{A}$. As everybody knows the longitudinal part of the electric field, to which $\vec{\nabla}\varphi$ contributes, is not an independent degree of freedom but instead is simply given in terms of particle coordinates by $\vec{\nabla} \cdot \vec{E}(r,t) = \rho(r,t)$ where

$$\rho(r,t) = \sum_i e^i \sum_a \delta(\vec{r} - \vec{r}_i^{\,a}(t)) \qquad (12.39)$$

is the charge density operator. However, the transverse components $\vec{E}^T$ and $\vec{B}^T$ obey the independent equal time commutation relations (Dirac 1958)

$$[E_i^T(r), A_j^T(r')] = i\hbar c[\delta_{ij}\delta(\vec{r} - \vec{r}')]^T , \qquad (12.40)$$

and they commute with all particle coordinates. $[\delta_{ij}\delta(\vec{r} - \vec{r}')]^T$ is the very non-local function whose Fourier transform is $\delta_{ij} - \hat{k}_i\hat{k}_j$. The fields are respectable operators, and can be used as usual to define correlation functions, e.g.

$$\chi''_{B_iB_j}(rt, r't') = \langle \frac{1}{2\hbar}[B_i(rt), B_j(r't')]\rangle \qquad (12.41)$$
$$= \int \frac{d\omega}{2\pi} \int \frac{d^3k}{(2\pi)^3} e^{i\omega(t-t') - i\vec{k}\cdot(\vec{r}-\vec{r}')} \chi''_{B_iB_j}(\vec{k}\omega).$$

This last equation implies the one restriction which we will maintain throughout: we will assume that we are dealing with a translationally invariant and, indeed, isotropic system, so that, for example,

$$\chi''_{B_iB_j}(\vec{k}\omega) = \chi''_{BB}(k\omega)(\delta_{ij} - \hat{k}_i\hat{k}_j) ,$$

$$\chi''_{E_iE_j}(\vec{k}\omega) = \chi''^{T}_{EE}(k\omega)(\delta_{ij} - \hat{k}_i\hat{k}_j) + \hat{k}_i\hat{k}_j \quad \chi''^{L}_{EE}(k\omega) . \qquad (12.42)$$

From Maxwell's equations it is apparent that, at $t = 0$, the operators $\dot{E}^T$ and $\dot{B}$ do not involve the particle interactions. Using the commutator (or classically, Poisson bracket) relations (12.40) we can therefore prove the two fundamental sum rules

$$\int \frac{d\omega}{\pi}\, \omega\, \chi''_{BB}(k\omega) = c^2k^2 , \qquad (12.43a)$$

$$\int \frac{d\omega}{\pi}\, \omega^3\, \chi''_{BB}(k\omega) = c^2k^2(c^2k^2 + \omega_p^{\,2}) , \qquad (12.43b)$$

where $\omega_p^{\,2} = \sum_i e_i^{\,2} n_i/m_i$.

And now we are ready to begin. In a charged system it is natural to consider the propagation and decay of electromagnetic fields since this is how many experiments are performed. We shall therefore consider the matrix of field correlation functions $C_{\mu\nu}(\vec{k}z)$, defined as usual by

$$C_{\mu\nu}(\vec{k}z) = \beta^{-1} \int \frac{d\omega}{\pi i} \frac{\omega^{-1}\chi''_{\mu\nu}(\vec{k}\omega)}{\omega - z} , \qquad (12.44)$$

and the associated memory matrix. At first we will discuss the 4 × 4 matrix of the transverse fields $\vec{E}^T$ and $\vec{B}$ which are uncoupled from all longitudinal fluctuations, in an isotropic system.

### 12.4.1 The Static Susceptibilities

Since we are aiming at applying the master formula (5.28) to this problem, we first consider the static susceptibilities $\chi_{\mu\nu}(k)$. The electric field is even under time reversal while the magnetic field is odd. Therefore the only non-vanishing terms are $\chi^T_{EE}(k)$ and $\chi_{BB}(k)$. Now from eq. (12.37) one sees that

$$\epsilon_{ik\ell} k_k \chi''_{E_\ell E_j} = \frac{\omega}{c} \chi''_{B_i E_j} \tag{12.45a}$$

and thus also

$$k^2 \chi''^T_{EE}(k\omega) = \left(\frac{\omega}{c}\right)^2 \chi''_{BB} . \tag{12.45b}$$

The sum rule (12.43a) therefore results in

$$\int \frac{d\omega}{\pi} \omega^{-1} \chi''^T_{EE}(k\omega) \equiv \chi^T_{EE}(k) = 1, \tag{12.46}$$

which holds for any k. What a nice result.

The magnetic susceptibility is a little more complicated. From Maxwell's equations we obtain the identity

$$k^4 \chi''_{B_i B_j} = \left(\frac{\omega}{c}\right)^4 \chi''_{B_i B_j} + \left(\frac{k}{c}\right)^2 \chi''^T_{J_i J_j} - \frac{2i\omega k^2}{c^2} \chi''^T_{E_i J_j} . \tag{12.47}$$

Now note that at equal times the only term in the current (12.36) which does not commute with the transverse electric field $E^T(r)$ is the vector potential A, and the relation (12.40) leads to

$$\int \frac{d\omega}{\pi} \chi''_{E_i J_j}(k\omega) = -i\omega_p^2 \delta_{ij} . \tag{12.48}$$

which is also true for the longitudinal part. This sum rule and (12.43b) therefore result in

$$\chi_{BB}(k) = 1 + \frac{1}{k^2 c^2} [\chi^T_{JJ}(k) - \omega_p^2] . \tag{12.49}$$

We shall cast this equation in a more useful form below. For now, let us hurry on.

### 12.4.2 The Frequency Matrix

Because of time reversal invariance the frequency matrix $\underset{\approx}{\omega}(k)$ has only E–B cross terms. These are, of course, easily calculated from the fundamental commutator (12.40). We obtain

$$\omega_{E_iB_j}(\vec{k}) = \omega_{B_jE_i}(\vec{k}) = \int \frac{d\omega}{\pi} \chi''_{E_iB_j}(\vec{k}\omega) = c\epsilon_{ijk}k_k . \tag{12.50}$$

At this point the reader may want to convince himself that in vacuum the correlation functions $C_{EE}(kz)$ etc. have poles when $z^2 = c^2k^2$. Indeed they do.

### 12.4.3 The Memory Function

The third object which we need to complete the formal representation of the field correlation functions is the matrix of memory functions, generally given by eq. (5.28). It involves a projector $Q = 1-P$ which we can formally write as

$$P = |E^T\rangle\beta^{-1}\chi^{-1}_{EE}\langle E^T| + |B\rangle\beta^{-1}\chi^{-1}_{BB}\langle B| . \tag{12.51}$$

The important property is that QA, where A is any operator, has no correlation with the transverse field components, i.e., $\langle E^T|Q = \langle B|Q \equiv 0$. Note that the definition of $P = 1-Q$ can include a sum over all wavevectors as in (5.45b), in a translationally invariant system. In an isotropic system we could also project out any longitudinal fluctuations without effecting the transverse results.

It follows immediately that the memory matrix element $\sigma_{\mu\nu}(\vec{k}z)$ vanishes when either of the indices refer to the magnetic field. This is because $\sigma_{B\cdot} \sim \langle\dot{B}|Q \ldots \sim \langle\nabla\times E|Q \equiv 0$, by Maxwell's equations. This leaves us with only one element of the transverse memory matrix,

$$\sigma^T_{E_iE_j}(\vec{k}z) = \bar{\sigma}^T(kz)\,(\delta_{ij} - \hat{k}_i\hat{k}_j). \qquad (12.52)$$

Noting further that $\sigma^T_{\dot{E}} \sim \langle \dot{E}|Q \sim \langle c\nabla \times B - J|Q \sim -\langle J|Q$, we can write $\bar{\sigma}^T$ in the form

$$\delta_{ij}\bar{\sigma}^T(kz) = \beta\langle J_i^T(k)|Q\frac{i}{z-QLQ}Q|J_j^T(k)\rangle . \qquad (12.53)$$

Note that, like all memory functions, $\bar{\sigma}^T(kz)$ is an analytic function of the frequency, for $\operatorname{Im} z \neq 0$, whose real and imaginary parts are therefore related by the usual Kramers-Kronig relations.

### 12.4.4 The Transverse Conductivity

We can now apply the general formula (5.28) and express the correlation functions in terms of $\bar{\sigma}$. One obtains

$$C^T_{EE}(kz) = \beta^{-1}\frac{iz}{z^2-[c^2k^2/\chi_{BB}(k)]+iz\bar{\sigma}^T(kz)}, \qquad (12.54a)$$

$$C_{BB}(kz) = \beta^{-1}\chi_{BB}(k)\frac{i(z+i\bar{\sigma}^T(kz))}{z^2-[c^2k^2/\chi_{BB}(k)]+iz\bar{\sigma}^T(kz)}. \qquad (12.54b)$$

What does it mean? In electrodynamics, for fields so small that the material response is linear, the local and frequency-dependent conductivity $\sigma_{ij}(\vec{k}\omega)$ is defined by $\delta\langle J_i(\vec{k}\omega)\rangle = \sigma_{ij}(\vec{k}\omega)\delta\langle E_j(\vec{k}\omega)\rangle$, or equivalently by

$$[\omega^2 - c^2k^2 + i\omega\sigma^T(k\omega)]\,\delta\langle E^T(k\omega)\rangle = \text{initial condition}, \qquad (12.55)$$

where $\langle E\rangle$ is the total field in the medium. In other words, $\sigma$ is defined by the propagation characteristics of electromagnetic fields. By comparison with (12.54) we can therefore identify the measured conductivity (or equivalently, the dielectric function $\epsilon^T$ since $\epsilon^T(kz) = 1 + (i/z)\sigma^T(kz)$) as

$$\sigma^T(kz) = \frac{c^2k^2}{iz}[1-\chi_{BB}^{-1}(k)] + \bar{\sigma}^T(kz) \ . \tag{12.56}$$

The first important result which follows from this identification is that, like the memory function $\bar{\sigma}^T$, the dynamic conductivity $\sigma^T(kz)$ is an analytic function of the frequency z except when Im z = 0, and therefore obeys the Kramers-Kronig relations rigorously. No, that is not a matter of course: the first really rigorous proof that this is so, equivalent to the one implicitly given here, has been given by Martin in 1967. (See also section 4.1 of Pines and Nozières, 1966.)

Equation (12.56) gives a very physical and intuitive separation of the measured conductivity. It is not $\sigma^T$ but $\bar{\sigma}^T$ which has the properties of a transport coefficient. $\bar{\sigma}^T(kz)$ can be expected to be regular as k and z go to zero. Moreover, the eq. (12.58) gives precise meaning to the usual, approximate, statement that the conductivity can be calculated by calculating the transverse current correlation function,

$$\bar{\sigma}^T(kz) = \beta\, C_{JJ}^{(n,m,i.)T}(kz), \tag{12.57}$$

for a system whose Hamiltonian contains the Coulomb but no magnetic interaction, since the projection operator Q in (12.53) removes the transverse electromagnetic fluctuations from the dynamics. With Q given by (12.51) this is seen to be true only if quadratic and higher field fluctuations (i.e., operator states like $|B(k-k')B(k')\rangle$ etc.) can be neglected. (By a somewhat strained analogy to the mode-mode coupling theory (Kadanoff and Swift, 1968), one would expect such terms to matter in an itinerant ferromagnet just above its critical point.)

### 12.4.5 Superconductors

The first term in (12.56) is a reversible, thermodynamic, contribution to the local conductivity. It becomes macroscopically important if the static local permeability $\mu(k) = \chi_{BB}(k)$ vanishes at $k = 0$, i.e., if $\mu = (\frac{\partial B}{\partial H}) = 0$. In superconductors as $k \to 0$,

$$\mu^{-1}(k) - 1 = (\omega_p^2/c^2k^2)\, n_s/n \tag{12.58}$$

defines the penetration depth $\lambda$ of (12.29) in the London limit, and thus gives an operational definition of the number $n_s$ of superconducting electrons. If $n_s \neq 0$ so that there is a Meissner effect, then it is also true that the measured conductivity as $k \to 0$,

$$\sigma^T(z) = \frac{i}{z}\,\omega_p^2 (n_s/n) + \text{regular terms} , \tag{12.59}$$

becomes infinite at low frequencies so that there are persistent macroscopic currents.

These definitions can be related to our microscopic considerations of section 12.2. As it stands, the identity (12.49) is not very useful since the static current correlation function $\chi^T_{JJ}(k)$ still contains all field fluctuations. Since, in our present language,

$$\chi^T_{JJ}(k) = \beta\, \langle J(k) | J(k) \rangle , \tag{12.60a}$$

we obtain its projected part by

$$\tilde{\chi}^T_{JJ}(k) = \beta \langle J(k) | Q | J(k) \rangle . \tag{12.60b}$$

Since $\tilde{\chi}^T$ has the transverse field fluctuations removed, it is much closer to the function which we computed in (12.17) or (12.22) for the equivalent uncharged system. Now since $\chi_{JE}(k)$ vanishes from time reversal, we have rigorously

$$\begin{aligned}[\chi^T_{JJ}(k) - \tilde{\chi}^T_{JJ}(k)](\delta_{ij} - \hat{k}_i\hat{k}_j) &= \chi_{J_iB_k}(k)\, \chi^{-1}_{BB}(k)\, \chi_{B_kJ_j}(k) \\ &= c^2k^2\, [\chi_{BB}(k) - 1]^2\, \chi^{-1}_{BB}(k)\, (\delta_{ij} - \hat{k}_i\hat{k}_j) .\end{aligned} \tag{12.61}$$

The second equation follows easily from Maxwell's equation,

$$\chi_{J_i B_j} = -\chi_{\dot{E}_i B_j} + c\,\chi_{(\nabla\times B)_i B_j} , \tag{12.62}$$

and the sum rule (12.50). If we insert (12.61) in our previous result (12.49) we obtain instead

$$\chi_{BB}(k) = \left[1 - \frac{\tilde{\chi}^T_{JJ}(k) - \omega_p^2}{c^2k^2}\right]^{-1} . \tag{12.63}$$

This is the rigorous equation which corresponds to the approximate result (12.27). Our present derivation may serve to elucidate what one does in a mean field approximation as given in section 12.2, and for those who are more comfortable with the latter, it again may elucidate the meaning of projectors when they appear as in (12.60).

[ Equation (12.63) has a curious consequence. Because $\chi_{BB}(k) > 0$ for all k, we find that as $k\to 0$, $\tilde{\chi}^T_{JJ}(k) - \omega_p^2 \leqq 0$. If we divide by $(e^2/m^2)$ and let $e\to 0$, the field fluctuations are removed automatically, and we obtain $\chi^T_{gg}(k\to 0) \leqq mn$ or $\rho_n \leqq \rho$ for an uncharged isotropic one component fluid. In chapter 10 this physically obvious result had been obtained only by reference to a longitudinal order parameter. Since as $c\to 0$ the transverse electromagnetic field does not fluctuate, we obtain further that $\chi^T_{gg}(k) < \rho$ for all k, if the particles have no magnetic moment. This simply shows that the equivalent charged system is diamagnetic. ]

12.4.6 Sum Rules

As $z \to \omega + i\epsilon$, $\sigma^T(k;\omega + i\epsilon) = \sigma'^T(k\omega) + i\sigma''^T(k\omega)$. We have already pointed out that $\sigma'^T$ and $\sigma''^T$ are connected by the usual Kramers-Kronig relations.

Moreover, since at large z, $\bar{\sigma}^T(kz) \to (i/z)\,\tilde{\chi}^T(k)$ we obtain the well-known sum rule, valid for all k,

$$\int \frac{d\omega}{\pi} \sigma'^T(k\omega) = \omega_p^2 , \tag{12.64}$$

for the measured conductivity. It is more physical to write it in terms of the regular, absorptive, part $\bar{\sigma}'^T$ as

$$\int \frac{d\omega}{\pi} \bar{\sigma}'^T(k\omega) = \omega_p^2 - c^2k^2[\chi_{BB}^{-1}(k) - 1] . \tag{12.65}$$

Further discussion of sum rules is given by Pines and Nozieres (1966).

### 12.4.7 Longitudinal Fluctuations

If we apply the memory function equation (5.28) to the single variable $E^L$, the longitudinal electric field, we obtain immediately the representation

$$C_{EE}^L(kz) = \frac{i\beta^{-1}\chi_{EE}^L(k)}{z + i\bar{\sigma}^L(kz)/\chi_{EE}^L(k)} , \tag{12.66}$$

where formally

$$\hat{k}_i\hat{k}_j\bar{\sigma}^L(kz) = \beta\langle J_i^L(k)|Q\frac{i}{z-QLQ}Q|J_j^L\rangle , \tag{12.67}$$

and where the projector Q now excludes the longitudinal field $E^L$ as well. (This, as we mentioned, does not affect the transverse results.) In (12.67) we have made use of the longitudinal microscopic Maxwell equation $\dot{E}^L + J^L = 0$. Altogether, our definition of the "regular" conductivity is thus summarized by

$$\bar{\sigma}_{ij}(kz) = i\beta\langle J_i(k)|Q[z-QLQ]^{-1}Q|J_j(k)\rangle . \tag{12.68}$$

To relate (12.65) and (12.68) to a measurement one would consider a system which, at $t = 0^-$, is held in a state of constrained equilibrium by means of some external force (i.e., an electric field) which induces a charge imbalance. If that force is turned off at $t = 0$, essentially instantaneously, then the return to full equilibrium is described by $\langle \dot{E}^L(r,t)\rangle + \langle J^L(r,t)\rangle = 0$, or by

$$-iz\langle E^L(kz)\rangle + \langle J^L(kz)\rangle = \langle E^L(k,t=0^+)\rangle, \tag{12.69}$$

for the longitudinal field, where

$$\langle E^L(kz)\rangle = \int_0^\infty dt\, e^{izt} \langle E^L(k,t)\rangle, \tag{12.70}$$

etc. Since for $t > 0$ there is no external field, one would define the measured local conductivity by the ratio $\langle J^L(kz)\rangle / \langle E^L(kz)\rangle$, and obtain

$$[z + i\sigma^L(kz)]\,\delta\langle E^L(kz)\rangle = i\delta\langle E^L(k,t=0^+)\rangle. \tag{12.71}$$

The same experiment is described generally by eq. (3.14) in linear response theory. We can therefore identify the measured longitudinal conductivity by comparing (12.71) and (12.66),

$$\sigma^L(kz) = \bar{\sigma}^L(kz)\, \chi_{EE}^{-1L}(k). \tag{12.72}$$

This microscopically well-defined quantity has all the characteristics of a generalized transport coefficient: $\sigma^L(kz)$ is an analytic function of the frequency $z$, for $\mathrm{Im}\, z \neq 0$, and if static field correlations are strong so that $\chi_{EE}^L(k)$ is large, $\sigma^L(kz)$ decreases correspondingly in accord with the general feature of "critical slowing down". The sum rule

$$\int \frac{d\omega}{\pi}\, \bar{\sigma}'^L(k\omega) = \omega_p^2, \tag{12.73a}$$

for the real part $\bar{\sigma}'^L(k\omega)$ of the complex function $\bar{\sigma}^L(k,\omega + i0)$, is valid at all values of k; the reader will recognize (12.73a) as the longitudinal f-sum rule. It is more physical, however, to write instead the equivalent sum rule

$$\int \frac{d\omega}{\pi} \sigma'^L(k\omega) = \omega_p^2 \, \chi_{EE}^{-1}(k) \tag{12.73b}$$

since $\sigma^L$ is more directly related to experiment. Note finally that, according to (12.66), the dispersion relation for charge fluctuations is

$$\omega + i\,\sigma^L(k\omega) = \omega + i\bar{\sigma}^L(k\omega)\chi_{EE}^{-1L}(k) = 0 \tag{12.74}$$

or $\epsilon^L(k\omega) = 0$, with the customary identity $\epsilon^L = 1 + i\sigma^L/\omega$. [The paper by Martin (1967) denotes by $\bar{\sigma}^\ell(kz)$ the quantity written as $\sigma^L(kz)$ in (12.70).]

Unfortunately, what we have described in the last several equations is not the most conventional definition of the local conductivity (Pines and Nozières 1966, Martin 1968). While eqs. (12.66) and (12.72) define the conductivity by

$$\begin{aligned} \epsilon^L(kz) &\equiv 1 + (i/z)\,\sigma^L(kz) \\ &= [1 - \frac{\chi_{EE}^L(kz)}{\chi_{EE}^L(k)}]^{-1} = [1 - \frac{\chi_{\rho\rho}(kz)}{\chi_{\rho\rho}(k)}]^{-1}, \end{aligned} \tag{12.75}$$

the customary definition is by (see section 4.1 of Pines and Nozières, 1966)

$$\epsilon^L(kz) \equiv 1 + (i/z)\,\sigma^L(kz) = [1 - \frac{\chi_{\rho\rho}(kz)}{k^2}]^{-1}, \tag{12.76}$$

where we have used Poisson's equation, $\vec{\nabla}\cdot\vec{E} = \rho$, to replace the longitudinal electric field by the charge density. Equation (12.76) is derived by considering the linear response to the total (external plus induced) electric field in the system. If we could prove that the identity (12.46), $\chi_{EE}^T(k) \equiv 1$, holds for the longitudinal fields as well, the two definitions (12.75) and (12.76) would coincide, and there would be no difficulty. We are not so lucky. However, if there are no field fluctuations of infinite range, the tensor $_{E_i E_j}(k)$ must be isotropic as k  0 so that

$$\lim_{k \;\; 0} \chi^{L}_{EE}(k) = \lim_{k \;\; 0} k^{-2} \int \frac{d\omega}{\pi} \omega^{-1} \chi''_{\rho\rho}(k\omega) = 1 \,. \qquad (12.79)$$

This sum rule reflects the screening of the electric field. It fails only in the hypothetical uniform perfect insulator in which compensating charge motion is inhibited, and potential correlations are of infinite range. In all other systems, including superconductors, (12.79) shows that, at least as k  0, (12.75) and (12.76) coincide. There is, however, a discrepancy at finite k. As a result the dielectric function, defined by (12.76), does not necessarily satisfy the Kramers-Kronig relations (Pines and Nozieres 1966, Martin 1967), although it will do so in almost all systems.

I will stop the discussion here. For more detail and many additional results (which could also be simply obtained within the formalism presented here) the reader should consult the literature quoted above. Even with this very brief and incomplete presentation I hope to have convinced the reader that charged systems, too, can be simply discussed and understood within the unifying framework of correlation and memory functions, sum rules, conservation laws, broken symmetry and all that, within which we have analyzed a rich variety of physical systems and phenomena.

## APPENDIX

## SCATTERING CROSS SECTIONS

In this Appendix we briefly recapitulate the standard expressions which relate the differential cross section, which is observed in inelastic scattering of neutrons, electrons or light from a many-particle system, to various correlation functions which describe the internal dynamics of these systems. This is a well-trodden path, and we make no attempt to critically review or discuss the assumptions on which the derivation is based. A more thorough discussion of scattering is given, for example, by Martin (1968).

### A.1 Electron Scattering

The simplest example is, in many ways, the scattering of electrons by the charges in a system with which they interact by the Coulomb potential

$$V = \sum_{\alpha} e\, e^{\alpha}\, V(R - r^{\alpha}) = e \int d^3r\, V(R-r)\, \rho(r), \qquad \text{(A. 1)}$$

where $e, \vec{R}$ are the charge and position of the external electron, $e^{\alpha}, \vec{r}^{\alpha}$ those of a target particle so that $\rho(r)$ is the charge density operator of the target system. If the incident electrons are sufficiently fast we can use the Born approximation; then the transition rate is given by Fermi's Golden Rule,

$$P_{i\to f} = \frac{2\pi}{\hbar} |V_{fi}|^2\, \delta(E_f - E_i - \hbar\omega). \qquad \text{(A.2)}$$

$E_i$ and $E_f$ are the initial and final energies of the target system so that $\hbar\omega$ is the energy loss of the electron. If we denote its momentum loss by $\hbar\vec{k}$, so that

$$\epsilon_f = \epsilon_i - \hbar\omega,$$

$$\vec{k}_f = \vec{k}_i - \vec{k}, \qquad (A.3)$$

the matrix element of V(R−r) between electron plane wave states is

$$\langle k_i | V(R-r) | k_f \rangle = V(k)\, e^{-i\vec{k}\cdot\vec{r}}, \qquad (A.4)$$

where

$$V(k) = \int d^3r\, V(r)\, e^{-i\vec{k}\cdot\vec{r}} = \frac{1}{k^2} \qquad (A.5)$$

for the Coulomb interaction. We therefore obtain

$$V_{if} = e\, V(k) \int d^3r\, e^{-i\vec{k}\cdot\vec{r}} \langle i | \rho(r) | f \rangle, \qquad (A.6)$$

with the last factor referring to the states of the target system. It is convenient to separate the δ-function in (A.2) in its Fourier components since

$$\langle i | \rho(r) | f \rangle\, \delta(E_f - E_i - \hbar\omega) =$$

$$(2\pi\hbar)^{-1} \int_{-\infty}^{\infty} dt\, e^{i\omega t} \langle i | \rho(\vec{r},t) | f \rangle, \qquad (A.7)$$

where $\rho(\vec{r},t)$ is the time dependent charge density, in Heisenberg representation. Therefore, putting the pieces into (A.2), and summing over all final states of the target and over those of the electron in $d^3k_f$, we obtain

$$\sum_f P_{i\to f} = \frac{d^3k_f}{(2\pi)^3} \cdot \frac{|eV(k)|^2}{\hbar^2} \int dr \int dr' \int dt\, e^{i\omega t - i\vec{k}(\vec{r}-\vec{r}')} \times$$

$$\times \langle i | \rho(\vec{r},t)\, \rho(\vec{r}',0) | i \rangle . \qquad (A.8)$$

Finally, averaging this transition rate over the equilibrium distribution of initial target states, we obtain the differential cross section for electron scattering,

$$\text{inc. flux} \times d\sigma = \langle \sum_f P_{i\to f} \rangle . \tag{A.9}$$

The incoming electron flux is $\hbar k_i/m_e$. And with $d^3k_f = k_f^2 dk_f d\Omega_f = (m/\hbar^2) k_f d\epsilon_f d\Omega_f$, we obtain the final result

$$\frac{d^2\sigma}{V d\epsilon_f d\Omega_f} = \frac{k_f}{k_i} \frac{m_e^2}{(2\pi)^3} \frac{e^2}{\hbar^5 k^4} S_{\rho\rho}(\vec{k}\omega), \tag{A.10}$$

where

$$S_{\rho\rho}(\vec{k}\omega) = \int d(t-t') \int d(\vec{r}-\vec{r}')\, e^{i\omega(t-t')-i\vec{k}(\vec{r}-\vec{r}')} \langle \rho(\vec{r},t)\rho(\vec{r}',t')\rangle$$

is the charge density correlation function. (In an electrically neutral system, $\langle\rho\rangle = 0$.) The reader should note that we have defined $\hbar\vec{k}$ to be the energy and momentum <u>loss</u> suffered by the scattered electron, see eq. (A.3). Many authors use, instead, the corresponding gains. The definition (A.3) is the convenient one with our sign convention for Fourier transforms. Since in thermal equilibrium

$$S_{\rho\rho}(k,-\omega) = e^{-\hbar\omega\beta} S_{\rho\rho}(k,\omega), \tag{A.12}$$

electrons can gain energy only at finite temperature when there are excitations in the system to nourish on, but they can always lose it.

## A.2 Neutron Scattering

Neutrons interact with the nuclei of the atoms or molecules in the system via a force of such short range that we can write

$$V = \sum_\alpha V(r_N - r^\alpha) = \sum_\alpha a\, \delta(r_N - r^\alpha). \tag{A.13}$$

where the pseudopotential a determines the strength of the contact interaction. We will omit the spin dependence of a. Then

$$V = a \int dr\, \delta(r_N - r)\, n(r) \tag{A. 14}$$

and we can apply our previous considerations without essential change. Replacing $e/k^2$ by a and $\rho(rt)$ by $n(rt)$ we obtain

$$\frac{d^2\sigma}{V\, d\epsilon_f d\Omega_f} = \frac{k_f}{k_i} \frac{m_N^2}{(2\pi)^3} \frac{|a|^2}{\hbar^5} S_{nn}(\vec{k}\omega). \tag{A. 15}$$

Note that $S_{nn}(k\omega)$ is defined via the fluctuations of the particle density. In addition to (A. 15), there is therefore also a term which comes from replacing $S_{nn}(k\omega)$ by

$$\int d(t-t') \int d(r-r')\, e^{i\omega(t-t') - i\vec{k}\cdot(\vec{r}-\vec{r}')} \langle n(\vec{r},t)\rangle \langle n(\vec{r}',t')\rangle. \tag{A. 16}$$

In a crystalline solid, this term describes elastic Bragg scattering. In a translationally invariant liquid this term contributes only in the forward direction.

Finally, the interaction a is generally spin-dependent. If this spin dependence is taken into account, and if it is assumed that the nuclear spins on different atoms are uncorrelated, one obtains a second contribution similar to (A. 15) but in which $S_{nn}(k\omega)$ is replaced by the self-correlation function

$$S_{self}(rt, r't') = \langle \sum_\alpha \delta(r - r^\alpha(t))\, \delta(r' - r^\alpha(t')) \rangle . \tag{A. 17}$$

This contribution is termed "incoherent"; it measures the self-diffusion of atoms. The contribution which we obtained in (A. 15) is termed "coherent".

## A.3 Magnetic Neutron Scattering

Neutrons have a magnetic moment, of magnitude $\mu^N$ = 1.91 $(e\hbar/m_N c)$, which interacts with the magnetization of the system via the magnetic dipole interaction. We can generally write the interaction in the form

$$V = \sum_{\alpha} 2\mu^N S_i^N V_{ij}(r^N - r^\alpha) \mu_j^\alpha$$

$$= \int dr\, 2\mu^N S_i^N V_{ij}(r^N - r) M_j(r). \qquad (A.18)$$

Then we can again compute a scattering cross section by the general procedure of section A.1. The dipole interaction is given by

$$V_{ij}(r) = \frac{-1}{4\pi} \nabla_i \nabla_j \frac{1}{r} \rightarrow \hat{k}_i \hat{k}_j . \qquad (A.19)$$

If we assume unpolarized neutrons, and sum over all final neutron polarizations, we obtain

$$\frac{d^2\sigma}{V\, d\epsilon_f d\Omega_f} = \frac{k_f}{k_i} \frac{(\mu^N m^N)^2}{(2\pi)^3 \hbar^5} \hat{k}_i S_{M_i M_j}(\vec{k}\omega) \hat{k}_j \qquad (A.20)$$

for the differential cross section per unit volume.

## A.4 Light Scattering

In a dielectric fluid light is scattered by local inhomogeneities of the dielectric constant which give rise, in the incident field, to a fluctuating polarization $\delta\vec{P}(rt)$. From Maxwell's equations, with $k^2 = (\omega^2/c^2)\epsilon$,

$$(\nabla^2 + k^2)\vec{E}(r,\omega) = -\frac{\omega^2}{c^2} \delta\vec{P}(r,\omega) , \qquad (A.21)$$

we obtain the scattered field

$$\vec{E}(r,\omega) = \frac{\omega^2}{4\pi c^2} \int d\vec{r}' \frac{e^{ik|r-r'|}}{|r-r'|} \delta\vec{P}(\vec{r}',\omega)$$

$$\approx \frac{e^{ikr}}{r} \cdot \frac{\omega^2}{4\pi c^2} \cdot \int d\vec{r}\; e^{-i\vec{k}\cdot\vec{r}'}\, \delta\vec{P}(\vec{r}',\omega)\,, \qquad \text{(A.21)}$$

where the second equation holds far from the scattering region, and $\vec{k} = k\frac{\vec{r}}{r}$ is in the direction of observation. The radiated power of frequency $\omega_f$ and polarization $\hat{e}_f$, observed outside the medium, is thus

$$\vec{N} = \hat{k}_f\, c\, \langle |\hat{e}_f \cdot \vec{E}(\vec{r},\omega)|^2 \rangle\, T^{-1}$$

$$= \hat{k}_f \cdot \frac{c}{r^2} \left(\frac{\omega_f^2}{4\pi c^2}\right)^2 \langle |\hat{e}_f \cdot \delta\vec{P}(k_f,\omega_f)|^2 \rangle\, T^{-1}\,, \qquad \text{(A.22)}$$

where $T = \int dt$ is the time during which scattering occurs. (Consideration of wave packets would give a better handle on this unsightly factor, and produce the same final result.)

Now in an incident field $\vec{E}^{in} = \hat{e}_i e^{i(\vec{k}_i\vec{r}-\omega_i t)}$, the induced polarization is

$$\delta P_i(r,t) = \delta\epsilon_{ij}(r,t)\, E_j^{in}(r,t)$$

or

$$\hat{e}_f \cdot \delta\vec{P}(k_f,\omega_f) = \delta\epsilon_{fi}(k_f-k_i,\omega_f-\omega_i)\,. \qquad \text{(A.23)}$$

If we write

$$S_{\epsilon\epsilon}(\vec{k},\omega) = \int d(t-t') \int d(r-r')\, e^{i\omega(t-t')-i\vec{k}\cdot(\vec{r}-\vec{r}')} \langle \delta\epsilon_{fi}(r,t)\,\delta\epsilon_{fi}(r',t')\rangle \qquad \text{(A.24)}$$

we obtain

$$\vec{N} = \hat{k}_f \cdot \frac{c}{r^2} \left(\frac{\omega_f^2}{4\pi c^2}\right)^2 V \cdot S_{\epsilon\epsilon}(\vec{k},\omega) \qquad \text{(A.25)}$$

where

$$\vec{k}_f = \vec{k}_i - \vec{k} ,$$

$$\omega_f = \omega_i - \omega . \tag{A.26}$$

If we therefore define the scattering cross section by

$$\text{inc. flux} \times d\sigma = N \cdot r^2 d\Omega_f \cdot \frac{d\omega_f}{2\pi} \tag{A.27}$$

and note that the incoming flux inside the medium is $c/\sqrt{\varepsilon}$, we obtain

$$\frac{d^2\sigma}{V d\Omega_f d\omega_f} = \frac{1}{4} \frac{\sqrt{\varepsilon}}{(2\pi)^3} \left(\frac{\omega_f}{c}\right)^4 S_{\varepsilon\varepsilon}(\vec{k}\omega). \tag{A.28}$$

With $S_{\varepsilon\varepsilon}$ given by (A.24), this is the standard light scattering formula. In its derivation, which is just the usual derivation of dipole radiation, we have treated the fluctuation $\delta\varepsilon_{ij}(r,t)$ of the dielectric tensor as a normal local variable. In a gas, with essentially isolated atoms or molecules, there is no problem; the dielectric properties are determined by the molecular polarizeability. In a liquid $\varepsilon(r,t)$ is a more correlated affair, particularly if the molecules contain permanent electric dipole moments as is the case for many liquid crystalline substances. In this case we have to assume that the range of those correlations which effectively determine $\varepsilon(r,t)$ is small on the scale of the wavelength of fluctuations which we probe.

In an isotropic liquid even the fluctuations of $\varepsilon$ are isotropic, $\delta\varepsilon_{ij} = \delta\varepsilon\,\delta_{ij}$ so that (A.28) leads only to unpolarized scattering. If we assume that $\delta\varepsilon$ fluctuates in response to fluctuations in the local particle density,

$$\delta\varepsilon(r,t) = \left(\frac{\partial\varepsilon}{\partial n}\right) \delta n(r,t) , \tag{A.29}$$

we obtain

$$\frac{d^2\sigma}{V d\Omega_f d\omega_f} = \left(\frac{1}{2}\frac{\partial \epsilon}{\partial n}\right)^2 \left(\frac{\omega_f}{c}\right)^4 (e_i \times \hat{k}_f)^2 \, S_{nn}(\vec{k}\omega) \frac{\sqrt{\epsilon}}{(2\pi)^3} \tag{A.30}$$

The contribution of temperature fluctuations, $\sim(\partial\epsilon/\partial T)^2$, is always negligible.

In liquid crystals, on the other hand, the largest contributions to $\delta\epsilon_{ij}$ come from fluctuations of the local alignment, as indicated in eq. (11.20), and depolarized scattering dominates. In this case even the equilibrium dielectric constant is uniaxial, and care must be taken to use the ordinary or extraordinary optical index for the incident and scattered beams, as the case may be. There are anisotropic fluctuations even in the isotropic pretransitional region of nematics. Their signature has also been seen in depolarized light scattering from many not-nematogen molecular liquids which are optically isotropic, but composed of anisotropic molecules.

## REFERENCES


Ailawadi, N. K., B. J. Berne, and D. Forster, 1971, Phys. Rev. A3, 1462.

Allen, J. F., and H. Jones, 1938, Nature 141, 243.

Ambegaokar, V., P. G. de Gennes, and D. Rainer, 1974, Phys. Rev. A9, 2676.

Andersen, H. C., and R. Pecora, 1971, J. Chem. Phys. 54, 2584.

Anderson, P. W., 1958, Phys. Rev. 112, 1900.

Anderson, P. W., and P. Morel, 1961, Phys. Rev. 123, 1911.

Anderson, P. W., 1964, "Concepts in Solids" (W. A. Benjamin, Inc., New York).

Anderson, P. W., 1964a, in "Lectures on the Many-Body Problem", vol. 2, ed. E. R. Caianello (Academic Press, New York), p. 113.

Anderson, P. W., 1966, in "Quantum Fluids", ed. D. F. Brewer (North-Holland Publishing Company, Amsterdam), p. 146.

Baym, G., 1969, in "Mathematical Methods of Solid State and Superfluid Theory", ed. R. C. Clark (Oliver and Boyd, Edinburgh), p. 121.

Bennett, H. S., and P. C. Martin, 1965, Phys. Rev. 138 A, 608.

Berne, B. J., J. P. Boon, and S. A. Rice, 1966, J. Chem. Phys. 45, 1086.

Berne, B. J., and D. Forster, 1971, Ann. Rev. Phys. Chem. 22, 563.

Berne, B. J., and G. D. Harp, 1970, Advan. Chem. Phys. XVII, 63.

Berne, B. J., and R. Pecora, 1975, "Dynamic Light Scattering with Applications to Chemistry, Biology and Physics", (John Wiley & Sons, New York, in press).

Bogoliubov, N., 1960, Physica 26, S1.

Callen, H. B., and T. A. Welton, 1951, Phys. Rev. 83, 34.

Chandrasekhar, S., 1943, Rev. Mod. Phys. 15, 1.

Chandrasekhar, S., 1961, "Hydrodynamic and Hydromagnetic Stability" (Oxford University Press, London).

Chatelain, P., 1948, Acta Cristallogr. 1, 315.

Chung, C. H., and S. Yip, 1969, Phys. Rev. 182, 323.

Collins, M. F., V. J. Minkiewcz, R. Nathans, L. Passell, and G. Shirane, 1969, Phys. Rev. 179, 417.

de Gennes, P. G., 1959, J. Phys. Chem. Solids 4, 223.

de Gennes, P. G., 1966, "Superconductivity of Metals and Alloys", (W. A. Benjamin, Inc., New York).

de Gennes, P. G., 1969, Phys. Letters 30A, 454.

de Gennes, P. G., 1974, "The Physics of Liquid Crystals",(Oxford University Press, London).

Dirac, P. A. M., 1958, "The Principles of Quantum Mechanics", (Oxford University Press, London).

Donnelly, R. J., 1967, "Experimental Superfluidity", (The University of Chicago Press, Chicago).

Dorfman, J. R., and E. G. D. Cohen, 1970, Phys. Rev. Letters 25, 1257.

Enz, Ch. P., 1974, Rev. Mod. Phys. 46, 705.
Ernst, M. H., E. H. Hauge, and J. M. J. van Leeuwen, 1970, Phys. Rev. Lett. 25, 1254.
Fano, U., 1957, Rev. Mod. Phys. 29, 74.
Fetter, A. L., and J. D. Walecka, 1971, "Quantum Theory of Many-Particle Systems", (Mc Graw Hill, New York).
Feynman, R. P., 1972, "Statistical Mechanics", (W. A. Benjamin, Inc., Reading, Mass.).
Forster, D., P. C. Martin, and S. Yip, 1968, Phys. Rev. 170, 155 and 160.
Forster, D., T. C. Lubensky, P. C. Martin, J. Swift, and P. S. Pershan, 1971, Phys. Rev. Letters 26, 1016.
Forster, D., 1974a, Ann. Phys. 84, 505.
Forster, D., 1974b, Phys. Rev. Letters 32, 1161.
Forster, D., 1975, "Theory of Liquid Crystals", to appear in Advan. Chem. Phys.
Frank, F. C., 1958, Discussions Faraday Soc. 25, 19.
Goldstone, J., 1961, Nuovo Cimento 19, 154.
Goldstone, J., A. Salam, and S. Weinberg, 1962, Phys. Rev. 127, 965.
Graham, R., 1974, Phys. Rev. Letters 33, 1431.
Halperin, B. I., and P. C. Hohenberg, 1969a, Phys. Rev. 188, 898.
Halperin, B. I., and P. C. Hohenberg, 1969b, Phys. Rev. 177, 952.
Hohenberg, P. C., and P. C. Martin, 1965, Ann. Phys. 34, 291.
Hohenberg, P. C., 1967, Phys. Rev. 158, 383.
Josephson, B. D., 1966, Phys. Letters 21, 608.
Kadanoff, L. P., and G. Baym, 1962, "Quantum Statistical Mechanics", (W. A. Benjamin, Inc., New York).
Kadanoff, L. P., and P. C. Martin, 1963, Ann. Phys. 24, 419.
Kadanoff, L. P., and J. Swift, 1968, Phys. Rev. 166, 89.
Kapitza, P. L., 1938, Nature 141, 74.
Katz, A.,and Y. Frishman, 1966, Nuovo Cimento 42A, 1009.
Kawasaki, K., 1970, Ann. Phys. 61, 1.
Kerr, W. C., K. N. Pathak, and K. S. Singwi, 1970, Phys. Rev. A2, 2416.
Khalatnikov, I. M., 1965, "An Introduction to Superfluidity", (W. A. Benjamin, Inc., New York.)
Kibble,T. W. B., 1966, in "Proceedings of the Oxford International Conference on Elementary Particles, 1965", (Rutherford High Energy Laboratory, Harwell).
Kirkwood, J. G., 1946, J. Chem. Phys. 14, 180.
Kristaponis, J.,and D. Forster, 1975, to be published.
Kubo, R., 1957, J. Phys. Soc. Japan 12, 570.
Kubo, R., 1959, in "Lectures in Theoretical Physics", vol. 1 ( Interscience, N.Y.).
Kwok, P. C., and P. C. Martin, 1966, Phys. Rev. 142, 495.
Landau, L. D., and E. M. Lifshitz, 1959, "Fluid Mechanics", (Addison-Wesley, Reading, Mass.).
Landau, L. D., and G. Placzek, 1934, Phys. Z. Sowjetunion 5, 172.
Lange, R. V., 1966, Phys. Rev. 146, 301.
Lebowitz, J. L.,1972, in "Statistical Mechanics", Proceedings of the Sixth IUPAP Conference, ed. S.A. Rice, (University of Chicago Press, Chicago), p. 41.
Lebowitz, J. L., and P. Résibois, 1965, Phys. Rev. A139, 1101.
Lubensky, T. C., 1971 , Ann. Phys. 64, 424.
Lubensky, T. C., 1970 , Phys. Rev. A2, 2497.
Ma, Shang-keng, 1975, "Critical Phenomena", (W. A. Benjamin, Inc., Reading, Mass.; to appear).

Martin, P. C., and J. Schwinger, 1959, Phys. Rev. 115, 1342.
Martin, P. C., 1965, in "Statistical Mechanics of Equilibrium and Non-Equilibrium" ed. J. Meixner (North-Holland, Amsterdam), p. 100.
Martin, P. C., 1967, Phys. Rev. 161, 143.
Martin, P. C., 1968, in "Many-Body Physics", eds. C. De Witt and R. Balian, (Gordon and Breach, New York).
Martin, P. C., 1969, in "Superconductivity", ed. R. D. Parks (Marcel Dekker, Inc., New York), p. 371.
Martin, P. C., O. Parodi, and P. S. Pershan, 1972, Phys. Rev. A6, 2401.
Martinoty, P., and S. Candau, 1971, Mol. Cryst. Liquid Cryst. 14, 243.
Mazenko, G. F., T. Y. C. Wei, and S. Yip, 1972, Phys. Rev. A6, 1981.
Mermin, N. D., 1966, J. Math. Phys. 8. 1061.
Minkiewcz, V. J., M. F. Collins, R. Nathans, and G. Shirane, 1969, Phys. Rev. 182, 624.
Mori, H., and K. Kawasaki, 1962, Progr. Theoret. Phys. (Kyoto) 27, 529.
Mori, H., 1965, Progr. Theoret. Phys. (Kyoto) 34, 423.
Nozières, P., 1964, "Theory of Interacting Fermi Systems", (W. A. Benjamin, Inc., New York).
Nozières, P., 1966, in "Quantum Fluids", ed. D. F. Brewer, (North-Holland Publishing Company, Amsterdam), p. 1.
Nozières, P., and D. Pines, 1964, "The Superfluid Bose Liquid", unpublished manuscript.
Nyquist, H., 1928, Phys. Rev. 32, 110.
Onsager, L., 1931, Phys. Rev. 37, 405 and 38, 2265.
Orsay Liquid Crystal Group, 1969, Phys. Rev. Letters 22, 1361.
Orsay Liquid Crystal Group, 1971, Mol. Cryst. Liquid Cryst. 13, 187.
Penrose, O., and L. Onsager, 1956, Phys. Rev. 104, 576.
Pines, D., 1965, in "Proceedings of the IXth International Conference on Low Temperature Physics", eds. J. G. Daunt, D. O. Edwards, F. J. Milford, and M. Jaqub, (Plenum Press, New York), p. 61.
Pines, D., and P. Nozières, 1966, "The Theory of Quantum Liquids", (W. A. Benjamin, Inc., New York.).
Puff., R. D., 1965, Phys. Rev. 137, A 404.
Rice, S. A., and P. Gray, 1965, "The Statistical Mechanics of Simple Liquids", (Interscience, New York).
Snyder, H., 1963, Phys. Fluids 6, 755.
Stanley, H. E., 1971, "Introduction to Phase Transitions and Critical Phenomena", (Oxford University Press, New York).
Stegeman, G. I. A., and B. P. Stoicheff, 1973, Phys. Rev. A7, 1160.
Stephen, M. J., and J. P. Straley, 1974, Rev. Mod. Phys. 46, 617.
Straley, J. P., 1972, Phys. Rev. A4, 675.
Tisza, L., 1938, Nature 141, 913.
Van Hove,L., 1954, Phys. Rev. 95, 249.
Wax, N., 1954, "Noise and Stochastic Processes", (Dover Publications, Inc., N.Y.).
Weidlich, W., 1971, Brit. J. Math. Statist. Psychol. 24, 251.
Weidlich, W., 1972, in "Collective Phenomena", (Gordon and Breach, N. Y.), p. 51.
Wilks, J., 1967, "The Properties of Liquid and Solid Helium", (Clarendon Press, Oxford).
Yang, C. N., 1962, Rev. Mod. Phys. 34, 694.
Zwanzig, R., 1961, in "Lectures in Theoretical Physics", vol. 3, (Interscience, New York).

Zwanzig, R. and R. D. Mountain, 1965, J. Chem. Phys. 43, 4464.
Zwanzig, R., 1972, in "Statistical Mechanics", Proceedings of the Sixth IUPAP Conference on Statistical Mechanics, eds. S. A. Rice, K. F. Freed, and J. C. Light, (University of Chicago Press, Chicago), p. 241.

INDEX